WITHDRAWN

Interfacial Phenomena in Metals and Alloys

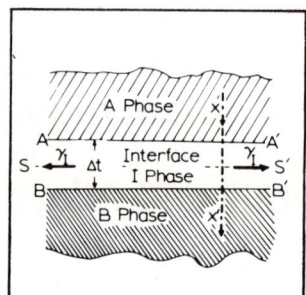

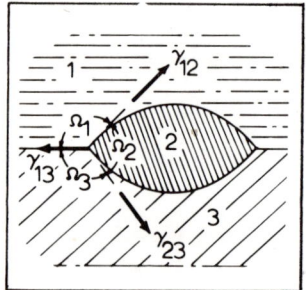

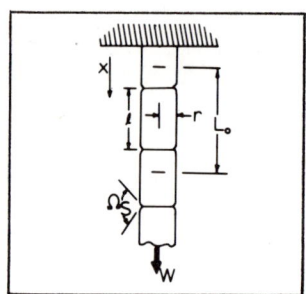

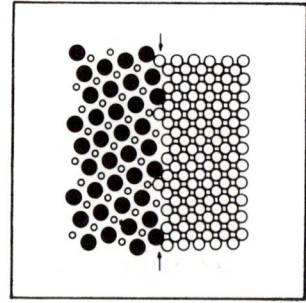

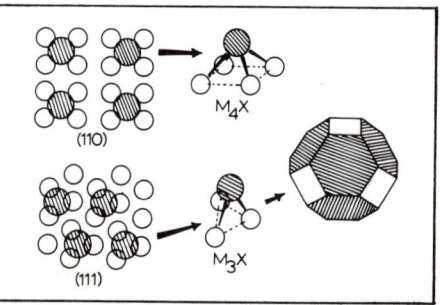

Interfacial Phenomena in Metals and Alloys

Lawrence E. Murr
Department of Metallurgical and Materials Engineering
New Mexico Institute of Mining and Technology

With a Foreword by Morris Cohen
Department of Materials Science and Engineering
Massachusetts Institute of Technology

 1975

Addison-Wesley Publishing Company
Advanced Book Program
Reading, Massachusetts

London • Amsterdam • Don Mills, Ontario • Sydney • Tokyo

Library of Congress Cataloging in Publication Data

Murr, Lawrence Eugene.
  Interfacial phenomena in metals and alloys.

  Includes bibliographies and index.
  1. Physical metallurgy.   2. Surfaces (Technology)
3. Metals.   4. Alloys.   I. Title.
TN690.M87   669'.95   74–30394

ISBN 0–201–04884–1
ISBN 0–201–04885–X pbk.

Copyright © 1975 by Addison-Wesley Publishing Company, Inc.
Published simultaneously in Canada

All rights reserved. No part of this publication may be reproduced, stored in a retrieval system, or transmitted, in any form or by any means, electronic, mechanical, photocopying, recording, or otherwise, without the prior written permission of the publisher, Addison-Wesley Publishing Company, Inc., Advanced Book Program, Reading, Massachusetts 01867, U.S.A.

Manufactured in the United States of America.

*Affectionately dedicated
to my wife
Pat
and to our children
Kim and Kyle*

CONTENTS

Foreword by Morris Cohen xi
Preface xiii

CHAPTER 1 THERMODYNAMICS OF SOLID INTERFACES 1
  1.1 Phenomenology of Solid Surface Free Energy 1
  1.2 Equilibrium Shape of Isolated Metallic Solids 3
      1.2.1 Multicomponent Systems 6
  1.3 The General Interface–Phase Concept 10
      1.3.1 Single-Component Systems: Pure Metals 13
      1.3.2 Multicomponent Systems: Alloys 15
      1.3.3 Interphase Interfaces in Heterogeneous Systems 21
  1.4 Summary 26
  References 27
  Problems 28

CHAPTER 2 INTERFACIAL EQUILIBRIA 31
  2.1 Interface Intersections; Polycrystal Equilibrium Configurations 31
  2.2 Grain Boundary Equilibria 33
      2.2.1 Grain Boundary Intersections in Random Sections 35
      2.2.2 Characterization of a Grain Boundary Section in Thin Films 39
      2.2.3 Concept of Grain Boundary Torque 41
      2.2.4 Grain Boundary Intersections with a Free Surface 43
  2.3 Twin Boundary Equilibria in FCC Metals and Alloys 45
      2.3.1 Twin Boundary–Free Surface Intersections 47
      2.3.2 Twin Boundary–Grain Boundary Intersections 54
  2.4 Multiphase Equilibria 66
      2.4.1 Equilibration of a Particle (Crystal) in Contact with a Solid Substrate . . 67
      2.4.2 Two- and Three-Phase Alloy Equilibria 71
  2.5 Summary 75
  References 75
  Problems 85

CHAPTER 3 INTERFACIAL FREE ENERGY 87
  3.1 Surface Energy of Liquid Metals and Alloys; The Concept of Surface Tension . 87
  3.2 Surface Tension Measurements 90
      3.2.1 The Sessile Drop Method 91
      3.2.2 The Pendant Drop Method 95
      3.2.3 Drop Weight Method 96
      3.2.4 Maximum Bubble Pressure Method 98

3.2.5 Surface Tensions at the Melting Point and Temperature Coefficients of Liquid Metals and Alloys 100
3.2.6 Surface Tension Isotherms for Alloy Solutions. 112
3.3 Surface Energy of Solid Metals and Alloys 116
    3.3.1 Zero-Creep Techniques for the Measurement of Solid–Vapor Surface Energies. 116
    3.3.2 Surface Free Energies and Temperature Coefficients of Solid Metals and Alloys 122
    3.3.3 Surface Energy Isotherms for Solid Alloys 127
3.4 Interfacial Free Energies in Solid Metals and Alloys 130
    3.4.1 Measurement of Grain Boundary Free Energy 131
    3.4.2 Effects of Temperature and Composition on Grain Boundary Free Energy 132
    3.4.3 Energetics of Twin Boundaries in Solid Metals and Alloys 138
    3.4.4 Measurement of Stacking-Fault Free Energies in Metals and Alloys 142
    3.4.5 Effects of Temperature and Composition on Stacking-Fault Free Energy. 150
    3.4.6 Measurement of Interphase Interfacial Free Energies in Metals and Alloys 153
3.5 Summary 157
References 158
Problems. 162

CHAPTER 4 STRUCTURE OF INTERFACES 165
4.1 Surface Structure in Liquid Metals and Alloys 165
4.2 Liquid Metal Interfacial Structure 167
4.3 Surface Structure of Solid Metals and Alloys 168
    4.3.1 Structure of Single Crystals 168
    4.3.2 Structure of the Solid–Liquid Interface 177
    4.3.3 Surface Structure Associated with Adsorption, Oxidation, and Related Surface Reactions 179
4.4 Calculation of Surface Tensions and Surface Energies 181
    4.4.1 Empirical Calculations of Metal Surface Energies 182
    4.4.2 Theoretical Calculation of Metal Surface Energies 184
    4.4.3 Theoretical Calculations of the Solid–Liquid Interfacial Free Energies . . 187
4.5 Interfacial Structure in Solid Metals and Alloys 187
    4.5.1 Grain Boundary Structure 188
    4.5.2 Interpretation and Description of Grain Boundary Structures 208
    4.5.3 Structure of Stacking-Fault and Twin Interfaces 219
4.6 Interphase Interface Structure 227
    4.6.1 Domain Boundary Structure 227
    4.6.2 Structure of Phase Boundaries 230
4.7 Adsorption, Absorption, Segregation, and Precipitation at Solid–Solid Interfaces 240
    4.7.1 Interfacial Adsorption and Absorption 241
    4.7.2 Solute Segregation and Precipitation at Interfaces 242
4.8 Calculation of Solid Interfacial Free Energies 244
    4.8.1 Theoretical Calculation of Grain Boundary Free Energies 245
    4.8.2 Theoretical Calculation of Twin Boundary and Stacking-Fault Free Energies in FCC Metals and Alloys 249
4.9 Summary 250
References 251
Problems. 256

CONTENTS

CHAPTER 5 PROPERTIES OF INTERFACES ... 259
  5.1 Adsorption and Reaction at a Free Surface ... 259
    5.1.1 Physisorption and Chemisorption ... 259
    5.1.2 Electronic Phenomena: Work Function Alteration ... 266
    5.1.3 Oxidation and Corrosion ... 267
    5.1.4 Reaction Kinetics and Catalysis ... 270
  5.2 Nucleation and Growth from the Vapor ... 276
    5.2.1 Homogeneous and Heterogeneous Nucleation ... 276
    5.2.2 Vacuum Vapor Deposition ... 280
  5.3 Metallic Friction and Adhesion ... 284
    5.3.1 Surface Structure and Friction ... 285
  5.4 Surface Mobility and Transport Phenomena ... 291
    5.4.1 Diffusion Phenomena ... 291
    5.4.2 Sintering Mechanisms and Sintering Diagrams ... 295
  5.5 Fracture of Metals and Alloys ... 303
    5.5.1 Surface Embrittlement by Segregation and Adsorption ... 306
    5.5.2 Grain Boundary Embrittlement and Intergranular Fracture ... 308
  5.6 Solid-Liquid Transitions: Solidification and Melting ... 314
    5.6.1 Nucleation from the Melt ... 314
    5.6.2 Growth Mechanisms and Kinetics ... 316
    5.6.3 Melting of Metals and Alloys ... 320
  5.7 Grain Boundary Migration and Sliding ... 322
    5.7.1 Boundary Migration during Recovery, Recrystallization, and Grain Growth ... 323
    5.7.2 Mechanisms of Grain Boundary Migration ... 330
    5.7.3 Mechanisms of Grain Boundary Sliding ... 331
    5.7.4 Grain Boundary Sliding and Creep in Polycrystals ... 335
    5.7.5 Grain Boundary Sliding and Superplasticity ... 336
  5.8 Influence of Grain and Interphase Boundaries on Mechanical Properties ... 338
    5.8.1 Grain and Interphase Boundaries as Sources for Dislocations ... 338
    5.8.2 Yield, Plastic Flow, and the Hall-Petch Relation ... 340
    5.8.3 Interphase Interface Strengthening ... 344
    5.8.4 Effect of Stacking-Fault Energy on Mechanical Properties ... 346
    5.8.5 Deformation-Induced Interfaces and Residual Strengthening ... 347
  5.9 Summary ... 350
  References ... 352
  Problems ... 359

Author Index ... 361
Subject Index ... 000

FOREWORD

Interfaces, both external and internal, constitute an integral part of the structure of materials, and so they enter directly into the structure/property/performance linkage which forms the central theme of materials science and engineering. These "boundary aspects" of bulk materials, because of their relatively discontinuous character, are not only difficult to study and elucidate, but they can also exert major influence on the behavior of materials and, hence, dominate their utility in many applications. Interfaces are often regarded as two-dimensional imperfections in materials, along with linear-defect dislocations and the point-defect vacancies and interstitialcies, but interfaces play an even deeper and more general role in nature's processes.

Two recent studies* by the National Academy of Sciences have independently called attention to the importance of interfaces (including external surfaces) in solid-state phenomena, and have placed high priority on research in this domain because of the challenging gaps in fundamental understanding and existing limitations in the performance of materials.

Fortunately, relatively powerful tools have now become available for probing the structure and behavior of interfaces. One need only mention such techniques as field-ion and atom-probe microscopy, Auger spectroscopy, high- and low-energy electron diffraction, scanning and transmission electron microscopy, quantitative stereology, high-vacuum and thin-film technologies, and a wide variety of photoelectric and semiconductor effects. External and internal surfaces are intimately involved in oxidation and corrosion, friction, wear and lubrication, plastic flow and fracture, intergranular embrittlement, nucleation and crystal growth, catalysis and reaction kinetics, solid-state transformations and resulting morphologies, recrystallization and grain growth, polymer degradation and flammability, adhesion and sintering, blood clotting and biocompatibility, integrated circuits and miniaturization, and virtually all types of materials processing.

The time has now come to bring this vital range of subject matter into the teaching core of materials science, alongside atomic structure, crystallography, ther-

---

*Physics in Perspective, 1972, National Academy of Sciences, Washington, D.C. *Materials and Man's Needs*, 1974, National Academy of Sciences, Washington, D.C.

modynamics, kinetics, and mechanics. The present book represents a significant step in that direction; it concentrates mainly on the metallic state for which the quantitative principles and relevant experimental data are available and thereby provides a unifying framework for materials more generally. Brought together here in pedagogical self-consistency are the thermodynamics, structure, and properties of the many types of interfaces which control both the bulk and surface behavior of materials. This is a pioneering effort which is likely to meet first success at the graduate level, but it is only a question of time before such an approach to interfacial phenomena will emerge in some undergraduate curricula as well.

Morris Cohen

PREFACE

This book was written with a number of objectives in mind. The foremost objectives were to produce a reasonably comprehensive text useful in conveying unifying characteristics of interfacial structure and the role played by interfaces in determining the properties of metals and alloys, and at the same time to provide a reference for students and researchers studying the structure and properties of interfaces. A particular effort was made to extensively tabulate experimentally determined values of interfacial free energies and energy ratios.

As a text, this book is intended to serve primarily as an introduction to interfacial phenomena at the graduate level in a materials science, metallurgy, or metallurgical and materials engineering curriculum, and it has been fully class-tested at this level. Because it surveys a wide range of areas in metallurgy and the materials sciences in which interfacial phenomena play an important role, this book could be used in a senior-level course where the necessary background in thermodynamics, atomic physics, solid-state physics, and physical metallurgy or ceramics has been acquired. Its fundamental approach and its attempt to document the development of experimental techniques and results in interfacial studies should make this book especially valuable for students and researchers in the fields of solid-state physics and chemistry, surface physics and chemistry, metallurgy, ceramics, and the broadest range of the materials sciences notwithstanding its explicit attention to metals and alloys. As an aid to the researcher and the student, original work has been cited where possible, and numerous additional references point the way for continued study or a more thorough description of theoretical or experimental details. In addition, practical problems are included at the end of each chapter to provide an opportunity for application and further clarification of the topics presented. Having covered the present text, the student should be fairly well equipped to critically read the many contributions, contemporary and future, dealing with interfacial phenomena, and to appreciate more fully the role played by interfacial phenomena in determining materials properties.

It was believed at the outset that a book was necessary that dealt not only with the structure and properties of surfaces, but more generally with the full range of solid-liquid-vapor interfaces in the context of real materials and the role played by interfaces in determining the properties of materials, particularly metals and alloys.

Although the material covered is, of course, biased to a large extent by the author's own research interests, a sincere effort has been made to present a balance in emphasis. The discussion of many topics will appear abbreviated only because of a lack of knowledge in the area.

The organization of this book into five long chapters is an attempt to present a more integrated chronology than is sometimes achieved with dozens of chapters, each treating a specific topic. The chronology begins with a basic overview of classical interfacial thermodynamics following the original concepts of Gibbs in Chapter 1; this is succeeded by a general treatment of interfacial equilibrium envisioned primarily as the balance of surface free energies from both a theoretical and an experimental viewpoint in Chapter 2. The experimental determination of interfacial free energies is presented in Chapter 3, followed by a general treatment of interfacial structure at the atomic level in Chapter 4. The final chapter illustrates some of the properties of metals and alloys specifically influenced by, or dependent on, interfacial energy and structure and provides for a kind of unification of the topics discussed in the preceding four chapters. This unification, more than any other concept, is one of the overriding purposes of this book.

I am indebted to the many authors and publishers who kindly provided me with photographs and other data, and for their permission to publish them. The various contributions are acknowledged in the appropriate figure captions. I am also grateful to my students and colleagues, who, over the years, have provided stimulating suggestions and comments for improvement, and who pointed out errors which persisted in various drafts of the manuscript. I am particularly grateful to Professor J. P. Hirth for having read the entire manuscript and for his comments and suggestions, many of which have been incorporated into the text; and to Professor Morris Cohen for his continued interest and encouragement in preparing the final manuscript. I am also grateful for the patient, competent typing and retyping of various portions of earlier manuscripts by Suzy Nevergold and Rosalie Anaya, and the final manuscript by Lorraine Valencia. The help of Elizabeth Fraissinet and my wife and daughter in the preparation and typing of the indexes is also gratefully acknowledged. I should also like to acknoweldge the support of research upon which a great deal of this book was based by the U.S. Office of Naval Research and the U.S. Atomic Energy Commission (now ERDA), the latter more recently through contracts with Sandia Laboratories, Albuquerque, New Mexico.

Finally, I am especially grateful to my wife and children for their constant support. The dedication of this work to them is indeed small compensation for their many sacrifices.

L. E. Murr

1
THERMODYNAMICS OF SOLID INTERFACES

1.1 PHENOMENOLOGY OF SOLID SURFACE FREE ENERGY

In order to study the structure and properties of a solid metallic (metal or alloy) interface, the nature and character of such an interface must first be established. One can initially characterize an interface (a planar interface) as a simple dividing surface: either a free surface—for example, a solid-vapor interface or a solid-liquid interface; or an internal dividing surface—a solid-solid interface.

The general case of a solid surface as a crystal surface (to which this monograph addresses itself) has been treated by Shuttleworth[1] as shown schematically in Fig. 1.1, which depicts the reversible deformation (at constant temperature) of an arbitrary crystal face of area A by very small amounts, dA_1 and dA_2, against surface stresses σ_1 and σ_2.

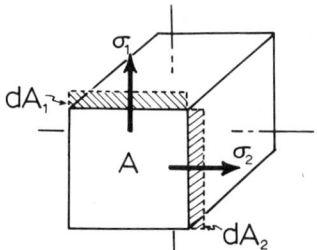

FIGURE 1.1 Reversible deformation of a crystal surface (area) element.

In order that no stresses occur in the volume of the crystal—that is, there is zero volume strain before deformation—the work done against the surface stresses must equal the increase in the Helmholtz surface free energy, F_A:

1

$$\sigma_1 \, dA_1 + \sigma_2 \, dA_2 = d(AF_A) \tag{1.1}$$

where F_A is the surface free energy per unit area defined thermodynamically by

$$F_A = E_A - TS_A \tag{1.2}$$

E_A being the total surface energy per unit area, T the absolute temperature (°K), and S_A the surface entropy per unit area. Shuttleworth[1] has shown that, for symmetrical crystals or a general isotropic solid, the surface stresses are equal and identical to a surface tension, γ, as in the case of a pure liquid. Thus, for $\sigma_1 = \sigma_2 \equiv \gamma$,* and $dA \equiv (dA_1 + dA_2)$, Eq. (1.1) can be written

$$\gamma \, dA = d(AF_A) \tag{1.3}$$

or

$$\gamma = F_A + A(dF_A/dA) \tag{1.4}$$

Note that γ will always be positive because work must always be expended in the creation of new surfaces. A negative surface tension, if it does in fact exist as an unstable surface, must inevitably lead to the formation of a new, stable surface. Otherwise, the surface must be in a state of dilation.

It is a well-established fact that, if the surface area of a pure liquid is increased by an isothermal deformation at constant volume, the second term on the right side of Eq. (1.4) will be zero because liquid atoms will be continuously transferred to the surface from the bulk, and, as a consequence, the number of atoms per unit area in the surface layer will remain constant. In addition, liquids are unable to support shear stresses. Therefore the surface tension (force per unit length) is numerically equal to the surface free energy (energy per unit area) for a liquid.

At relatively low temperatures in metallic solids, the atoms are relatively immobile. As a result, after low-temperature deformation the total number of atoms in a surface remains effectively unchanged, and dF_A/dA is not zero. However, at elevated temperatures where an interface is able to migrate—that is, where diffusion is appreciable—the number of atoms per unit area of surface or interface layer could tend to be constant, and at a sufficiently high temperature it might be assumed that $dF_A/dA \simeq 0$ for a metallic solid. Under these conditions, the solid surface tension, γ, will be approximately equal to the surface free energy, F_A, in Eq. (1.4).*

---

*As recently shown by F. V. Nolfi and C. A. Johnson [*Acta Met.*, **20**, 769 (1972)], this assumption is physically impossible. For example, $\sigma \simeq 0$ for copper. This feature is implicit in the application of Eq. (1.4) to solids where, as discussed in the text, $\gamma \simeq F_A$ for a solid is only approximately true for certain conditions. In addition, P. R. Couchman, W. A. Jesser, D. Kuhlmann-Wilsdorf, and J. P. Hirth [*Surface Sci.*, **33**, 429 (1972)] point out that surface stress is not simply related to surface strain. If the excess energy per surface atom does not

Since, in the case of solid surfaces, one is concerned with the reversible work of creating new surface, the term specific surface free energy or simply surface free energy is a more appropriate thermodynamic description for an interface than surface tension when $\gamma \simeq F_A$ [Eq. (1.4)] at elevated temperatures. This characterization will be utilized in the development of the interface concept, and the measurement of the associated reversible work of formation of homogeneous and heterogeneous phase boundaries.

It must be cautioned at the outset that the assignment of a surface tension to a solid interface (either solid-vapor or solid-solid) is valid only when a reversible equilibrium is established. This feature is, in the broad spectrum of metals and alloys, doubtful in many instances. We must note in particular that surface tension, as defined for a liquid, probably obtains only for metals and alloys when they melt and are maintained in the liquid phase. The quantity measured as interfacial free energy in a solid metal or alloy must therefore be carefully considered in light of these facts, and the particular assumptions made.

1.2 EQUILIBRIUM SHAPE OF ISOLATED METALLIC SOLIDS

The thermodynamics of isolated particles (as small crystals) was set forth originally by Gibbs,[2] Curie,[3] and Wulff,[4] and expanded upon in the later works of Semenchenko,[5,6] Herring,[7] Mullins,[8] and Cabrera,[9] to name only a few. Among other things, the equilibrium (ignoring gravitational effects) of isolated particles provides a means of determining the relative values of anisotropy of (specific) surface free energies, a feature embodied in the well-known Wulff theorem which has its origin in the following treatment.

Let us initially consider for simplicity an isolated, single-component crystalline (centrosymmetric) metal (a pure metal particle) and the thermodynamic equilibrium of this metal crystal in relation to its surface morphology (shape). The total surface free energy of a crystalline particle, F^c, will be given by

$$F^c = \sum_i \gamma_i A_i \tag{1.5}$$

where γ_i is the (specific) surface free energy of the ith crystal face having an area A_i. Since the crystal shape will be determined by the facet or crystal face morphology, equilibrium shape must demand a specific minimum in the crystal surface free energy according to the so-called Gibbs-Curie theorem,[2,3,*]

$$(dF^c)_V = 0; \quad (d^2 F^c)_V > 0 \tag{1.6}$$

---

vary with infinitesimally small changes of surface atom density, the surface stress vanishes. Stresses at interfaces and free surfaces can be eliminated by spontaneous generation and motion of dislocations. It should also be emphasized that the treatment presented is implicit for planar interfaces.

*The derivative d is sometimes replaced by δ or Δ to denote an infinitesimal change. It should be understood that in this treatment changes are small.

for constant volume, V. Although Eq. (1.6) is not specifically related to crystal morphology, Wulff[4] added a condition relating the crystal volume to the possible changes in specific surface area (that of the ith interface). He regarded the crystal volume, V^c, as the sum of the volumes of the pyramids constructed on the faces A_i whose vertices converge in the center of the crystalline particle and wrote

$$V^c = \frac{1}{3} \sum_i A_i \lambda_i \tag{1.7}$$

where λ_i denotes the length of the normal to the ith face of area A_i extended into the crystal center. Or, simply stated, λ_i is the height of the ith pyramid constructed inside the crystal volume as stipulated above.

Following Semenchenko,[5] the free energy of the total system of a pure metal particle surrounded by its saturated vapor or melt can be written

$$F = (F^c + U^c) + U^{\widetilde{V},L} \tag{1.8}$$

where U^c denotes the total (volume) free energy of the crystalline phase, and $U^{\widetilde{V},L}$ denotes the surrounding vapor ($\widetilde{V}$)- or liquid (L)-phase free energy. If the number of moles in the crystalline (particle) phase is n^c, and correspondingly $n^{\widetilde{V},L}$ denotes the number of moles in the vapor or liquid phase enveloping the particle, then for equilibrium conditions,

$$n = n^c + n^{\widetilde{V},L} = \text{constant} \tag{1.9}$$

and additionally, since we demand that the volume of the system remain constant,

$$V = V^c + V^{\widetilde{V},L} = \text{constant} \tag{1.10}$$

Since the total work done on the system is equal to the change in free energy at constant temperature, we invoke Eq. (1.6) for equilibrium conditions and obtain[6]

$$\sum_i \gamma_i \, dA_i + \left(\frac{\partial F}{\partial n^c}\right)_{T,V} dn^c + \left(\frac{\partial F}{\partial V^c}\right)_{T,V} dV^c + \left(\frac{\partial F}{\partial n^{\widetilde{V},L}}\right)_{T,V} dn^{\widetilde{V},L}$$

$$+ \left(\frac{\partial F}{\partial V^{\widetilde{V},L}}\right)_{T,V} dV^{\widetilde{V},L} = dF = 0 \tag{1.11}$$

By definition,[2]

$$\mu^p = (\partial F/\partial n^p)_{T,V} \tag{1.12}$$

and

$$P^p = -(\partial F/\partial V^p)_{T,V} \tag{1.13}$$

where μ^p and P^p are the chemical potential and pressure of phase p. Substituting Eqs. (1.12) and (1.13) and the derivatives of Eqs. (1.9) and (1.10) into Eq. (1.11) then results in

$$\sum_i \gamma_i A_i + (\mu^c - \mu^{\widetilde{V},L}) dn^c - (P^c - P^{\widetilde{V},L}) dV^c = 0 \tag{1.14}$$

Since from Eq. (1.7)

$$dV^c = \frac{1}{3}(\sum_i A_i d\lambda_i + \sum_i \lambda_i dA_i) \tag{1.15}$$

and considering any (small) change in volume of the crystal to simply reflect a displacement of A_i through a distance, $d\lambda_i$, in the form

$$dV^c = \sum_i A_i d\lambda_i \tag{1.15a}$$

then

$$\sum_i A_i d\lambda_i = \frac{1}{3}(\sum_i A_i d\lambda_i + \sum_i \lambda_i dA_i) \tag{1.15b}$$

or

$$dV^c = \frac{1}{2}\sum_i \lambda_i dA_i \tag{1.15c}$$

Substituting Eq. (1.15c) into Eq. (1.14) then results in

$$\sum_i [\gamma_i - (\lambda_i/2)(P^c - P^{\widetilde{V},L})] dA_i + (\mu^c - \mu^{\widetilde{V},L}) dn^c = 0 \tag{1.16}$$

If one considers in Eq. (1.16) that small variations in dA_i and dn^c are essentially independent, then the corresponding coefficients must be independently zero. Consequently

$$2\gamma_i/\lambda_i = (P^c - P^{\widetilde{V},L}) \tag{1.17}$$

and

$$\mu^c = \mu^{\widetilde{V},L} \tag{1.18}$$

for thermodynamic equilibrium. Equation (1.18) expresses the well-known fact[2] that the chemical potential of the solid (pure) crystal will be equal to that in its saturated vapor or liquid. On the other hand, the pressure difference, $P^c - P^{\widetilde{V},L}$ in Eq. (1.17), will be constant for this system in equilibrium so that Eq. (1.17) can be rewritten in the original form of the Wulff theorem as

$$\gamma_1/\lambda_1 = \gamma_2/\lambda_2 = \ldots = \gamma_i/\lambda_i = \text{constant} \qquad (1.17)$$

or

$$2\gamma_i/\lambda_i = K_W \qquad (1.17a)$$

where K_W, the Wulff constant, is given by the Gibbs-Thomson theorem.[10] Equations (1.17) and (1.17a) express the fact that the ratio of the (specific) surface free energy, γ_i, of the ith crystal face or planar surface facet to the height of the pyramid, λ_i, constructed within the particle on the area A_i as shown in Fig. 1.2, as the base, will be a constant. It should be noted, however, that this constant, K_W, applies only for the equilibrium shape of a particular crystal; that is, the equilibrium shape will include only those crystal faces for which the ratio $2\gamma_i/\lambda_i$ is equal to the Wulff constant, K_W. While the equilibrium shape for "particles" will generally be a polyhedron, the equilibrium shape may also be bound partially or completely by "curved" surfaces. This latter feature was treated by C. Herring [*Phys, Rev.*, **82**, 87 (1951)].

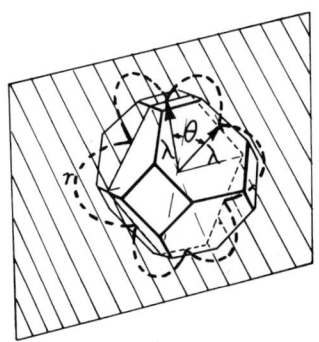

FIGURE 1.2   Idealized Wulff construction for equilibrium crystal shapes. The planar projection of the polar relationships is referred to as a γ-plot. Note that energy minima (cusps) occur in the crystallographic planes.

1.2.1 Multicomponent Systems

We might consider more generally the treatment of multicomponent (alloy) systems. Here, according to Gibbs,[2] the crystal surface free energy is expressed by

$$F^c = \sum_{ij} A_i(\gamma_i + \mu_{ij}\Gamma_{ij}) \qquad (1.19)$$

where μ_{ij} denotes the chemical potential of the jth component at the ith crystal face (of area A_i), and Γ_{ij} is defined as the surface excess of the jth component, characterized as the Gibbs adsorption isotherm[2] (at the ith crystal face)—that is:

$$\Gamma_{ij} \equiv n_j^c/A_i = -(\partial\gamma_i/\partial\mu_{ij})_T \quad (1.19a)$$

Semenchenko[5,6] has treated the two-component system (j = 1, 2) in detail by considering Eq. (1.19) to replace F^c in Eq. (1.8). Although the situation is considerably more complex, it has been established generally that Eq. (1.17) is valid for a multicomponent crystalline particle.

The Wulff theorem, implicit in Eq. (1.17), has been generalized somewhat by Herring,[7] and its theoretical implications have been treated by Frank[11] and Mullins.[12] In addition, Miller and Chadwick,[13] for example, have illustrated that various Wulff constructions (corresponding to specific equilibrium shapes) provide a means to relate polar plots of surface free energy, γ_i, γ-plots, and the equilibrium shapes by considering an orientation parameter as illustrated in Fig. 1.2.

It should be apparent that a truly minimum free energy shape of a particle in the absence of external forces will approximate a sphere. This feature is particularly true of homogeneous nucleation where a solid cluster forms by a vapor-to-liquid supercooling as shown, for example, in Fig. 1.3(a). In such instances, γ_i [Eq. (1.17a)] varies infinitesimally with crystallographic orientation (for example, θ in Fig. 1.2) so that λ_i is essentially constant, the shape being distorted somewhat by gravitational effects. On the other hand, crystalline systems at elevated temperature can deviate noticeably from spherical shapes because γ_i can in fact vary significantly with surface orientation as a consequence of the specific crystallographic arrangements of atoms near (and composing) different surfaces. The resulting equilibrium shapes typically attain a faceted appearance as shown in Fig. 1.3(b)–(d). Note in Fig. 1.3(b) and (d) that the equilibrium shapes are polyhedra composed of $\{111\}$ and $\{001\}$ planes as described by K. H. Behrndt [*Thin Solid Films,* 7, 415 (1971)]. In addition to facets, the equilibrium body also exhibits sharp edges and curved edges (or faces), caused by the adjustments of atoms over specific areas to make γ_i minimum. These features are recognizable in Fig. 1.3, particularly in Fig. 1.3(c).

In dealing with crystalline solids, it must be recognized that surfaces, regardless of their apparent smoothness due to resolution limitations in observing them, are usually not atomically flat. That is, when one is dealing, for example, with a hemispherical or spherical crystalline particle shape [as in Fig. 1.3(a)], the surface will be composed of a multitude of crystallographic facets, having higher-order indices as the surface approaches a spherical (or hemispherical) shape. Ultimately, a perfectly "smooth" metal–crystal surface will be composed of atomic steps where the atomic planes conform to the associated radius of curvature. This feature is illustrated in Fig. 1.4, which shows an electrochemically etched iridium wire tip which has been smoothed to perfection by field evaporation of the end form in the field-ion microscope. It is particularly notable in Fig. 1.4(a) that the shank area

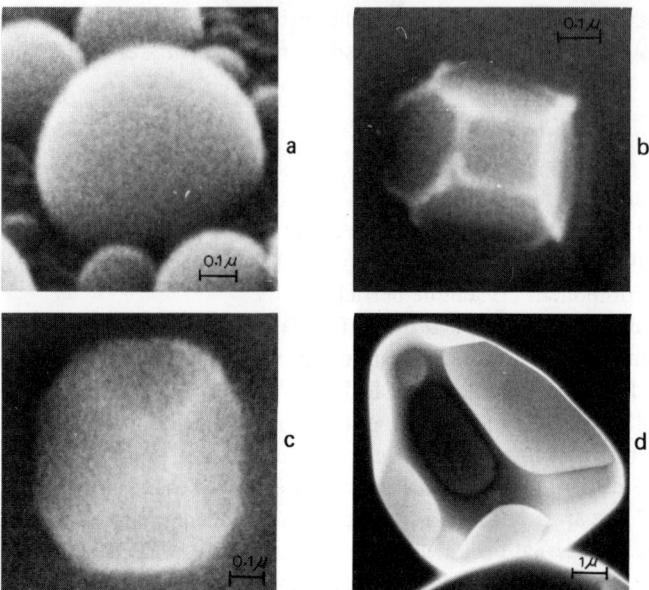

FIGURE 1.3 Equilibrium shapes of crystalline solids observed in the scanning electron microscope. (a) Tin particles nucleated by vapor deposition onto an NaCl substrate at $0.8T_m$ (T_m = melting point in °K). (b) Nickel crystal on ThO_2 substrate after 50 hours at 1200°C in H_2 (electron beam nearly normal to substrate). (c) NiCr (80 : 20) crystal on ThO_2 substrate after 50 hours at 1700°C in H_2 (electron beam nearly normal to substrate). (d) Iridium crystal after 50 hours at 1700°C in helium observed in the SEM.

away from the tip surface possesses a recognizable crystallographic (faceted) morphology. Figure 1.4(b) illustrates the atomic surface structure of an iridium end form similar to that in Fig. 1.4(a), but with a considerably smaller average radius of curvature. Each "ring" of atoms constitutes the edge of an atom layer, and consequently a step into (or out of) the surface. Several prominent crystal planes (facets) are indexed in Fig. 1.4(b) according to the standard stereographic projection along [001].

It should be pointed out that, although the Wulff theorem presupposes interfacial free energy to be minimized when particles exist as single crystals, Fullman[14] has shown that this need not be true for systems possessing a set of low-energy faces which do not constitute a closed form. This feature has important implications in the treatment of several interfaces—for example, grain boundaries—in subsequent chapters.

The detailed study of surface (solid-vapor) structure and surface interactions will not be pursued here, but will be discussed with regard to specific particle-matrix (solid-solid) systems in Chapter 4. The interested reader is referred to the

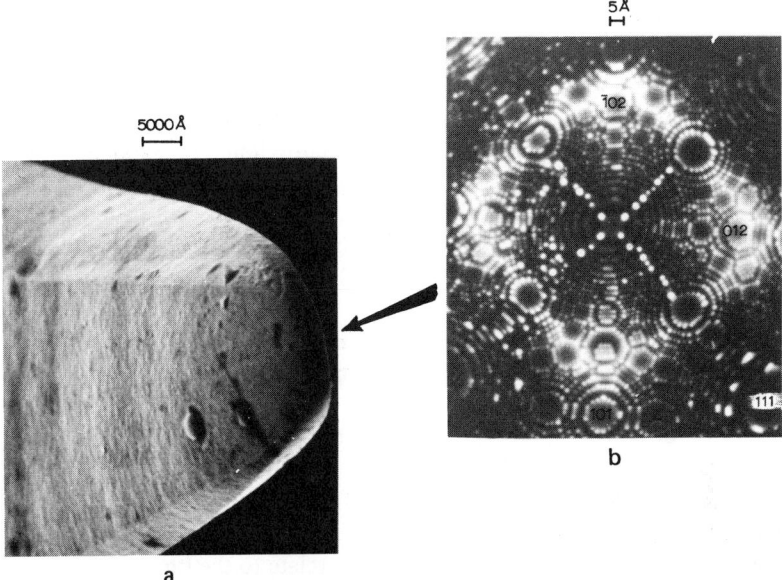

FIGURE 1.4 Atomically "smooth" surface features. (a) Scanning electron microscope view of an electropolished iridium end form extensively blunted by field evaporation in the field-ion microscope. (b) Field-ion microscope image of an iridium end form similar to that in (a) viewed essentially end-on along the axis of the wire as shown by the arrow in (a). The wire axis is parallel to [001].

treatments by Winterbottom,[15] Gjostein and Winterbottom,[16] Dunning,[17] and Basterfield et al.[18] In addition, we shall not be specifically concerned with the energetics of crystalline particles or their equilibrium shapes. These features have been studied in detail by Sundquist,[19] Winterbottom,[20] and Moore.[21] It is to be noted that changes in the γ-plot of a surface and the surface structure itself will be expected to be influenced by adsorption thermodynamics [Eq. (1.19)]. These features have been treated for bulk surfaces by Gjostein[22] and by Robertson and Shewmon;[23] and recent work by Burmeister[24] has demonstrated the applicability to variations in surface topography.

It might be instructive at this point to make one final comment concerning particle shapes, particularly where they relate to thermodynamic equilibrium. Semenchenko[6] has shown that, since[2]

$$(\partial \mu / \partial P)_T = V_j (\partial \mu / \partial T)_P = -S \tag{1.20}$$

where S is the effective entropy for one gram-molecule, the thermal dependence on shape can be expressed by

$$\lambda_i^2 (\partial T/\partial \lambda_i)_P = (2\gamma_i/S^{\widetilde{V},L} - S^c) V^c \tag{1.21}$$

where $S^{\widetilde{V},L}$ and S^c are the saturated vapor- or liquid-phase entropy and crystal phase entropy, respectively. Equation (1.21) expresses the variation in the temperature of the phase transition [solid crystal-to-liquid (or vapor)] with crystal dimensions, V^c, at constant pressure. The important thermodynamic implication in Eq. (1.21) is that the melting point must always decrease with a reduction in the particle (crystal) size at constant pressure. This feature can phenomenologically influence the thermodynamics of polycrystalline aggregates and dispersed and precipitated phases to some extent.

An alternative means for expressing thermodynamic variations of crystals with their size—in a form similar to Eq. (1.21)—can be written

$$RT \ln (P^c/P^{\widetilde{V}}) = 2\gamma_i V^c/\lambda_i = K_W V^c \tag{1.22}$$

where R is the well-known gas constant, P^c is the pressure associated with the crystal, and $P^{\widetilde{V}}$ is the saturated vapor pressure at a temperature T, assuming ideal gas behavior, and considering these features to relate to the ith crystal face—that is, a plane area of area A_i. Equation (1.22) illustrates the dependence of vapor pressure on crystal size. Ideally, P^c in Eq. (1.22) can be replaced by P^L if the "crystal" is presumed to be a "liquid" drop. It might also be observed with reference to Eq. (1.22) that those crystal faces for which $2\gamma_i/\lambda_i \neq K_W$ will either dissolve or evaporate if $2\gamma_i/\lambda_i > K_W$, or grow out of existence if $2\gamma_i/\lambda_i < K_W$.

1.3 THE GENERAL INTERFACE–PHASE CONCEPT

In the treatment of a metallic particle in the preceding discussions, the surface was considered to represent, or was implied to be, the interface separating a solid-phase particle from its saturated vapor or liquid phase. In a more general sense, an interface can be described as any dividing surface that interrupts a periodic lattice arrangement, or separates homogeneous phases of varying crystal structure, or heterogeneous phases of varying composition and crystal structure, or both, or any combination of these possibilities.

The original treatment of an interface by Gibbs[2] was characterized by the introduction of a mathematical surface called the dividing surface. Gibbs' treatment allowed for a thermodynamic description of an interface without a detailed knowledge of the structure of the transition region. In contrast to Gibbs' rather abstract treatment, van der Waals and Bakker[25] and Verschaffelt[26] contributed to the theory of a general interface by considering it to be a region of small but uniformly finite thickness. Guggenheim[27] clarified this concept by treating the interface as a planar phase separating two homogeneous bulk phases.

Following Gibbs[2] and Guggenheim,[27] a general plane interface separating two homogeneous bulk phases, A and B, can be represented as shown in the schematic

THERMODYNAMICS OF SOLID INTERFACES

of Fig. 1.5. Note in Fig. 1.5 that the Gibbs dividing surface, SS', represents a special case of the interface phase when $\Delta t = 0$, where Δt is the interface thickness or the limits of the transition region between phases A and B. The properties of the interface phase, I, are assumed to be uniform (homogeneous) in a direction parallel to the dividing surface, SS', but not in a direction perpendicular to SS' (along xx' in Fig. 1.5). Figure 1.6 illustrates the most general examples of the possible variation in volume concentration of a component c_j of the system in Fig. 1.5 with reference to the dividing surface, SS'.

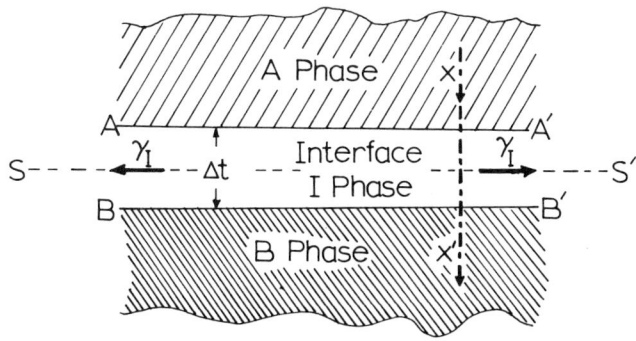

FIGURE 1.5 Thermodynamic representation of a general interface. AA' and BB' represent the limits of the interface phase as parallel planes of area A. The arbitrary dividing surface, SS', is an identical plane of area A.

We now consider the Gibbs-Duhem equation for a homogeneous plane-surface system (phases A and B are the same):

$$S\,dT - V\,dP + \sum_j n_j\,d\mu_j = 0 \qquad (1.23)$$

Applied to the interface phase, any consideration of changes which occur with pressure at constant volume and area must realize that the pressure across SS' in Fig. 1.5 (along xx') will be influenced by the resolved interfacial free energy, γ_I, shown as heavy vectors along SS'. As a consequence, $V\,dP$ in Eq. (1.23) must be replaced by $V^I\,dP - A\,d\gamma_I$, where V^I is the interfacial volume equal to $\Delta t A$, and A is the interfacial area. We can then rewrite Eq. (1.23) for the interface phase, I, as

$$S\,dT - \Delta t A\,dP + A\,d\gamma_I + \sum_j n_j\,d\mu_j = 0 \qquad (1.24)$$

which on division by A and rearrangement gives

$$d\gamma_I = -S_I\,dT + \Delta t\,dP - \sum_j \Gamma_j\,d\mu_j \qquad (1.25)$$

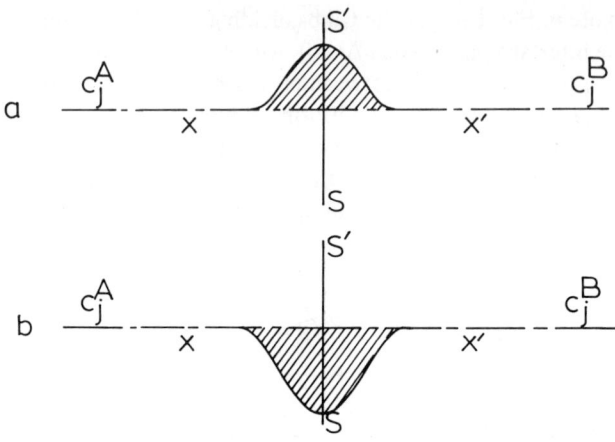

FIGURE 1.6 Idealized variations in volume concentration, c_j, of the jth component in the general two-phase (multicomponent) system of Fig. 1.5 for $c_j^A = c_j^B$. (a) Positive adsorption. (b) Negative adsorption.

where γ_I is the specific interfacial free energy or simply the interfacial free energy (per unit area of interface), S_I is the interfacial entropy (per unit area of interface), and Γ_j [from Eq. (1.19a)] is the surface excess of the jth component (the number of moles of component j per unit area of interface).

The significance of the $\Delta t\, dP$ term in Eq. (1.25) will be expected to vary according to the physicochemical nature of the interface. For example, if the interface is essentially a physical dividing surface between crystallographically or geometrically different phases—that is, where $c_j^A = c_j^B$ and $\Gamma_j = 0$—then Δt might be expected to be on the order of an atomic spacing—that is, approaching zero as an ideal Gibbs dividing surface. On the other hand, if Δt is significantly less than A (the interfacial area under consideration), then $\Delta t\, dP$ would be expected to approach zero, since dP would also vary insignificantly (it will be identically zero if P is constant). Finally, for systems where $\Gamma_j \neq 0$, the properties of the interface might indeed be described in terms of a broad transition zone where Δt is of the order of several hundred or even several thousand atom diameters. In this case, $\Delta t\, dP$ may still be zero if the pressure is constant. It is apparent from these options that an interface may be considered generally either as a physical dividing surface ($\Delta t \cong 0$) or as a chemical transition zone ($\Delta t > 0$), or as a combination of both. These features are illustrated in Fig. 1.7, which shows an atomically resolved grain boundary in the surface of an iridium bicrystal, and a precipitate-free zone characterizing a grain boundary in an Al-Zn-Mg alloy. It should also be noted that, to specify the composition of an interface, it will be necessary to assign some value to Δt.

THERMODYNAMICS OF SOLID INTERFACES 13

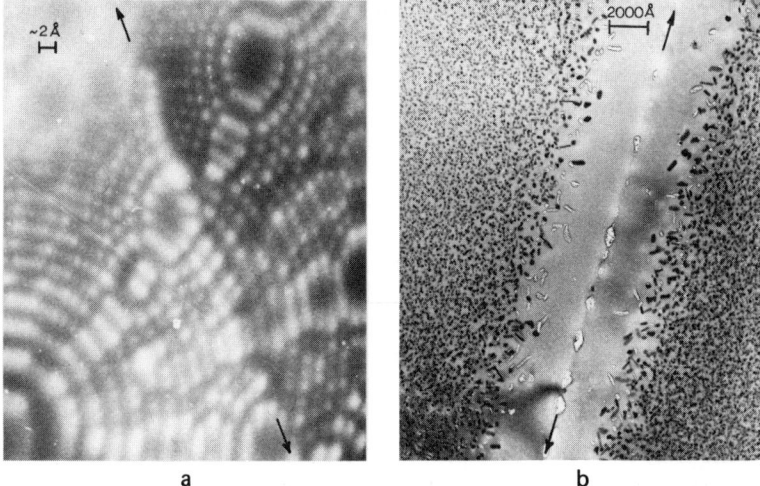

FIGURE 1.7 Grain boundary interface. (a) Field-ion image showing atomically resolved iridium bicrystal portions separated by a planar (straight) grain boundary. Planar grain boundary segment in an Al–Zn–Mg alloy. Δt might be considered to be the width of the precipitate-free zone, or that of the "physical" dividing surface in the middle of the zone. [(a) is courtesy of Dr. O. T. Inal; (b) is courtesy of Dr. G. W. Lorimer from ref. 28.]

1.3.1 Single-Component Systems: Pure Metals

Temperature Dependence of Interfacial Free Energy. For a single-component system, it is generally valid to assume $\Delta t\, dP$ to be negligible in Eq. (1.25). If, in addition, we assume that there is no nonequilibrium concentration of vacancies, then μ will be constant, and Eq. (1.25) can be written simply as

$$(d\gamma_I/dT)_V = -S_I \cong (dF_A/dT)_V \tag{1.26}$$

Equation (1.26) shows that the surface free energy of an idealized (nonadsorbing) interface in a pure metal at elevated temperature will have a negative temperature coefficient (entropy will always be positive). Equation (1.26) is a perfectly general assumption for any pure metal interface—free surface (solid–vapor interface), grain boundary, twin boundary, or stacking fault—provided no vacancy adsorption occurs at the interface.

The Gibbs Adsorption Isotherm for a Pure Metal. In many experiments, it is physically possible to confirm the absence of a nonequilibrium vacancy concentration at or near an interface. In addition, at elevated temperatures, vacancies would be expected to be annihilated. Nevertheless, a moving grain boundary or other interface might be expected to be a suitable sink for vacancies or vacancy clusters (see Fig. 1.8), and vacancy adsorption to an interface might occur under certain conditions—for example, Suzuki adsorption to a stacking fault.[29-31] Consequently, in the perfectly general case of a single-component system, Eq. (1.25) might be written in the form

$$d\gamma_I = -S_I \, dT - \Gamma_1 \, d\mu_1 - \Gamma_v \, d\mu_v \qquad (1.27)$$

where Γ_v and μ_v are the vacancy excess and chemical potential, respectively. If we introduce the common designation of mole fraction (or atomic fraction) of the jth component, identified by

$$x_j = n_j/\Sigma n_j; \quad \Sigma x_j = 1 \qquad (1.28)$$

where the mole fraction, x_j, is a pure number, and consider that $d\mu_1$ and $d\mu_v$ in the interface must equal those in the bulk phases A and B in Fig. 1, then we obtain

$$d\mu_1 = -S_1 \, dT + (d\mu_1/dx_v) \, dx_v \qquad (1.29)$$

and

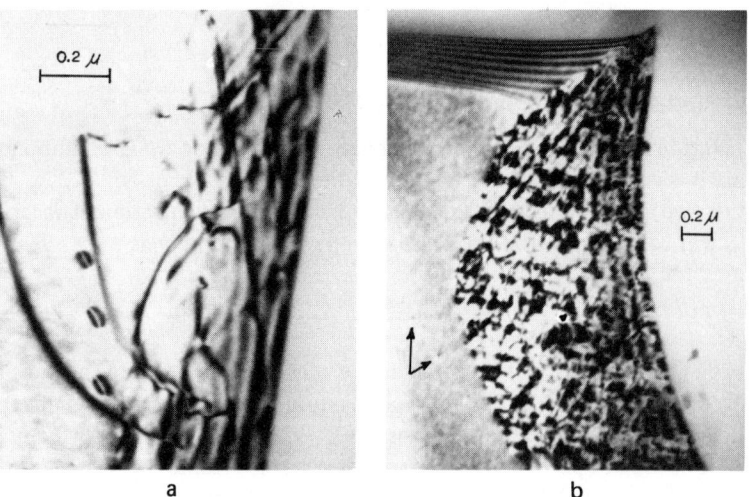

FIGURE 1.8   Grain boundaries in pure (99.999%) aluminum. (a) Clearly defined vacancy clusters in the vicinity of a grain boundary. (b) Complex contrast features at a specific grain boundary section which suggest high vacancy (cluster) concentration. Individual small clusters in the matrix are indicated by arrows. Observations in (a) and (b) were thin film (bright-field) electron transmission microscopy.

THERMODYNAMICS OF SOLID INTERFACES

$$d\mu_v = -S_v\, dT + (d\mu_v/dx_v)\, dx_v \tag{1.30}$$

so that Eq. (1.27) becomes

$$d\gamma_I = -(S_I - \Gamma_1 S_1 - \Gamma_v S_v)\, dT - \left(\Gamma_1 \frac{d\mu_1}{dx_v} + \Gamma_v \frac{d\mu_v}{dx_v}\right) dx_v \tag{1.31}$$

where S_1 and S_v denote the partial molar entropies of the component (matrix) atoms and vacancies in the interface, respectively. Since the chemical potentials (the μ's) cannot vary independently, but are related by the Gibbs-Duhem equation in the form,

$$x_v\,(d\mu_v/dx_v) + (1 - x_v)(d\mu_1/dx_v) = 0, \tag{1.32}$$

then Eq. (1.31) simplifies to

$$\left(\frac{d\gamma_I}{dT}\right)_V = -(S_I - \Gamma_1 S_1 - \Gamma_v S_v) - \left(\Gamma_v - \frac{x_v}{x_1}\Gamma_1\right)\frac{d\mu_v}{dx_v}\frac{dx_v}{dT} \tag{1.33}$$

The first term in parentheses on the right-hand side of Eq. (1.33) gives the direct effect of temperature upon the interfacial free energy; the second term in parentheses accounts for the indirect effects due to variation of vacancy segregation (adsorption) at the interface with temperature. Both terms are invariant with respect to the position of the interface planes in Fig. 1.5. It is observed that, even if Γ_1 is negligible as might safely be assumed for a pure metal, Γ_v might not be negligible. In addition, it should be noted that μ_v can be varied independently when excess vacancies are quenched in and maintained in metastable equilibrium with small vacancy clusters as might be suggested in Fig. 1.8(b).

For the case of constant temperature and pressure, Eq. (1.31) can be written in the following form of the Gibbs adsorption isotherm for a pure metal interface:

$$\left(\frac{d\gamma_I}{dx_v}\right)_{T,P} = -\left(\Gamma_v - \frac{x_v}{x_1}\Gamma_1\right)\frac{d\mu_v}{dx_v} \tag{1.34}$$

or

$$\left(\frac{d\gamma_I}{d\mu_v}\right)_{T,P} = -\left(\Gamma_v - \frac{x_v}{x_1}\Gamma_1\right) \tag{1.34a}$$

Since Γ_1 in Eq. (1.34) will generally be small or negligible for a pure metal, the adsorption of vacancies to an interface will lower the specific surface free energy.

1.3.2 Multicomponent Systems: Alloys

Binary Alloys. In the simplest alloy system, the homogeneous binary solid-solution systems where $c_j^A = c_j^B$ (or $x_j^A = x_j^B$) as in Fig. 1.6(a) and (b), the general form of the Gibbs-Duhem equation

$$(x_\nu \, d\mu_\nu + \sum_j x_j \, d\mu_j) = 0 \tag{1.35}$$

allows us to write

$$x_\nu \, d\mu_\nu + x_1 d\mu_1 + x_2 \, d\mu_2 = 0 \tag{1.36}$$

Considering, as implicit in Eq. (1.29), that $d\mu_\nu$, $d\mu_1$, and $d\mu_2$ in the interface must equal those in the bulk phases it separates then allows Eq. (1.25) to be written in the expanded form:

$$d\gamma_I = -(S_I - \Gamma_\nu S_\nu - \Gamma_1 S_1 - \Gamma_2 S_2) dT - (\Gamma_\nu - \frac{x_\nu}{x_1} \Gamma_1) d\mu_\nu$$

$$- (\Gamma_2 - \frac{x_2}{x_1} \Gamma_1) d\mu_2 \tag{1.37}$$

assuming $\Delta t \, dP$ to be negligible, where S_ν, S_1, and S_2 are the partial molar entropies of vacancies and components 1 and 2 in the interface, respectively, and Γ_ν, Γ_1, and Γ_2 represent the corresponding adsorptions of vacancies, solvent, and solute.

As indicated earlier by McLean[32] for weakly segregating solute atoms (or vacancies), the excess-entropy term in Eq. (1.37) would tend to be dominant. However, for an appreciable segregation of solute or vacancies, the second term in Eq. (1.37) would be the dominant one. Depending on the nature of the segregation, and the possible interaction of vacancies with solute atoms, the interfacial free energy could either increase or decrease with temperature, in a regular or nonlinear fashion.

Kinoshita and Eguchi[33] have discussed the arrangement of atoms and vacancies in alloys in thermal equilibrium where ordering or clustering can take place, and they concluded that, in binary alloys with short range order or clusters, the probability that a nearest-neighbor site of a vacancy is occupied by an atom of any particular kind does not necessarily vary monotonically with temperature. They conclude that in some alloys—for example, α-Cu–Al, this probability increases after decreasing or decreases after increasing. Similarly, the origin of the grain boundary precipitate-free zone [Fig. 1.7(b)], which occurs prominently in Al–Cu and Al–Zn–Mg alloys, has been described by Unwin et al.[28] to be related to the vacancy and solute concentration profile at the grain boundary, or diffusion-controlled loss of excess vacancies at the grain boundary.

It is apparent that, under certain conditions, the mechanical properties of alloys (or even pure metals) can be significantly influenced by vacancy adsorption to an interface—that is, to the extent that γ_I influences the particular property in question. In addition, the interaction of solute with vacancy concentrations can also contribute to these phenomena. However, there is no available information about Γ_ν, and only very little about x_ν.[34] As a consequence, Γ_ν and x_ν have been assumed to be negligible in the past. While there is no fundamental basis for this assumption aside from the fact that one may not be able to see any evidence of vacancies or vacancy clusters in the vicinity of an interface (with particular reference here to internal or solid-solid interfaces), we shall accept it for the moment in order to simplify Eq. (1.37) to the form

$$d\gamma_I = -(S_I - \Gamma_1 S_1 - \Gamma_2 S_2)\, dT - \left(\Gamma_2 - \frac{x_2}{x_1}\Gamma_1\right) d\mu_2 \qquad (1.37a)$$

Hirth[31] has pointed out that Eq. (1.37a) is probably a fairly accurate description of an interface in a concentrated solid solution where the vacancy terms would be negligible. It is again noted in Eq. (1.37a) that, for a strongly segregating solute (component 2 in this notation), the second parenthesized term on the right-hand side can cause the interfacial free energy to increase with temperature.

The Gibbs Adsorption Isotherm for a Binary Alloy. At constant temperature and pressure, Eq. (1.37a) becomes the familiar Gibbs adsorption equation for a binary alloy:

$$\left(\frac{d\gamma_I}{d\mu_2}\right)_{T,P} = -\left(\Gamma_2 - \frac{x_2}{x_1}\Gamma_1\right) \qquad (1.38)$$

or

$$-\frac{d\gamma_I}{dx_2}\bigg|_{T,P} = \Gamma_{2(1)} \left(\frac{d\mu_2}{dx_2}\right) \qquad (1.39)$$

where

$$\Gamma_{2(1)} = \left(\Gamma_2 - \frac{x_2}{x_1}\Gamma_1\right) = \left(\Gamma_2 - \frac{x_2}{1-x_2}\Gamma_1\right) \qquad (1.40)$$

The advantage of Eq. (1.39) over Eq. (1.38) is that $\Gamma_{2(1)}$ (the original notation of Gibbs[2]) is a natural and convenient measure of the interfacial adsorption of component 2 relative to component 1. Consequently $\Gamma_{2(1)}$ is ideally the relative adsorption density at the interface located so that Γ_1 vanishes, and it can be calculated from experimental measurements as shown, for example, by Cahn and Hilliard,[35] while Γ_1 and Γ_2 cannot be derived individually without knowing the associated changes in molar volumes across the interface. It should be noted also that $\Gamma_{2(1)}$ is invariant with respect to the placement of the dividing surface.

On examining Eq. (1.39) one observes that, if $d\gamma_I/dx_2$, $\Gamma_{2(1)}$, and $d\mu_2/dx_2$ are known independently for a particular system, it can be immediately determined whether the observed excess (or deficiency) is consistent with equilibrium segregation. Unfortunately, these complete data are not available for any alloy system, although measurements have been made of solute excess at free surfaces[36] and at grain boundaries in alloys which were characteristically dilute (low solute concentrations).[37-39] Inman et al.[40] have also measured the variation in surface (solid-vapor) free energy and grain boundary free energy in copper with small additions of antimony and have concluded that the concomitant energy reduction with solute addition resulted because of interfacial adsorption, consistent with Eq. (1.39).

Cahn and Hilliard[35] have argued that if the chemical potential dependence on composition in an alloy obeys Henry's law:

$$d\mu_2/dx_2 = kT/x_2 \tag{1.41}$$

then substitution of this assumption in Eq. (1.39) results in an estimate of the upper bound for the solute excess which can be present as an equilibrium adsorbate at a grain boundary in a binary alloy in the form

$$\Gamma^0_{2(1)} < \frac{\gamma_{gb}}{kT(1 + \ln x_e/x_0)} \tag{1.42}$$

where $\Gamma^0_{2(1)}$ is the measured value of $\Gamma_{2(1)}$ at an atomic fraction x_0, x_e is the solubility limit, γ_{gb} is the grain boundary free energy, and k is Boltzmann's constant. Equation (1.42) additionally assumes $d\gamma_{gb}/dx_2$ to be linear up to x_0 and logarithmic thereafter. Bitler[41] has shown that these assumptions are not generally valid for solids, and that indeed considerable equilibrium solute excess can occur at a grain boundary—that is, without forming a precipitate (a new phase). In addition, it must be noted that Cahn and Hilliards' assumptions are specific to an interface which is ideally a Gibbsian dividing surface. This is, in the physical sense, unrealistic in view of Fig. 1.6.

On examining Eq. (1.39) one observes that, if there is an excess of component 2 at the interface ($\Gamma_2 > \Gamma_1$) [Fig. 1.6(a)], $\Gamma_{2(1)} > 0$ and γ_I decreases with an increase in x_2. This is commonly referred to as positive adsorption. Conversely, for negative adsorption, $\Gamma_2 < \Gamma_1$ [Fig. 1.6(b)], and $\Gamma_{2(1)} < 0$; γ_I will increase with a decrease in the solute concentration, x_2, in the interface. From measurements of γ_I and μ (sometimes referred to as activity) over a range of concentrations, x_2, $\Gamma_{2(1)}$ may be calculated as a function of x_2. Several possible variations of interfacial free energy with composition (x_2) in a binary system are illustrated in Fig. 1.9, and superimposed on this plot are several possible variations of μ_2 (solute activity) with composition. The condition for a maximum or minimum in $\Gamma_{2(1)}$ to be present is expressed by the following identity:

$$\left(\frac{d^2\gamma_I}{dx_2^2}\right)\frac{d\mu_2}{dx_2} = \left(\frac{d^2\mu_2}{dx_2^2}\right)\frac{d\gamma_I}{dx_2} \tag{1.43}$$

Note that this adsorption characteristic, depicted as the turning point in the magni-

THERMODYNAMICS OF SOLID INTERFACES

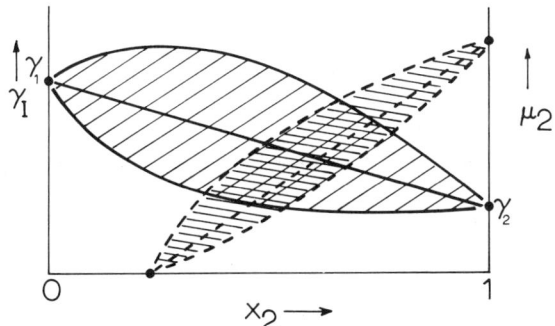

FIGURE 1.9 Possible variations of γ_I (solid curves) and μ_2 (broken curves) with composition in a binary system; γ_1 and γ_2 represent the interfacial free energies of the pure metals 1 and 2. The alloy systems correspond to $c_j^A = c_j^B$ in Fig. 1.6(a) or (b). The shading represents the ranges of variations of the sets of curves.

tude of adsorption, will be dependent on the interaction of the components with one another.

Interstitial and Substitutional Alloys. For conditions of constant temperature and pressure, and on the assumption that vacancies can be ignored, or their effect incorporated into the adsorption properties of the substitutional or interstitial component (component 2), Eq. (1.38) will be applicable. For an interstitial alloy, $\Gamma_1 = 0$, so that Eq. (1.38) becomes simply

$$(d\gamma_I/d\mu_2)^{\text{interstitial}} = -\Gamma_2 \tag{1.44}$$

In the case of a substitutional alloy, $\Gamma_1 = -\Gamma_2$; and, since $(x_1 + x_2) = 1$, Eq. (1.38) becomes

$$(d\gamma_I/d\mu_2)^{\text{substitutional}} = \Gamma_2/x_1 \tag{1.45}$$

It must again be cautioned that there is no a priori basis for assuming x_v and Γ_v to be negligible in Eqs. (1.44) and (1.45). In fact, vacancies can be important even in a concentrated interstitial solid solution if a binding energy exists between the vacancies present and the solute (interstitial) atoms. We might note in this regard that the general conditions corresponding to equilibrium in an interface in a simple interstitial or substitutional alloy are expressed by

$$d\gamma_I = -(S_I - \Gamma_v S_v - \Gamma_2 S_2)\, dT - \Gamma_2\, d\mu_2 \tag{1.46}$$

and

$$d\gamma_I = -[S_I - \Gamma_v S_v + \Gamma_2(S_1 - S_2)] \, dT - \left(\frac{\Gamma_2}{x_1}\right) d\mu_2 \quad (1.47)$$

respectively. For an ideal (dilute) solid solution, we can also assume Henry's law to apply in the form

$$d\mu_2 = -kT \, d \, (\ln x_2) \quad (1.41a)$$

Ternary and Multicomponent Solid-Solution Alloys. It might be expected that adsorption phenomena and variations in interfacial free energy with concentration and temperature will differ markedly in multicomponent alloys by comparison with two-component systems because of the complex chemical processes that can occur. However, although very few experimental data are available for two-component systems, there are essentially none for ternary or more complex multicomponent alloys. Therefore, it would seem somewhat presumptuous to attempt a detailed description of interfacial thermodynamics in these more complex systems. On the other hand, many of the technologically important and most useful alloys are multicomponent systems, and some consideration must be given to these systems.

We shall consider the case of a three-component system in which components 2 and 3 are considered additions to component 1 as a basis. For simplicity, we shall consider only the conditions where temperature and pressure are constant and write the adsorption equation as

$$d\gamma_I = -\Gamma_{v(1)} \, d\mu_v - \Gamma_{2(1)} \, d\mu_2 - \Gamma_{3(1)} \, d\mu_3 \quad (1.48)$$

where $\Gamma_{v(1)}$, $\Gamma_{2(1)}$, and $\Gamma_{3(1)}$ are the relative adsorptions (excesses) of vacancies, component 2 and component 3, respectively, with reference to the base component (component 1), having the form of Eq. (1.40). It is observed from Eq. (1.48) that the interfacial free energy could increase or decrease, depending on the action of the components separately, or their reaction or interaction.[32] Vacancy-coupled processes such as those described by Aust et al.[42] would be expected to be particularly important in multicomponent systems. These features are in fact suggested in the recent radiation-enhanced alloy decomposition experiments of Shriver and Wuttig.[43] Semenchenko,[6] for example, has theoretically treated several cases, including surface-active components which lower the interfacial concentration and adsorption of another component, and various possibilities conducive to positive (component excess) and negative (component desorption deficiency, or depletion) adsorption at an interface. Several cases in a ternary system have also been treated mathematically by deBruyn.[44]

As an interesting example, one might consider the case of a three-component system in which component 2 is systematically depleted at a grain boundary by reacting with an excess of component 3, where at the outset $\Gamma_3 < \Gamma_2$. If large precipitates form in equilibrium with the interface, then it is possible that both components 2 and 3 exhibit negative adsorption in the regions between the precipitates. Consequently, the interfacial free energy in the depleted regions could be increased appreciably. The precipitates formed, assuming that they grow to appreciable sizes, would be considered a new phase. If the precipitate-phase interface interconnected with that of the grain boundary, forming an interfacial equilibrium, and if the new interphase interfacial free energy were appreciably above that of the unreacted alloy grain boundary, then the entire grain boundary free energy would be effectively enhanced.

It might be possible, with regard to the previous discussion, to promote situations of phase equilibria where a component could be made to deplete (or desorb) from an interface (particularly a grain boundary) through a systematic diffusion reaction in which precipitation would occur only in the matrix phases, away from the grain boundary. Such a process would enhance the mechanical properties by strengthening the grain boundary, while precipitation hardening the matrix. Systems of this type might be devised from a systematic solution of the adsorption equation, and a full knowledge of the existing phase equilibria data, and would be particularly desirable where grain boundary precipitation would otherwise weaken the interface.

It should be noted that the general interface discussed in this section—whether in regard to single or multicomponent systems—has been a planar dividing surface. Additionally, it was implicit in our thermodynamic derivations that the interface was stationary. The effects of adsorption, in particular, would be enhanced by any motion of the interface. This is particularly true of grain boundary motion during annealing. Kasen[45] has in fact demonstrated this feature even for ultrapure aluminum, where segregation to moving grain boundaries was shown to depend on the distance and rate at which the interface moved regardless of the purity, suggesting segregation contributions from both equilibrium and nonequilibrium vacancy-coupled processes.[32,42] Such a process will ideally account for the appearance of the grain boundary in Fig. 1.8(b). It is therefore to be noted in retrospect that vacancy contributions to interface thermodynamics must be carefully considered.

1.3.3 Interphase Interfaces in Heterogeneous Systems

The previous section, while developing the concept of a general interface separating two phases A and B, was primarily concerned with homogeneous systems where phases A and B (Fig. 1.5) were identical—that is, where $x_j^A = x_j^B$. This concept can be utilized in the treatment of a solid-vapor (free surface) interface in a homogeneous alloy (or pure metal) and solid-solid interfaces such as grain boundaries, twin boundaries, or stacking faults in homogeneous metals and alloys.

In cases where phases A and B in Fig. 1.5 are chemically different—that is, where $x_j^A \neq x_j^B$—the interface is referred to as an interphase interface—for example, an interphase grain boundary for a dividing surface between two solid grains having different compositions. Other special cases could also involve phase transformations where a dividing surface separates the new phase from the matrix, or even precipitates or eutectic phases where the dividing surface at certain points could be treated as a planar interface.

While the thermodynamic quantities characterizing the state of the system are expected to change abruptly at the interface, the condition for internal equilibrium in a heterogeneous system requires that the temperature and pressure be constant throughout, and that the chemical potential of individual components be the same in the phases separated by the dividing surface; that is,

$$\mu_j^A = \mu_j^B \tag{1.49}$$

in Fig. 1.5, and A and B are homogeneous phases. The corresponding Gibbs-Duhem equation for these conditions is then written

$$(c_\nu^A - c_\nu^B) d\mu_\nu + \sum_j (c_j^A - c_j^B) d\mu_j = 0 \tag{1.50}$$

and the corresponding relative adsorption density for a multicomponent homogeneous two-phase system becomes

$$\Gamma_{j(1)} = \Gamma_j - \left(\frac{c_j^A - c_j^B}{c_1^A - c_1^B}\right) \Gamma_1 \tag{1.51}$$

which is invariant with respect to placement of the dividing surface.* Several of the more obvious cases expressed in volume concentrations, c_j, of the jth component referenced to the dividing surface (SS') in Fig. 1.5 for $c_j^A \neq c_j^B$ are shown in Fig. 1.10. While Eq. (1.51) ignores any appreciable interface thickness, Δt, this can be taken into account in some respects as shown for planar interfaces by Ono and Kondo[46]

$$\Gamma'_{j(1)} = \Gamma_{j(1)} + \Delta t \left(\frac{c_1^A c_j^B - c_1^B c_j^A}{c_1^A - c_1^B}\right) \tag{1.52}$$

which in the homogeneous case ($c_j^A = c_j^B = c_j$) becomes

$$\Gamma'_{j(1)} = \Gamma_{j(1)} + c_j \Delta t \tag{1.52a}$$

equal to Eq. (1.40) for a binary alloy where Δt might justifiably be assumed to be negligibly small.

*The reader should note that the position of the dividing surface itself is not unique, so that in a binary system Γ_2 and Γ_1 are variable even though $\Gamma_{2(1)}$ is not.

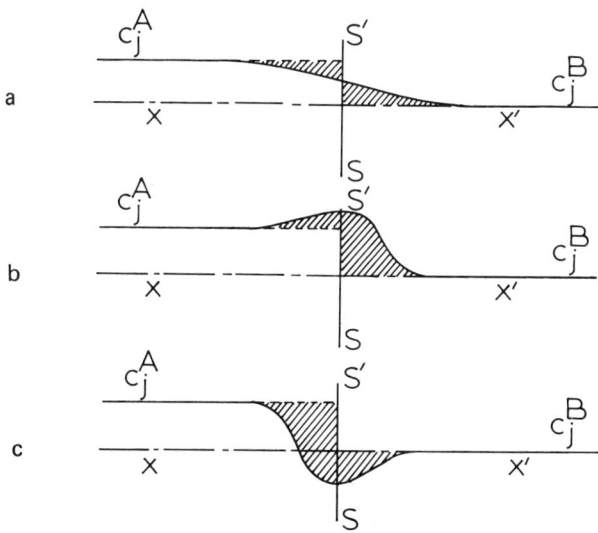

FIGURE 1.10 Idealized variations in volume concentration, c_j, of the jth component in the general two-phase (multicomponent) system of Fig. 1.5 for $c_j^A \neq c_j^B$. (a) No excess for a symmetric distribution about the dividing surface. (b) Effective position adsorption. (c) Effective negative adsorption.

The detailed treatment of interphase interfaces will, of course, require a detailed knowledge of the associated phase equilibria. We are in general unable to deal adequately with conditions where temperature is not constant because of the expected variations of interface composition with temperature. However, in situations where certain assumptions can be made concerning composition (phase stability) within a certain temperature range, properly modified forms of Eq. (1.24) will be applicable. Variations in interphase-interface energy for two-component heterogeneous systems at constant temperature could be expected to fall within the range ideally depicted in Fig. 1.9, and variation in phase boundary free energy would be extremely complex, conceivably increasing or decreasing in a complex fashion.

Case Examples of Interphase Interfaces. Various general examples of heterogeneous systems in a simple two-phase situation are illustrated in Fig. 1.11. Note that the phase boundaries are considered to be a plane section in the system, and that the compositions are essentially fixed at specific temperature and pressure. We are not concerned in this treatment with the phase equilibria so much as with the thermodynamics of the interface. The situations implicit in Fig. 1.11 also ignore crystallographic variations in the phases, as well as variations of the interfacial free energy with the orientation and geometry of the phase boundary.

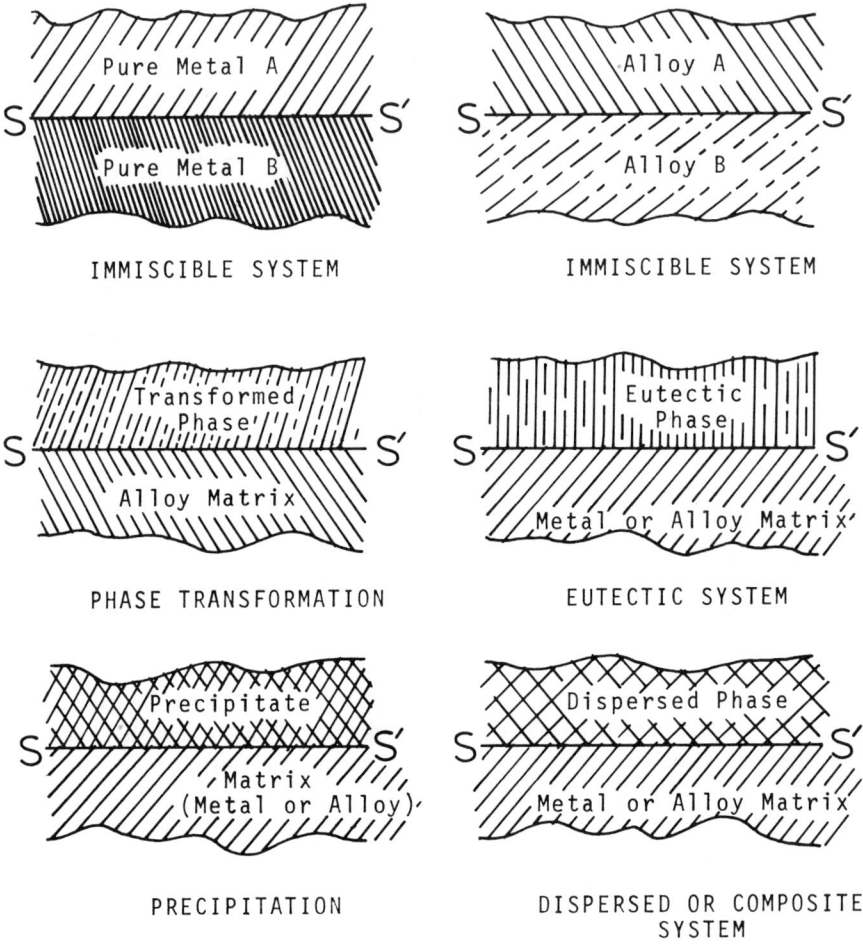

FIGURE 1.11   Idealized cases giving rise to interphase interfaces. In all cases the phases may have the same or different crystal structure—usually different lattice parameters if both are solids. One phase may be liquid.

In the specific case of coherent precipitates and dispersoids within a metal or alloy matrix, the shapes of particles, particularly in solid-solid systems, will depend on the associated elastic strain energies in addition to the interfacial free energies; in general there will be anisotropy of the associated elastic constants of the particle and matrix phases, demonstrated for example in the work of Tien and Copley.[47] In addition, large coherent particles will tend to lose coherency so that the strain energy contribution will tend to decline with the relative size of a precipitate phase.[48] For incoherent particles or fibers, it is generally thought that strain energy contributions are unimportant.[49,50] This is also assumed to be true for eutectic systems. Ignoring the coherency properties and the particle shapes, the interfacial thermodynamics are specified as outlined previously. An example of this implication is shown in the sequence of electron micrographs in Fig. 1.12.

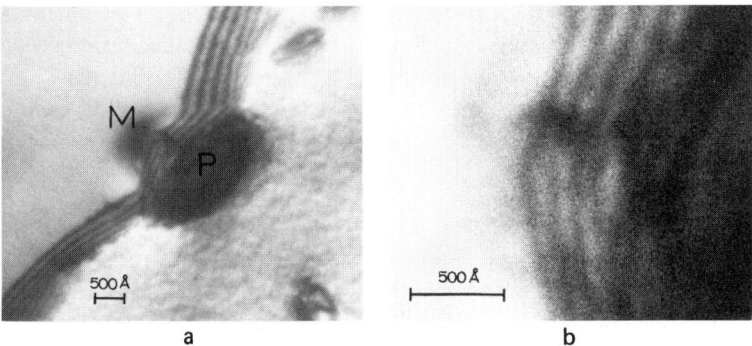

FIGURE 1.12 Grain boundary precipitate treated as a two-phase system at the interphase interface. (a) Electron transmission image of carbide precipitate in the grain boundary of a Ni-Cr-Fe alloy (Inconel 600). (b) Magnified view of (a) showing the interphase interface separating phase P (precipitate) from the matrix phase, M. The interface plane is inclined to the surfaces of the thin section and gives rise to the fringe pattern.[51]

Curved Interfaces. An important feature particularly prominent in the case of precipitates or similar inclusions or phases having highly nonplanar shapes is that of curvature of the interface [apparent to some extent in Figs. 1.8(b) and 1.12]. Gibbs[2] has shown that the fundamental formulas for planar interfaces may be applied in practice to curved interfaces; however, in most cases a straightforward application must assume the curvature not be too great. The treatment is clarified somewhat by introducing the condition that the thickness of the interface must be small compared to its radii of curvature—that is, $\Delta t_i << R_i$, where i refers to the ith interface or the ith portion of an area of interface. Curved surfaces, as these relate primarily to liquid-liquid interfaces, have been treated in detail by Guggenheim[27] and deBruyn[44] among others; and it has been shown that, when the curvature is not large in comparison with the thickness of the interface, the concept of interfacial (surface) tension becomes ambiguous. Consequently, for solids, the conditions must demand that the radius of curvature be large, or some account must be made of the variation in interfacial free energy with crystallographic orientation. In the particular case of precipitates or other situations where the interface exhibits noticeable curvature, the analysis might simply be approximated according to the scale of examination as shown in Fig. 1.12. Even in the case of Fig. 1.12, it can be noted that, if the interface thickness is taken to be as much as several atom layers (~10 Å), it is observed that $\Delta t/R \cong 0.007$ ($R \cong 1500$ Å).

In most engineering applications the energetics of a specific interface—such as that of one particular precipitate or dispersoid particle—will serve no useful purpose. Consequently, assumptions made concerning the planarity of an interface when treated statistically over a large number of observations or measurements in obtaining an average interfacial free energy, or observations or measurements of systematic changes in this average, will surely bear some metallurgical significance. It will become apparent in the succeeding chapters (particularly Chapter 2) that, when equilibrium situations are measured at interface intersections, for example, the planarity conditions become increasingly accurate as the scale of measurement becomes increasingly fine.

1.4 SUMMARY

An attempt has been made in this chapter to construct a perfectly general thermodynamic treatment of solid interface energetics applicable to stacking faults, twin boundaries, grain boundaries, phase boundaries, and the solid–vapor or solid–liquid surface in metals and alloys. The notable features of this treatment have been the establishment of generalized equations relating changes of interfacial free energy with temperature and composition, and the variation of interfacial free energy with crystallographic orientation implicit in the development of the Wulff criterion for the equilibration of crystal surface shape. The analysis has shown that vacancy concentration can contribute markedly to the interfacial free energy variations either by direct segregation to the interface, or by interaction with solute atoms.

It must be cautioned, once again, that any assumptions made regarding a specific interface in a metal or alloy must be carefully considered in establishing a final thermodynamic description. This is particularly true in the neglect of vacancy concentrations, and in the treatment of the interface as a dividing surface of negligible thickness; that is, $\Delta t\, dP \cong 0$. It must be also noted that the thermodynamic treatment presented has tacitly assumed the interface to be a plane section. This assumption is only approximately valid if the thickness of the interface, Δt_i, is much smaller than the local radius of curvature, R_i—that is, $\Delta t_i << R_i$; $i \equiv i$th interface or ith specific interfacial area.

We should note, in conclusion, that any thermodynamic description of a solid interface must of necessity be approximate where it proposes to deal specifically with interfacial free energy, particularly solid–solid interfaces. It is difficult to imagine a surface tension acting in a grain boundary, for example, which has exactly the same significance as that at a liquid surface or interface. Certainly, this approximation will hold at or near the melting point, but there is no information on the degree of error as the equilibrium temperature applicable to a solid-solid interface deviates from the melting point. For the most part, the thermodynamic relations obtained in the direct or modified Gibbs' form are accurate only if a reversible equilibrium has been established, a feature that certainly cannot be assured at low temperatures, particularly below the recrystallization temperature in metals and alloys where interfaces are generally static. In the succeeding treatment of interfacial free energy—particularly its proposed measurement—these assumptions will be tested and evaluated.

REFERENCES

1. R. Shuttleworth, *Proc. Phys. Soc., Ser. A*, **63**, 444 (1950).
2. J. W. Gibbs, "The Scientific Papers of J. Willard Gibbs," Vol. I, Dover Publications, New York (1961).
3. P. Curie, *Z. Kristallogr.*, **12**, 651 (1897).
4. G. Wulff, *Z. Kristallogr.*, **34**, 449 (1901).
5. V. K. Semenchenko, *Zh. Fiz. Khim.*, **19**, 350 (1945).
6. V. K. Semenchenko, "Surface Phenomena in Metals and Alloys," Addison-Wesley Publishing Co., Reading, Massachusetts, p. 272 (1962).
7. C. Herring, in "Structure and Properties of Solid Surfaces," R. Gomer and C. S. Smith (eds.), University of Chicago Press, Chicago, Chapter I (1953).
8. W. W. Mullins, in "Metal Surfaces," ASM, Metals Park, Ohio, Chapter 2 (1963).
9. N. Cabrera, *Surface Sci.*, **2**, 320 (1964).
10. C. A. Johnson, *Surface Sci.*, **3**, 429 (1965).
11. F. C. Frank, in "Metal Surfaces, Structure, Energetics, and Kinetics," ASM, Metals Park, Ohio, p. 1 (1963).
12. W. W. Mullins, in "Metal Surfaces, Structure, Energetics, and Kinetics," ASM, Metals Park, Ohio, p. 17 (1963).
13. W. A. Miller and G. A. Chadwick, *Proc. Roy. Soc., Ser. A*, **312**, 257 (1969).
14. R. L. Fullman, *Acta Met.*, **5**, 638 (1957).
15. W. L. Winterbottom, in "Surfaces and Interfaces I," Syracuse University Press, Syracuse, New York, p. 133 (1967).
16. N. A. Gjostein and W. L. Winterbottom, in "Fundamentals of Gas-Surface Interactions," Academic Press, New York, p. 42 (1967).
17. W. J. Dunning, in "The Solid-Gas Interface," Vol. I, Marcel Dekker, New York, p. 271 (1967).
18. J. Basterfield, W. A. Miller, and G. C. Weatherly, *Can. Met. Quart.*, **8**, 131 (1970).
19. B. E. Sundquist, *Acta Met.*, **12**, 67 (1964).
20. W. L. Winterbottom, *Acta Met.*, **15**, 303 (1967).
21. A. J. W. Moore, *J. Phys. Chem. Solids*, **23**, 907 (1962).
22. N. A. Gjostein, *Acta Met.*, **11**, 957 (1963).
23. W. M. Robertson and P. G. Shewmon, *J. Chem. Phys.*, **39**, 2330 (1963).
24. J. Burmeister, *J. Crystal Growth*, **11**, 131 (1971).
25. J. D. van der Wahls and G. Bakker, in "Handbuch der Experimentalphysik," Vol. 6, W. Wien and F. Horms (eds.), Akademische Verlagsgesellschaft, Leipzig (1928).
26. I. E. Verschaffelt, *Acad. Roy. Belg. Bull. Classe Sci.*, **22**, 402 (1936).
27. E. A. Guggenheim, *Trans. Faraday Soc.*, **36**, 397 (1940).
28. P. N. T. Unwin, G. W. Lorimer, and R. B. Nicholson, *Acta Met.*, **17**, 1363 (1969).

29. H. Suzuki, *J. Phys. Soc. Japan,* **17,** 322 (1962).
30. J. P. Hirth and J. Lothe, "Theory of Dislocations," McGraw-Hill Book Co., New York (1968).
31. J. P. Hirth, *Met. Trans.,* **1,** 2367 (1970).
32. D. McLean, "Grain Boundaries in Metals," Clarendon Press, Oxford, p. 148 (1957).
33. C. Kinoshita and T. Eguchi, *Acta Met.,* **20,** 45 (1972).
34. F. C. Fraikor and J. P. Hirth, *J. Appl. Phys.,* **38,** 2312 (1967).
35. J. W. Cahn and J. E. Hilliard, *Acta Met.,* **7,** 219 (1959).
36. H. P. Bonzel and H. B. Aaron, *Scripta Met.,* **5,** 1057 (1971).
37. M. C. Inman and H. R. Tipler, *Acta Met.,* **6,** 73 (1958).
38. M. A. Fortes and B. Ralph, *Acta Met.,* **15,** 707 (1967).
39. D. F. Stein, A. Joshi, and R. P. Laforce, *Trans. ASM,* **62,** 776 (1969).
40. M. C. Inman, D. McLean, and H. R. Tipler, *Proc. Roy. Soc., Ser. A,* **273,** 538 (1963).
41. W. R. Bitler, *Scripta Met.,* **5,** 1045 (1971).
42. K. T. Aust, R. E. Hanneman, P. Niessen, and J. H. Westbrook, *Acta Met.,* **16,** 291 (1968).
43. B. C. Shriver and M. Wuttig, *Acta Met.,* **20,** 1 (1972).
44. P. L. deBruyn, in "Fundamental Phenomena in the Materials Sciences," Vol. 3, Surface Phenomena, L. J. Bonis, P. C. deBruyn, and J. J. Duga (eds.), Plenum Press, New York, p. 20 (1966).
45. M. B. Kasen, *Acta Met.,* **20,** 105 (1972).
46. S. Ono and S. Kondo, in "Encyclopedia of Physics," Vol. X, S. Flügge (ed.), Springer-Verlag, Berlin, p. 144 (1960).
47. J. K. Tien and S. M. Copley, *Met. Trans.,* **2,** 543 (1971).
48. A. Kelly and R. B. Nicholson, *Progr. Mater. Sci.,* **10,** 151 (1963).
49. H. I. Aaronson, in "Decomposition of Austenite by Diffusional Processes," Wiley-Interscience, New York, p. 387 (1962).
50. J. W. Christian, "Transformations in Metals and Alloys," Pergamon Press, New York (1965).
51. L. E. Murr, "Electron Optical Applications in Materials Science," McGraw-Hill Book Co., New York (1970).

PROBLEMS

(1.1) What are the thermodynamic implications of the statement: Iridium melts at 2455°C? Write out the complete statement of thermodynamic equilibrium at the melting point in terms of the Gibbs free energy, on the one hand, and the Helmholtz free energy, on the other.

(1.2) Show, using simple two-dimensional sketches, the shapes of crystals where $\gamma_{(001)} = \gamma_{(111)}$; $\gamma_{(111)} = 1.1\gamma_{(001)}$. Make a three-dimensional sketch for $\gamma_{(011)} = 0.8\gamma_{(001)}$. Make a polar ($\gamma$) plot of the surface tension for the case where $\gamma_{(110)} = 2\gamma_{(100)}$, and explain why the equilibrium form of such a crystal should consist only of $\{100\}$ planes.

(1.3) At equilibrium, the pressure inside a liquid drop exceeds that outside the drop by an amount ΔP given by

$$\Delta P = 2\sigma/R$$

where σ is an isotropic surface stress, and R is the radius of the liquid drop. The increase in chemical potential which results from the pressure increase noted above is given by

$$\Delta\mu = (2\sigma/R) V_a$$

where V_a denotes the atomic volume. If the vapor is treated as an ideal gas for which the chemical potential is equal to

$$kT \ln P + C$$

where C is a constant, then the vapor pressure, P, in equilibrium with the spherical drop is related to that over a flat surface, P_0, by

$$\Delta\mu = kT \ln (P/P_0)$$

Derive the Gibbs-Thompson equation for a liquid drop having a form similar to that of Eq. (1.22), and show the conditions necessary for this equation to be identical to Eq. (1.22). Calculate the Wulff constant for a mercury droplet having a diameter of 100 Å at 40°C. Explain, with reference to the Gibbs-Thompson equation for a liquid drop derived, what will happen in a system of mercury drops having a distribution of sizes (a distribution of R values).

(1.4) J. W. Gibbs has stated that, when a system's mole number is treated as an independent variable in the same sense as its entropy and volume, "we shall have for the complete value of the differential of E,

$$dE = T\,dS - P\,dV + \mu\,dn$$

where

$$T \equiv (\partial E/\partial S)_{V,n}; \quad P \equiv -(\partial E/\partial V)_{S,n}; \quad \mu \equiv (\partial E/\partial n)_{S,n}"$$

Show that, when n is variable,

$$\mu = (\partial G/\partial n)_{T,P}$$

Show also the derivation of Eq. (1.23).

(1.5) Considering logical assumptions concerning the excess and mole fraction of component 2 at a binary alloy grain boundary, show mathematically, and with the use of graphs, the increase of grain boundary free energy with temperature. Assume Henry's law to be applicable in the form expressed by Eq. (1.41a).

(1.6) Describe an interstitial solid and compare it with a substitutional solid. Describe the possible interfacial thermodynamics at the interface formed by surface welding of one to the other.

(1.7) Discuss, as either a liquid-phase or a solid-phase system, a possible chemical situation in which a component would be systematically depleted from an interface.

(1.8) Considering the work reported in ref. 43, discuss the importance of vacancy-coupled processes and their effect on interfacial free energy using Eq. (1.48). Which specific interface is involved in the particular situation described in ref. 43?

(1.9) A large ThO_2 particle in a pure nickel matrix can be considered to approximate a planar two-phase system along small areas of its morphology where $\Delta t_i \ll R_i$. Write out the complete expression for the relative adsorption density of oxygen at the interface. Discuss the applicability of the planar interface thermodynamics for a particle size distribution ranging from 50 Å to 5000 Å (diameter).

(1.10) Consider the case of pure iron containing 0.15 w/o phosphorus (corresponding to an atomic fraction of 0.0027). If the solubility limit is known to be 0.02 (atomic fraction of phosphorus), calculate the interfacial excess of phosphorus (atoms/cm^2) to be expected for equilibrium segregation at grain boundaries whose mean interfacial free energy has been measured to be 600 ergs/cm^2 at $T = 1620°K$. Estimate the thickness of the adsorbed layer.

2
INTERFACIAL EQUILIBRIA

2.1 INTERFACE INTERSECTIONS; POLYCRYSTAL EQUILIBRIUM CONFIGURATIONS

It was pointed out in Section 1.1 that in crystalline solids the work involved in forming a surface and the work involved in stretching it are not, as in the case of liquids, identical [Eq. (1.4)]. This distinction was originally pointed out by Gibbs,[1] in arguing that no simple relationship is to be expected between stress and surface tension of solid crystals. Herring[2] has also indicated the difference between a liquid and a solid to lie in the fact that the number of surface atoms will decrease in comparison to those in the interior for a liquid drop experiencing an internal compression, whereas for a solid crystal the state of strain and the number of surface atoms are unrelated. It has been shown, however, in numerous diffraction studies of uncontaminated crystal surfaces, that the surface atoms are indeed arranged differently from those in the interior, and that the surface atom density is at least several percent less than that of the bulk.[3-5] Consequently, as argued in Section 1.1, it is not unreasonable, in considering interfacial equilibria at elevated temperatures, to assume the surface free energy and the surface tension to be similar, and to treat an interface as if a surface tension or surface stress were resolved within it as shown ideally in Fig. 1.5.

Herring[2] has described the problem of finding the equilibrium shape of a solid system—a polycrystalline solid—as being similar in principle to the problem of finding the equilibrium shape of a single-crystal particle; that is, its general solution is assumed to involve an extension of the Wulff theorem outlined in Section 1.2. In fact, it can be observed (as illustrated in Fig. 2.1) that an idealized polycrystalline solid can be constructed of a continuous array of regular polyhedra, ideally the tetrakaidecahedron shown previously in Fig. 1.3(b) and (d). Most polycrystalline metals and alloys do not, however, consist of regular polyhedra, but are composed of phase or grain structures having irregular shapes which topologically satisfy the Euler-Poincaré relation[6]

$$n_0 - n_1 + n_2 - n_3 + \ldots + (-1)^k n_k = 1 \qquad (2.1)$$

31

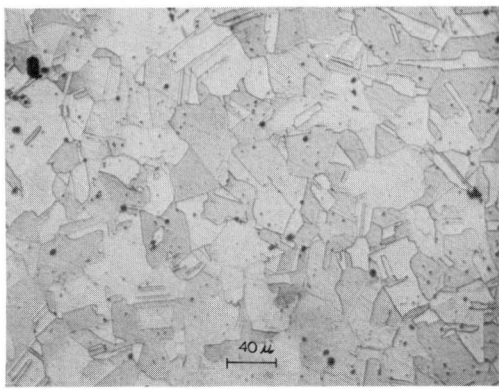

FIGURE 2.1 Polycrystalline metal grain structure. (Copper sample courtesy of D. Huo.)

where n_0, n_1, n_2, and n_3 denote the number of grain vertices, edges, faces, and volumes, respectively, in configurations of any number of dimensions and features $(k + 1)$, where $k = 0, 1, 2$. In a three-dimensional system (a bulk polycrystalline metal or alloy) the number of grains or volume phases, n_3, is given by

$$n_3 = \sum_{\Delta=3}^{9} (6 - \Delta) n_2(\Delta) - 1 \qquad (2.2)$$

where Δ represents the number of edges of a face of a polyhedron, and $n_2(\Delta)$ is the number of two-dimensional polygons (grain faces) having Δ edges. In typical metals and alloys, the average value of Δ is slightly greater than 5. The average number of grain faces is about 13; and the average number of vertices, 23. For a two-dimensional microstructure as commonly observed by optical metallography on a polished metal or alloy surface, Eq. (2.2) becomes simply

$$\sum_{\Delta=3}^{9} (6 - \Delta) n_2(\Delta) = 0 \qquad (2.3)$$

This feature is illustrated in Fig. 2.1.

Since the work of Harker and Parker[7] and the classic treatment of metallurgical microstructures by Smith,[8] the interfaces of grains or phases have been considered to possess a surface free energy or an associated surface tension which acts in all directions in the plane of the interface (Fig. 1.5), exerting a "pull" upon the grain edges, and opposed to that of other interfaces along their mutual line of intersection. The equilibrium configurations that develop in a three-dimensional polycrystalline material can therefore be treated as a balancing of surface free energy vectors resolved in the plane of the interfaces, meeting at a common point of reference.

INTERFACIAL EQUILIBRIA

Herring[2] has treated the general case of intersecting interfaces which applies to solid–solid intersections in polycrystalline solids, and to the intersection of internal solid interfaces with a free surface (solid–vapor interface), by considering the equilibrium condition at a triple junction of resolved surface free energy vectors (Fig. 2.2) to be expressed by an equation of the form

$$\sum_{i=1}^{3} [\gamma_i \sigma_i + (\sigma_i \times z)(\partial \gamma_i / \partial \Omega_i')] = 0 \tag{2.4}$$

where γ_1, γ_2, and γ_3 are the specific surface free energies resolved in the associated interfaces as vectors (σ_i), z is a unit vector coincident with the line of intersection of the interfaces (directed into the page at 0 in Fig. 2.2), and $\partial \gamma_i / \partial \Omega_i'$ accounts for the dependence of the ith interfacial free energy on the orientation of its surface (an implicit inclusion of the Wulff condition), which is presumed to be related to Ω or some other geometric or crystallographic parameter. The terms in $\partial \gamma_i / \partial \Omega_i'$ account for the fact that each interface will ideally strive to attain a minimum energy configuration by rotating to some orientation of lower γ_i, with an effort measured by a so-called torque, $\partial \gamma_i / \partial \Omega_i'$, per unit area of interface. As a conse-

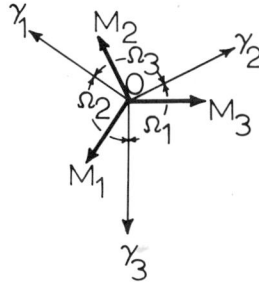

FIGURE 2.2 Intersection equilibrium of three general interfaces. The heavy vectors denote the associated torques, M_j, assuming a right-hand convention in Eq. (2.4).

quence of Eq. (2.4) any boundary whose orientation is not that of a cusp in the polar γ-plot (Fig. 1.2) will generally be a curved surface. Conversely, any plane boundary or interface might be expected to lie at or near a cusp in the corresponding γ-plot—that is, to be a low-energy boundary.

2.2 GRAIN BOUNDARY EQUILIBRIA

In many metallurgical systems it is tacitly assumed that interfacial free energy is negligibly influenced by crystallographic orientation, and that interfacial intersections represent the true equilibrium conditions—that is, that the angles between the interface traces at the intersection are the dihedral angles. This situation is ideally maintained only when the interface planes meet perpendicular to a reference surface (so that the intersection line is a surface normal) and the reference surface coincides with the phase or grain surfaces composing the intersecting system. Such a two-dimensional grain structure is only likely to equilibrate at high temperatures, as indicated in Fig. 2.3. For the assumption that $\partial \gamma_i/\partial \Omega_i$ is zero in Eq. (2.4) (which is particularly valid at or near the melting point because of the relaxation in rigid crystal periodicity) as depicted in Fig. 2.3(b), the equilibrium condition becomes simply the Dupré equation:

$$\gamma_1/\sin \Omega_2 = \gamma_2/\sin \Omega_3 = \gamma_3/\sin \Omega_1 \tag{2.5}$$

where, if $\gamma_1 = \gamma_2 = \gamma_3 \equiv \gamma_{gb}$ (a constant grain boundary free energy), the dihedral angles, Ω_i, are 120°. Unequal dihedral angles in the ideal situation of Fig. 2.3(a) must therefore indicate a difference in the specific grain boundary free energies.

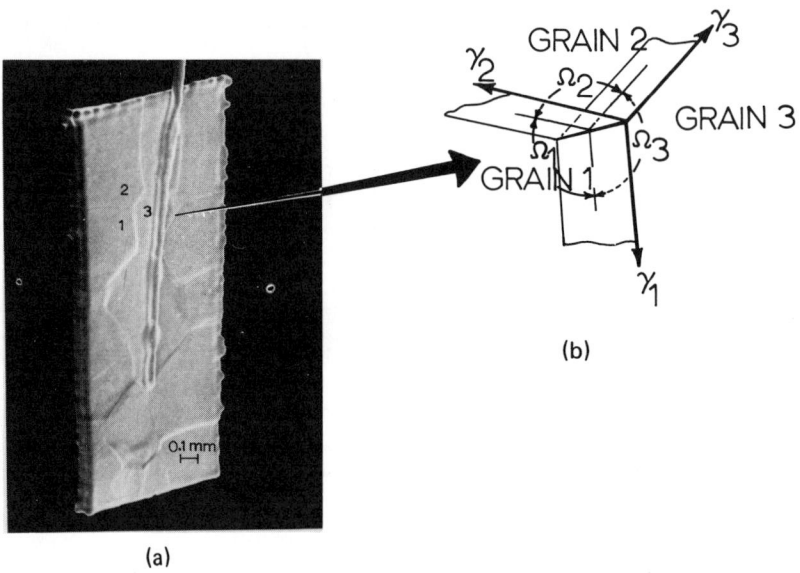

FIGURE 2.3 Two-dimensional grain boundary structures at high temperature. (a) Scanning electron micrograph of a small 304 stainless-steel sheet sample suspended in a helium atmosphere for 50 hours at 1370°C (~0.95T_m, where T_m is the absolute melting temperature). Grain boundaries are observed to penetrate normal to the sheet surfaces. (b) Simple energy balance at a junction of grains in (a).

It should be noted that any interfacial equilibrium situation, particularly where it relates specifically to polycrystalline grain structure, is a conditional or pseudo-

INTERFACIAL EQUILIBRIA

equilibrium. True equilibrium as defined by Eq. (1.5) occurs only when the solid becomes a single crystal with the annealing out of all the internal interfaces.

2.2.1 Grain Boundary Intersections in Random Sections

In the general case of grain boundaries in random solid sections or in metal or alloy sheets or plates, the equilibrium conditions may not approximate those of a two-dimensional grain structure, and the grain boundary planes will not be perpendicular to the specimen or section surfaces as shown in Fig. 2.4. Since the angles

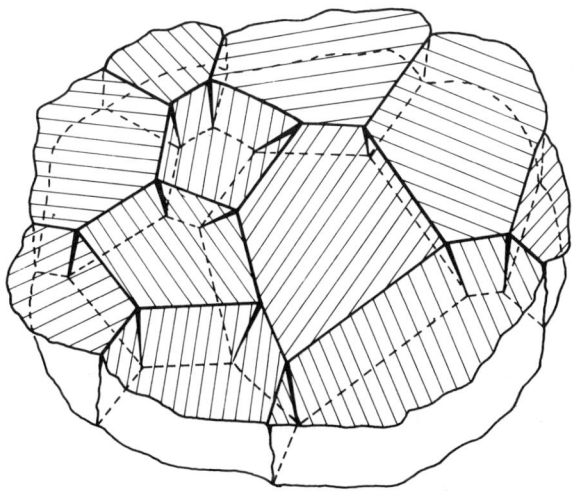

FIGURE 2.4   Grain structure in a random slice or sheet of bulk metal or alloy. (From Murr.[9])

measured on the specimen surface by optical metallography will not in general be the true dihedral angles, a means must be devised to measure the approximate angle of inclination of the grain boundaries (treated as planes if they penetrate the slice from surface to surface as in Fig. 2.4), or to confirm that the inclination angles are indeed approaching 90°. One method for accomplishing this is to treat the section geometrically with respect to some reference system, as outlined, for example, in the work of Fullman.[10] As a simple illustration, consider that a random metallographic section polished on both sides is indexed near a triple junction in a jig with indentation markers as shown in Fig. 2.5. The distance, d, perpendicular to a particular boundary trace composing the triple junction is then measured, and the section is inverted. The distance, d', perpendicular to the boundary trace on the opposite face is then measured. Treating the boundary as a planar interface, and knowing the section thickness, then allows the inclination angle, θ, to be approxi-

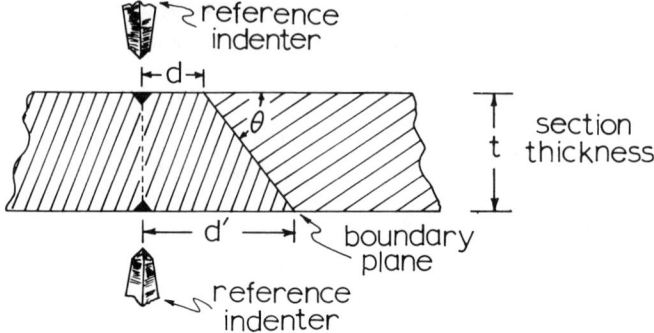

FIGURE 2.5   Measurement of boundary plane inclination in a solid section which it penetrates.

mately determined from

$$\theta = \tan^{-1}(t/w); \quad w = (d' - d) \tag{2.6}$$

The true dihedral angles of intersection can then be determined by measuring the trace intersection angles, w_1, w_2, and w_3, and considering the spherical trigonometric cosine law in the general (cyclic) form:

$$\cos \Omega_1 = -\cos \theta_1 \cos \theta_2 + \cos w_1 \sin \theta_1 \sin \theta_2 \tag{2.7}$$

In studies of opaque sections where the grain or phase size is too small to penetrate from surface to surface, it is usually not possible to determine interfacial geometries with any accuracy. However, techniques of stereometric microscopy (quantitative metallography) can be employed in obtaining information on the spatial structure by analyzing the two-dimensional microstructures.[11,12]

The techniques of thin-film transmission electron microscopy[13-16] are unprecedented in their accuracy in analyzing grain boundary equilibria because of the fact that electropolished thin sections are in many cases ideally random slices of a three-dimensional bulk solid[9] and are transparent, allowing the internal structure and associated geometry (Fig. 2.4) to be observed directly. In addition, although grain surface orientations can be determined by optical metallography using stereographic techniques,[10,17,18] the selected-area electron diffraction facility associated with transmission electron microscopy permits the grain or phase orientations to be obtained, and in many cases the crystallographic orientation of the specimen surface can be uniquely established.

Figure 2.6 illustrates the determination of the true intersection equilibria from the associated projection geometry for a thin metal film in the electron microscope.

INTERFACIAL EQUILIBRIA 37

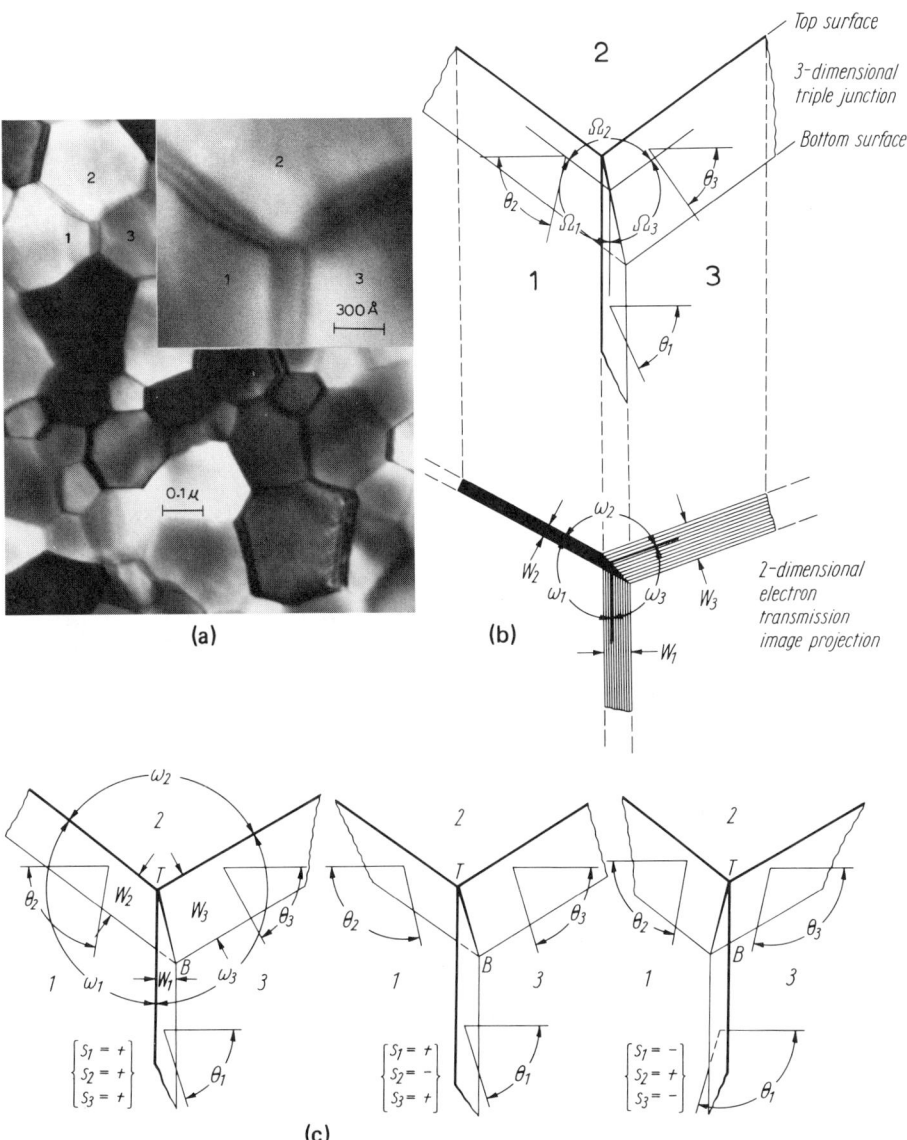

FIGURE 2.6 Triple junctions and junction conventions for grain boundary equilibria in thin electron-transparent metal films. (a) Vapor-deposited tin film ~1200 Å thick. Insert shows enlarged triple junction. (b) Schematic view of image projection of triple junction in (a). (c) Junction conventions for projection geometries in the electron microscope. The top surface of the specimen is denoted by heavy lines along T. B denotes bottom surface. S_1, S_2, and S_3 simply denote the first (+1) or second (−1) quadrant conventions on θ_1, θ_2, and θ_3, respectively, viewed with reference to grain 1. [(b) and (c) are from Murr.[19]]

To facilitate the calculations of the intersection geometry at a triple junction, a system of conventions is established as illustrated in Fig. 2.6(c) to systematically account for sign changes as a result of variations in the inclination angles of the boundary planes composing the junction.[9,19] Although Fig. 2.6(c) illustrates only three of the eight possible intersection geometries, the concept is basically indicated in the sign convention for the boundary inclinations as designated S_1, S_2, and S_3. Note that, in Fig. 2.6(c), $S_1, S_2, S_3 = +1, -1, +1$ and $-1, +1, -1$ result simply by viewing the same triple junction from opposite sides of the foil section. Considering Eq. (2.7), the appropriate expression for the true dihedral angles can be determined for a general triple junction in a thin section in the electron microscope from

$$\Omega_1 = \cos^{-1}(S_1 S_2 \cos\theta_1 \cos\theta_2 + \cos\omega_1 \sin\theta_1 \sin\theta_2)$$
$$\Omega_2 = \cos^{-1}(-S_2 S_3 \cos\theta_2 \cos\theta_3 + \cos\omega_2 \sin\theta_2 \sin\theta_3) \quad (2.8)$$
$$\Omega_3 = \cos^{-1}(-S_1 S_3 \cos\theta_1 \cos\theta_3 + \cos\omega_3 \sin\theta_1 \sin\theta_3)$$

where

$$\theta_1 = \tan^{-1}(t/w_1); \quad \theta_2 = \tan^{-1}(t/w_2); \quad \theta_3 = \tan^{-1}(t/w_3) \quad (2.9)$$

and t is the section thickness in the vicinity of the junction; ω_1, ω_2, and ω_3 are the angles measured in the projected image, and w_1, w_2, and w_3 are the projected boundary widths.

In cases where thin films are made by vapor deposition or a similar growth process where the section thickness is uniform, the thickness, t, can be determined by several techniques, including optical interferometry.[20] However, in electropolished sections (which are statistically more meaningful in examining bulk materials) the thickness varies considerably from point to point, and it is necessary for an accurate analysis to determine the local specimen thickness at or near the junction system of interest. This can be accomplished by the analysis of diffraction contrast phenomena such as extinction fringes at boundary planes,[14,16] or by the direct measurement of the projection widths of interfaces of known crystallography (and inclination) with respect to that of the surface orientation. A typical example of this latter feature is the measurement of twin boundary, stacking-fault, or slip trace projections in face-centered cubic metal and alloy films[15,16] as illustrated in Fig. 2.7. In many cases, stacking-fault or dislocation slip (and cross-slip) traces can be generated locally at an interface (grain boundary) plane by focusing the electron beam onto the area. Thermal and contamination stresses then cause dislocations to propagate within the associated grains or to emanate from grain boundary sources, thereby creating a diffraction contrast effect which projects the trace.[9,14,16] Since, for cubic metals and alloys (fcc and bcc in particular) dislocation slip traces and related planar crystallographic features will occur in identifiable (hkl) planes, a determination of the surface orientation (HKL) will allow the section thickness to

INTERFACIAL EQUILIBRIA

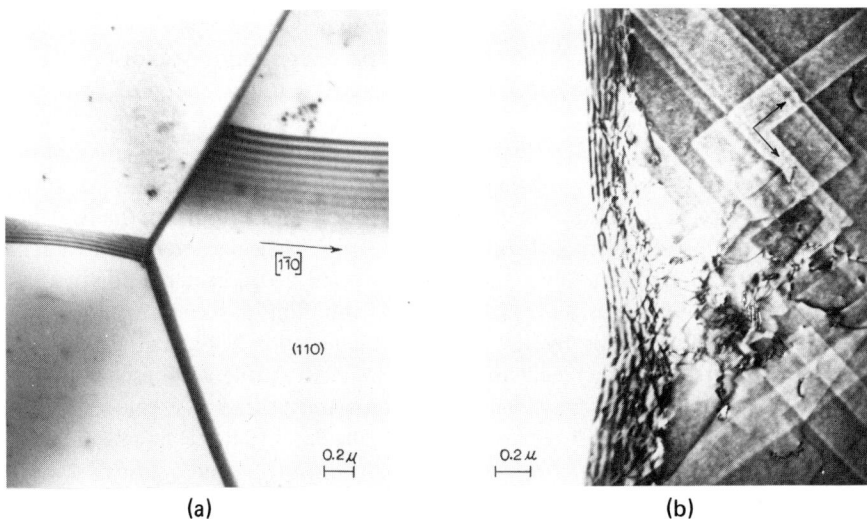

FIGURE 2.7 Planar crystallographic (111) features in fcc thin sections from which the local thickness can be determined in order to measure grain boundary-plane inclinations. (*a*) Twin boundary plane in stainless steel near a triple junction. (After Murr.[16]) (*b*) Dislocation slip and cross-slip traces in pure aluminum generated by the action of the electron beam in order to measure the local boundary inclination.

be determined from

$$t = W_T \tan \theta_T \qquad (2.10)$$

where W_T and θ_T are the projected width and inclination angle of the twin, stacking fault, or slip plane:

$$\cos \theta_T = \frac{hH + kK + lL}{\sqrt{h^2 + k^2 + l^2}\sqrt{H^2 + K^2 + L^2}} \qquad (2.11)$$

For fcc metals and alloys, $\{hkl\} = \{111\}$. Consequently, simply determining the associated grain surface orientation and the trace direction [Fig. 2.7(*a*)] will allow θ_T to be uniquely determined. In most instances for cubic materials, θ_T can be obtained from tables.[16]

2.2.2 Characterization of a Grain Boundary Section in Thin Films

In the generalized case of a grain boundary or any segment of grain boundary as a plane section, the interface can be treated as a dividing surface separating two similar crystals of arbitrary (relative) orientation joined together at the interface. To completely describe a general bicrystal, five parameters or degrees of freedom are necessary: Two degrees of freedom are associated with the amount of misorientation, and three degrees of freedom are associated with the direction of the grain boundary normal. Goux[20] has described a graphical method for determining the five bicrystal parameters utilizing a Laue back-reflection x-ray technique, and Lange[21] has given a generalized mathematical analysis for determining the five degrees of freedom of a bicrystal section without any particular reference to a grain boundary structural model. Murr *et al.*[22] have treated the characterization of a grain boundary interface as a bicrystal section in thin films in the electron microscope as depicted schematically in Fig. 2.8. In the notation of Fig. 2.8, θ denotes the (positive, first-quadrant) inclination of the boundary plane with respect to grain surfaces $(hkl)_A$ and $(hkl)_B$ coincident with the specimen surfaces, Θ characterizes the misorientation of grain A relative to B as an angular difference along a direction common to the orientations of A and B [$(hkl)_A$ and $(hkl)_B$], and ϕ positions the trace of the interface plane as it intersects the surfaces with reference to the bisector of the misorientation angle, Θ ($\Theta/2$, in Fig. 2.8); ϕ is commonly referred to as the asymmetry angle.

In the perfectly general case, $(hkl)_A \neq (hkl)_B$, and there are five degrees of freedom characterizing a boundary plane section, as shown, for example, in Fig. 2.7(b). However, for a situation where $(hkl)_A = (hkl)_B$, the boundary plane can be characterized by only three parameters: θ, Θ, and ϕ. Since ϕ is defined with respect to the bisector of Θ, and since θ is related to Ω at a triple junction [Eq. (2.8)], the torques implicit in Eq. (2.4) as defined in terms of Ω' can, in principle, be related to any of these parameters. However, when $\theta = 90°$, the torque associ-

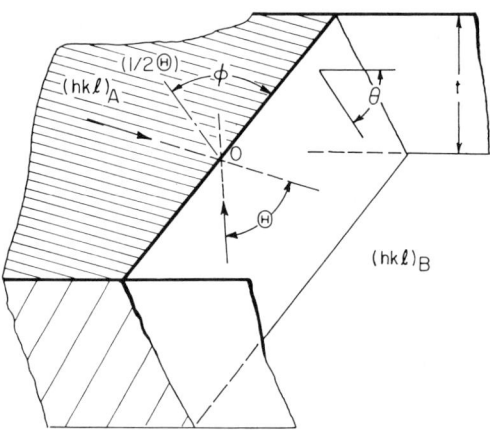

FIGURE 2.8   Geometrical conventions for characterizing a grain boundary plane section in a thin electron-transparent foil. (From Murr *et al.*[22])

ated with any boundary can be uniquely characterized in terms of the misorientation, Θ, when $(hkl)_A = (hkl)_B$.

When crystals of the same orientation are joined in such a way that the interface (grain boundary) plane is perpendicular to the surfaces of the crystals, and the boundary plane is symmetrically disposed with respect to crystals 1 and 2 (or grains A and B)—that is, $\phi = 0°$ or $90°$—then the grain boundary is uniquely described by only one degree of freedom, Θ (the misorientation parameter). Such idealized systems are rarely observed in bulk metals or alloys except where exaggerated grain structures may develop through extremely long anneals [as in Fig. 2.3(a)], and symmetry conditions may arise by coincidence. Several interesting features related to boundary structure can be investigated in such idealized systems; they will be discussed in detail in Chapter 4.

2.2.3 Concept of Grain Boundary Torque

The energy derivatives, or the so-called interfacial torques in Herring's general equilibrium equation [Eq. (2.4)], are considered to represent a tendency of a boundary to rotate to a minimum energy position; or conversely, in the case of a boundary in a low-energy position, the torque opposes any tendency to deviate from this position. Because, as we shall see in a later section, the solution of Eq. (2.4) is not possible without certain assumptions, the derivative terms or torques cannot be described unambiguously. Since the torque represents an anisotropy of grain boundary energy, it should be possible to detect torque components or resultant torques resolved along a boundary by observations of deviations in the asymmetry or inclination of a boundary. Furthermore, the occurrence of a resolved torque at an internal (solid–solid) interface must, in an equilibrium situation, be balanced by some appropriate torque in an adjacent boundary region. If this were not the case, grain growth at elevated temperatures would be characterized by prominent grain rotations as a result of unbalanced torques; this is generally not observed. Where an unbalanced torque does exist, the associated grain structure will be unstable and anneal out easily.

Hartt et al.[23] and Basterfield et al.[24] have shown evidence for the existence of anisotropic grain boundary energies as serrated (faceted) grain boundaries in zinc, and Weinberg[25] has observed rotations of whole triple junction configurations in zinc tricrystals grown with parallel basal planes. Figure 2.9 illustrates a similar phenomenon in a thin foil of austenitic stainless steel observed in the electron microscope. The unique feature in Fig. 2.9 is that the grain boundary separates grains of the same orientation; consequently variations occurring in the boundary plane are characterized by a change in boundary asymmetry (ϕ in Fig. 2.8). It is to be noted in Fig. 2.9 that, if a resultant torque or net effective torque (NET) is resolved at each boundary serration as a means to balance the resolved interfacial free energies, the NET's occur in a balanced array, which is a necessary condition for an equilibrium configuration. If the serrations in the boundary of Fig. 2.9 are considered as a special case of an equilibrated junction with a symmetrically

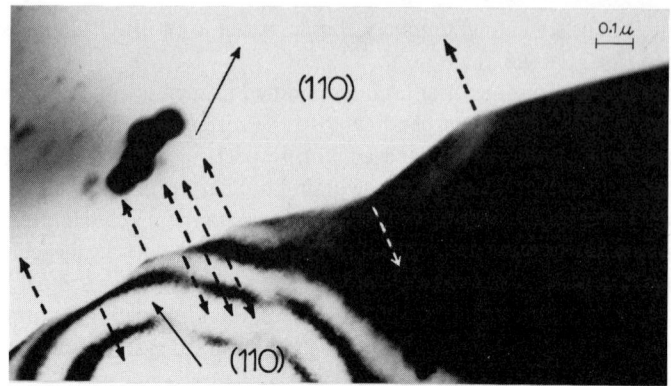

FIGURE 2.9 Grain boundary serrations in a grain boundary in 304 stainless steel, annealed at 1000°C. The arrows in the grains separated by the boundary indicate the associated [110] directions. The heavy broken arrows suggest resultant or net effective torques which act to "equilibrate" the serrations.

resolved NET at the serrations, Eq. (2.4) could be written in the simplified form

$$\Sigma M = 2\gamma_{gb} \cos \Omega/2 \qquad (2.12)$$

where ΣM represents the NET (heavy broken arrows in Fig. 2.9), γ_{gb} is the average grain boundary free energy, and Ω is the dihedral angle at the serration (angle opposite the resolved torque). A measurement of the dihedral angles associated with the serrations in Fig. 2.9 shows that $\Sigma M/\gamma_{gb}$ varies from about 0.30 to 0.50. Similar ratios of grain boundary torque to average grain boundary free energy have been determined by Hess[26] and by Miller and Williams[27] in aluminum rod and sheet specimens, respectively, and have been found to range from 0 to 0.30, with a mean value of about 0.15. G. H. Bishop, W. H. Hartt, and G. A. Bruggeman [*Acta Met.*, **19**, 37 (1971)] have observed similar faceting for grain boundaries in zinc where $\Sigma M/\gamma_{gb}$ varied between 0.1 and 0.5. Horylev and Murr,[28] in addition, have measured $\Sigma M/\gamma_{gb}$ associated with twin grain boundary intersections (see Section 2.3.2) in 304 stainless steel and found a mean value of about 0.10.

It can be observed in retrospect that, for a torque to exist, a consistent change in boundary morphology with respect to some specific or implicit reference must occur. Consequently, it is difficult to believe a torque to be associated with any particular grain boundary at a triple junction unless some recognizable variation in interfacial morphology occurs, and the assumption that torques are negligible at triple junctions is generally valid when the interfaces meeting at the junction are essentially planar. It will be shown in Chapter 4 that this generally is not true on an atomic level.

2.2.4 Grain Boundary Intersections with a Free Surface

When metals or alloys are annealed at sufficiently high temperature in vacuum or inert gas atmosphere as shown in Fig. 2.3(a), the grain boundaries will equilibrate with the surface and form a groove at the point of intersection of the grain boundary with the surface. Chalmers et al.[29] were among the first to describe this thermal grooving at the intersection of a solid-solid interface with the free (solid-vapor) surface as a condition of interfacial free energy minimization. Mullins[30,31] has treated the conditions of groove formation theoretically on the basis that solid surface energy is independent of crystallographic orientation. Gjostein[32] and Robertson[33] determined the mechanisms of grain boundary grooving based on Mullins' theories for copper and liquid lead, respectively; these mechanisms have been shown to be essentially the same in chromium, molybdenum, and tungsten as well as their alloys with rhenium in the work of Allen.[34]

The four possible mechanisms that can contribute to the dimensions and shape of groove profiles can be summarized as (1) evaporation-condensation, (2) volume diffusion in the solid, (3) volume diffusion in the vapor (or liquid) phase at the solid surface with which the grain boundary equilibrates, and (4) surface diffusion. Figure 2.10 illustrates schematically the equilibrium groove profiles which result by the transport of matter at the intersection by these mechanisms. Since it has been shown that torques are negligible in the equilibration of grain boundary-surface intersections, and since we may assume that the symmetry of intersection is as depicted in Fig. 2.10, the balance of the associated interfacial free energies is expressed by

$$\gamma_{gb} = 2F_S \cos(\Omega_S/2) \tag{2.13}$$

where γ_{gb} is the grain boundary free energy, F_S is the surface free energy at the

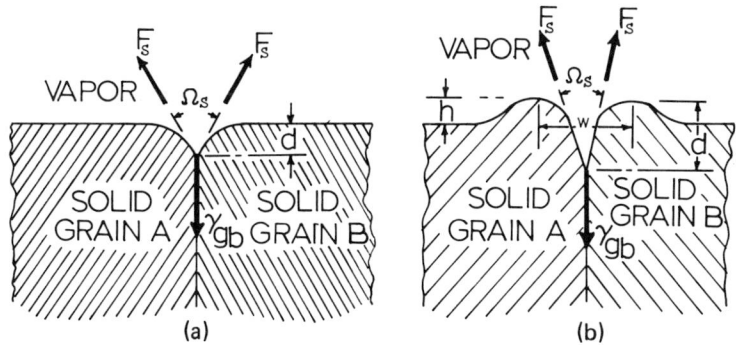

FIGURE 2.10 Thermal groove profiles for grain boundary-surface equilibria. (a) Evaporation-condensation-induced profile. (b) Diffusion-induced profile.

solid–vapor surface (in the case of the solid–liquid interface, F_S should be replaced by the notation γ_{SL}), and Ω_S is the dihedral angle defining the groove (Fig. 2.10).

For evaporation–condensation, the groove profile [Fig. 2.10(a)] is governed by an equation of the form[31]

$$d = 1.13 \sqrt{At} \cot (\Omega_S/2) \qquad (2.14)$$

where t is the time of thermal anneal, and

$$A = P_0 F_S V_\alpha^2 / (2\pi M)^{\frac{1}{2}} (kT)^{\frac{3}{2}} \qquad (2.14a)$$

with P_0 = vapor pressure of the flat surface.

V_α = atomic volume.

M = molecular weight of metal or alloy vapor.

$T \equiv °K$.

Conversely, for volume diffusion, the groove profile [Fig. (2.10(b))] is governed by[31]

$$\cot (\Omega_S/2) = 4.95 \, (d/w); \quad (h/d) \cong 0.15 \qquad (2.15)$$

and

$$w = 5.0 \, (A't)^{\frac{1}{3}} \qquad (2.16)$$

where

$$A' = C_0 V_\alpha^2 F_S / kT \qquad (2.16a)$$

and C_0 is the equilibrium concentration of diffusing metal atoms which, for volume diffusion through the vapor phase, is equal to P_0/kT. The corresponding theoretical relations for surface diffusion governing the groove profile in Fig. 2.10(b) are

$$\cot (\Omega_S/2) = 4.73(d/w); \quad (h/d) \cong 0.21 \qquad (2.17)$$

and

$$w = 4.6 \, (A''t)^{\frac{1}{4}} \qquad (2.18)$$

where

$$A'' = D_S \, V_\alpha^{\frac{4}{3}} \, F_S/kT \tag{2.18a}$$

and D_S is the surface diffusion coefficient.

Several common examples of grain boundary-surface groove equilibria and techniques for measuring and observing groove profiles are illustrated in Fig. 2.11. Groove geometries can be measured interferometrically using an optical interference microscope as described by numerous researchers[35-37] and illustrated in Fig. 2.11(a) and (b). It has been shown [S. Amelinckx, N. Bennendijk, and W. Dekeyser, *Physica,* **19**, 1193 (1953)] that, for interference patterns such as those in Fig. 2.11(a) and (b),

$$\frac{\Omega_S}{2} \simeq \tan^{-1} \frac{2X}{\lambda M \tan(\omega/2)} \tag{2.13a}$$

where X is the fringe spacing, λ is the wavelength, M the linear magnification, and ω the groove angle in the fringe pattern. More accurate measurements are possible by the direct observation of groove geometries in the scanning[38] or transmission[38-41] electron microscopes as illustrated in Fig. 2.11(c)-(g). It is possible, as shown in Fig. 2.11(f), to align wire samples in the specimen holder of the transmission electron microscope with the wire axis perpendicular to the electron beam. This technique, originated by Inman and Tipler,[39] has the particular advantage that the projection shadowgraph of a grain boundary groove is the true or idealized geometry as depicted schematically in Fig. 2.10. In this mode of analysis, the groove angle can be measured at very high magnifications in order to ensure that the groove root is being considered at the point where the resolved grain boundary energy and surface free energy vectors meet.

It must be pointed out that the equilibrium conditions described by Eqs. (2.13) through (2.18) are valid only for symmetrical intersections, or approximations of such situations. As a consequence, annealed bulk metals and alloys may not exhibit symmetrical intersections where thermal equilibrium has not been established, in which case, additionally, large unbalanced torques may exist. Since Drechsler and Nicholas[42] have found that surface energy anisotropy decreases with increasing temperature, the neglect of torque contributions would be expected to be most valid near the melting point.

2.3 TWIN BOUNDARY EQUILIBRIA IN FCC METALS AND ALLOYS

It is well known that annealing twins in fcc metals and alloys have an associated interfacial free energy which is very small by comparison with a grain boundary of the high-angle variety—usually characterized by a misorientation, Θ (Fig. 2.8), of greater than roughly 30°; these features are treated in detail in Chapter 3. In

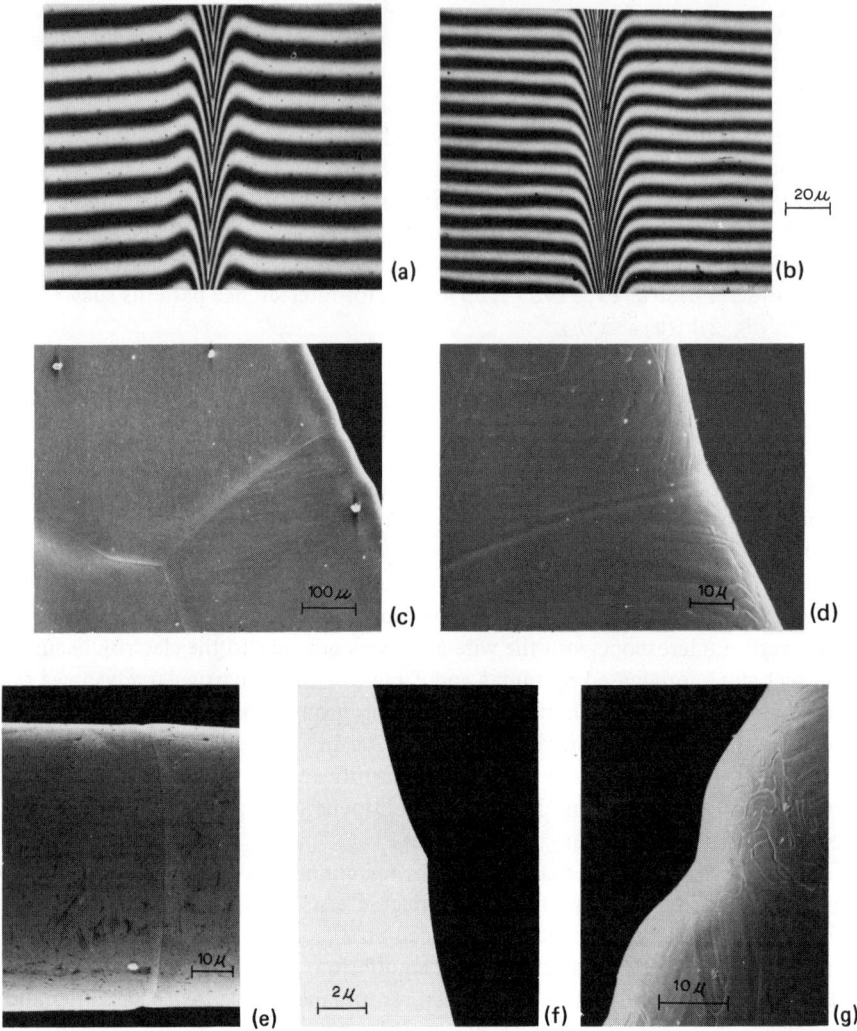

FIGURE 2.11 Grain boundary grooves formed at metal and alloy surfaces. (a) Optical interferograms of a grain boundary groove in a chromium surface during annealing for 6 hours at 1550°C in argon. (b) Same as (a) except annealed in vacuum. (c) Austenitic stainless-steel plate [left portion of Fig. 2.3(a)] observed in the scanning electron microscope. (d) Enlarged view of (c) showing well-defined groove at the thermally rounded edge. (e) Grain boundary traversing nickel wire annealed 40 hours at 1060°C in helium. (f) Shadowgraph of grain boundary groove in stainless-steel wire annealed as in (e), observed in the transmission electron microscope with the electron beam perpendicular to the wire axis. (g) Scanning electron micrograph of groove profile in stainless-steel wire similar to (f) after 50-hour anneal at 1360°C in helium. [(a) and (b) are courtesy of B. C. Allen.]

addition, the intersection of a twin boundary with another interface—for example, a grain boundary or the free surface—would be expected to give rise to grain boundary torques at the intersection because of the fact that, depending upon the crystallography of the twinning operation, the propagation of a single twin boundary or a twin band into a grain boundary, or the encounter with a free surface, will cause an abrupt orientation change at the junction of a twin boundary with another interface. This is tantamount to saying that an interface of energy γ_0 before encountering a twin boundary will have an energy γ_0 on one side of the intersection and an energy γ' on the other. As a consequence, the interfacial anisotropy could be characterized by $\Delta\gamma_I = \gamma_0 - \gamma' \neq 0$ or $\gamma_0/\gamma' \neq 1$.

Twin boundaries in fcc materials are of particular interest because for coherent boundaries the twin interface—that is, the dividing surface separating the matrix and its twin—coincides with the $\{111\}$ plane; it is a true planar interface. Non-coherent (non-$\{111\}$ plane) twin interfaces in fcc metals and alloys are also coincident with a crystal plane which may be unique to either the matrix or twin lattices, or a plane common to both lattices. These boundaries, as will be observed in Chapter 3, have associated surface free energies greater than the coherent twin boundaries, but generally less than the average high-angle grain boundary free energy.

Twins in fcc lattices, whether possessing the coherent or noncoherent boundary characteristic, are sometimes classified as first order, second order, or n order, depending on the number of repeated twinning operations about different $\{111\}$ planes, the order of the twin indicating the number of twinning operations. Friedel[43] showed that twins of the first order ($\{111\}$ annealing twins) possess a coincidence lattice in which one lattice site in every three is common to both the matrix grain and the twin grain separated by the twin interface. Friedel used a twin index, Σ, to express the number of lattice sites to coincidence lattice sites, with $\Sigma = 3$ for the first-order twins. Consequently, crystals in a higher-order twin relationship are considered to possess coincidence lattices whose twin index is given by $\Sigma = (3)^n$, where n is the number of twinning operations. The coincidence lattice notation for general crystal interfaces—that is, grain boundaries—has been treated by Kronberg and Wilson[44] and modified by Brandon,[45] Pumphrey and Bowkett,[46] and others. We shall defer until Chapter 4 a detailed treatment of the coincidence lattice concept as it relates to grain boundary structure in general.

2.3.1 Twin Boundary-Free Surface Intersections

In dealing with the concept of the γ-plot in Section 1.2, it was phenomenologically demonstrated that in the case of the solid–vapor interface it is expected that the specific surface free energy will be orientation-dependent; this feature has been discussed in detail by Herring.[47] In addition, as we discussed above, the occurrence of appreciable torques can be expected at certain twin boundary intersections with a free surface or a grain boundary because of the systematic crystallographic transformation which must occur between the matrix (reference) grain and its twin. While there have been no direct determinations of the orientation dependence of the absolute magnitude of interfacial free energies, relative values of γ_I for different crystal planes can in principle be obtained from an analysis of the dihedral angles formed at the intersection of a grain boundary with the solid-vapor (free surface) interface as illustrated in Fig. 2.11.

Mykura[48] first reported on the intersection of twin boundaries with a free surface in a pure metal, and demonstrated that the slope of the γ-plot can be determined by applying Herring's equation [Eq. (2.4)] for the equilibrium angles at twin boundary-surface junctions. In dealing with the specific case of first-order annealing twin bands intersecting the free surface, Mykura[49] considered the two junctions as a system in equilibrium, and the corresponding surface groove profiles as shown schematically in Fig. 2.12. The associated equilibrium equations for this

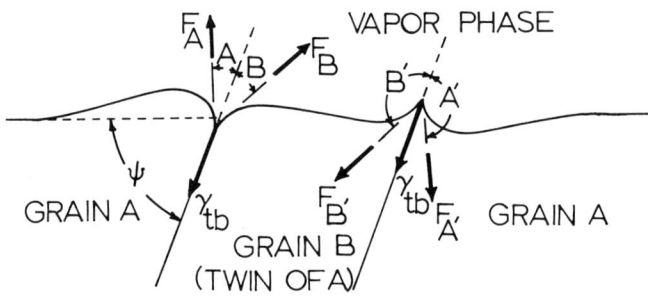

FIGURE 2.12   Surface profile schematic for twin band-free surface intersection system.

system are written as follows (with reference to Fig. 2.12)[2]:

$$\gamma_{tb} = F_A \cos A + F_B \cos B - (\partial F_A/\partial A) \sin A - (\partial F_B/\partial B) \sin B$$

$$\gamma_{tb} = F_{A'} \cos A' + F_{B'} \cos B' + (\partial F_{A'}/\partial A') \sin A'$$

$$+ (\partial F_{B'}/\partial B') \sin B' \qquad (2.19)$$

It is readily recognized in Eq. (2.19) that each equilibrium equation contains five unknown quantities—$\gamma_{tb}, F_A, F_B, \partial F_A/\partial A$, and $\partial F_B/\partial B$ in the first, etc., assuming of course that the angles A and B as well as A' and B' can be accurately measured from the associated groove profiles. It is further apparent from Fig. 2.12 that the groove and ridge intersection profiles unique to twin band intersections with a free surface as illustrated in Fig. 2.13 essentially demand the resolution of a net effective torque in the twin boundary. The net effective torques have equal magnitudes

INTERFACIAL EQUILIBRIA 49

FIGURE 2.13 Interferogram of twin boundary groove and ridge configuration for twin band intersecting a free surface in pure gold annealed for several days in hydrogen at 1030°C (0.27 μ/fringe). (Courtesy of N. A. Gjostein.)

but an opposite sense at the two junctions, similar in principle to the scheme previously depicted in Fig. 2.9. Mykura pointed out that a unique torque value can be obtained under certain circumstances from the symmetry conditions that occur when $\psi = 90°$ in Fig. 2.12, and in addition made several assumptions in order to solve Eq. (2.19). In the case where $\psi \cong 90°$ and the torque components are assumed to be small or negligible, it can be argued, as discussed by Mukura, that $F_A = F_B = F_S$, and $A = B = \Omega_S/2$. Equation (2.19) then assumes the symmetrical groove form of Eq. (2.13):

$$\gamma_{tb} = 2F_S \cos(\Omega_S/2) \qquad (2.20)$$

where Ω_S is the groove (or ridge) angle. Mykura[49] measured a relative energy of $\gamma_{tb}/F_S = 0.015$ for nickel at 1000°C in vacuum. If the torques are not zero when $\psi \cong 90°$, Eq. (2.19) defines a single torque value—that is, $\partial F_A/\partial A = \partial F_B/\partial B$. Generally, however, Mykura assumed that torques acting in the same direction were equal, and the γ surface was taken to be symmetrical over some unspecified range of orientations about (100) and (111) orientations. By graphical integration of graphs of the derivative terms in Eq. (2.19) measured from twin boundary groove profiles and plotted against orientation, Mykura obtained the orientation dependence of the surface tension for pure nickel.

Winslow and Shewmon[50] have developed a variation of Mykura's method for measuring torques which consists in measuring twin boundary groove angles at the same twin boundary on different faces of a symmetric bicrystal using the relation

$$\frac{1}{F_S}\left(\frac{\partial F_S}{\partial \theta}\right)_a \sin \theta_a + \frac{1}{F_S}\left(\frac{\partial F_S}{\partial \theta}\right)_b \sin \theta_b = \cos \theta_a - \cos \theta_b \qquad (2.21)$$

where it has been assumed that $F_A = F_{A'} = F_B = F_{B'} = F_S$ in Eq. (2.19); a and b refer to the two faces of the bicrystal on which the measurements are made; and θ is a surface orientation parameter as described generally in the γ-plot of Fig. 1.2. By properly picking the orientation of one of the faces, one of the torque terms in Eq. (2.21) can be made equal to zero. This technique was employed by Robertson and Shewmon[51] in studies of the crystallography of impurity adsorption on copper surfaces.

Relation between Torque, Chemical Potential, Surface Excess, and Orientation. Robertson and Shewmon[51] assumed that, while $\partial F_S/\partial \theta$ has jump discontinuities (cusps in the γ-plot, Fig. 1.2), F_S is continuous between these cusps. In addition, as noted from Eq. (1.19a) for $F = F_S$, Γ_j can be identified with $\partial F_S/\partial \mu_j$ for an alloy (multicomponent) system. Consequently,

$$\frac{\partial}{\partial \mu_j}\left(\frac{\partial F_S}{\partial \theta}\right) = \frac{\partial}{\partial \theta}\left(\frac{\partial F_S}{\partial \mu_j}\right) \qquad (2.22)$$

and for a dilute solid-solution alloy system

$$-(\partial \Gamma_j/\partial \theta)_{\mu_j} = (\partial M_S/\partial \mu_j)_\theta \qquad (2.23)$$

where

$$M_S \equiv (\partial F_S/\partial \theta)$$

Since Mykura's method as described above allows relative values of torque to be determined as M_S/F_S, Eq. (2.23) can be employed to determine values for $(\partial \Gamma_j/\partial \theta)/F_S$. Gjostein[52] has also developed a phenomenological theory governing the shape alterations of the γ-plot due to adsorption from the gas phase.

Mathematically Consistent Solution of the Twin Boundary-Free Surface Equilibrium Equations. Winterbottom and Gjostein[53] have argued that Mykura's analysis is not rigorous because his solution of Eq. (2.19) requires assumptions to be made about the γ-plot prior to its determination. Consequently, such a procedure might be expected to yield a γ-plot that must conform to the assumed boundary conditions, and may not be a true γ-plot. In the analysis of Winterbottom and Gjostein,[53] a mathematically consistent solution of the equations of equilibrium at twin boundary-free surface intersections [Eq. (2.19)] was developed that required no assumptions other than that $\gamma(\theta, \phi) \equiv \Gamma_S(\theta, \phi)$ and its derivatives be piece-wise continuous functions of θ and ϕ, defined in Fig. 2.14(a), with a finite number of discontinuities. Note that θ and ϕ in Fig. 2.14(a) simply define the γ-plot in spherical coordinates, and θ is the same as that depicted previously in

INTERFACIAL EQUILIBRIA

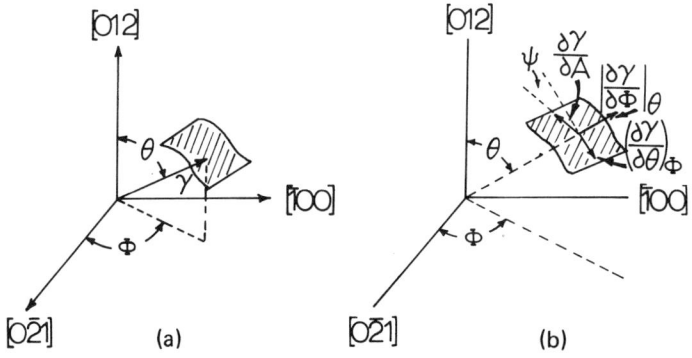

FIGURE 2.14 Definition of the coordinates of the γ-plot (a), the general directional derivative, $\partial F_S/\partial A$, its orthogonal components $(\partial F_S/\partial \phi)_\theta$ and $(\partial F_S/\partial \theta)_\phi$, and the angle ψ (b). (After Winterbottom and Gjostein.[53])

Fig. 1.2. By considering a normalization of Eq. (2.19) (with reference to Fig. 2.12) as

$$1 = \left(\frac{F_A}{\gamma_{tb}}\right)\cos A + \left(\frac{F_B}{\gamma_{tb}}\right)\cos B - \frac{1}{\gamma_{tb}}\left(\frac{\partial F_A}{\partial A}\right)\sin A$$
$$- \frac{1}{\gamma_{tb}}\left(\frac{\partial F_B}{\partial B}\right)\sin B \qquad (2.24)$$

and by expressing the directional derivatives, $\partial F_S/\partial A$, etc., in terms of the angular coordinates, θ and ϕ, as illustrated in Fig. 2.14(b), we obtain

$$\frac{\partial F_S}{\partial A} = \frac{\partial F_S}{\partial \theta} \cdot \frac{\partial \theta}{\partial A} + \frac{\partial F_S}{\partial \phi} \cdot \frac{\partial \phi}{\partial A}$$

where

$$\partial\theta/\partial A = \cos\psi$$
$$\partial\theta/\partial A = \sin\psi/\sin\theta \qquad (2.25)$$

Modifying Eq. (2.24) accordingly results in a completely general stability relation:

$$1 = \left(\frac{F_A}{\gamma_{tb}}\right)\cos A + \left(\frac{F_B}{\gamma_{tb}}\right)\cos B$$

$$-\frac{1}{\gamma_{tb}}\left(\frac{\partial F_A}{\partial \theta_A}\cos\psi_A + \frac{\partial F_A}{\partial \psi_A}\frac{\sin\psi_A}{\sin\theta_A}\right)\sin A$$

$$-\frac{1}{\gamma_{tb}}\left(\frac{\partial F_B}{\partial \theta_B}\cos\psi_B + \frac{\partial F_B}{\partial \phi_B}\frac{\sin\psi_B}{\sin\theta_B}\right)\sin B \quad (2.26)$$

Taking $F_A/F_S = F_B/F_S \cong 1$ (where F_S represents the average surface free energy), Eq. (2.26) can be rewritten as

$$\cos A + \cos B = \frac{\gamma_{tb}}{F_S} + \frac{1}{F_S}$$

$$x\left[\frac{\partial F_S}{\partial \theta}\bigg|_A \cos\psi_A + \frac{\partial F_S}{\partial \phi}\bigg|_A \cdot \left(\frac{\sin\psi}{\sin\theta}\right)_A\right]\sin A + \frac{1}{F_S}$$

$$x\left[\frac{\partial F_S}{\partial \theta}\bigg|_B \cos\psi_B + \frac{\partial F_S}{\partial \phi}\bigg|_B \cdot \left(\frac{\sin\psi}{\sin\theta}\right)_B\right]\sin B \quad (2.27)$$

This problem, as treated by Winterbottom and Gjostein, is one of interpolation where a Fourier series interpolation function expanded about the [012] pole in the form[52]

$$\gamma(\theta,\phi)/\gamma_{tb} \equiv F_S(\theta,\phi)/\gamma_{tb} = \sum_{m,n=0}^{\infty} \lambda_{mn}\,[a_{mn}\cos(m2\theta)\cos(n\phi)$$
$$+ b_{mn}\sin(m2\theta)\cos(n\phi)] \quad (2.28)$$

where

$$\lambda_{mn} = \begin{cases} \frac{1}{4} & \text{for } m = n = 0 \\ \frac{1}{2} & \text{for } m > 0, n = 0, \text{ or } m = 0, n > 0 \\ 1 & \text{for } m > 0, n > 0 \end{cases}$$

is substituted into Eq. (2.27) to define an array of equations linear in the unknown coefficients a_{mn} and b_{mn}, which for simplicity can be denoted x_i:[53]

$$\left.\begin{array}{l} 1 = e_{11}x_1 + \ldots + e_{1m}x_m \\ 1 = e_{21}x_1 + \ldots + e_{2m}x_m \\ \ldots \\ 1 = e_{n1}x_1 + \ldots + e_{nm}x_m \end{array}\right\} \quad (2.29)$$

The x's in Eq. (2.29) represent the unknown series coefficients, while the e_{ij}'s represent the constants determined from the measurements of the groove and ridge angle twin band-free surface intersections as depicted in Fig. 2.12 (and illustrated typically in Fig. 2.13). The reader is referred to the original work of Winterbottom and Gjostein[53] for the complete details of the derivations, and to a subsequent paper describing the orientation dependence of the surface free energy of gold using the twin boundary grooving technique.[54]

Experimental γ-Plots of FCC Metals. Measurements of twin boundary-free surface intersection equilibria incorporating the solutions and assumptions outlined above allow for the determination of the extent of surface energy anisotropy in fcc metals (or alloys). Any assumptions made, however, as was pointed out by Winterbottom and Gjostein,[53] generally affect the accuracy of the resulting γ-plot. Figure 2.15 illustrates several experimentally determined γ-plots based on measurements of twin boundary groove and ridge profiles.[55] Shown for comparison in Fig. 2.15 are two theoretical γ-plots. The existing experimental results can be accounted for with some degree of success by the pairwise bonding model coupled with the use of appropriate potential functions to account for interatomic potentials.[58] Several discrepancies persist, however—for example, the fact that γ_{max} does not fall on the [121] zone for the experimental γ-plots as predicted theoretically. Additionally, it should be pointed out that the copper γ-plot in Fig. 2.15 has been obtained by McLean and Gale[56] by measuring the dihedral angles of grain boundary grooves for

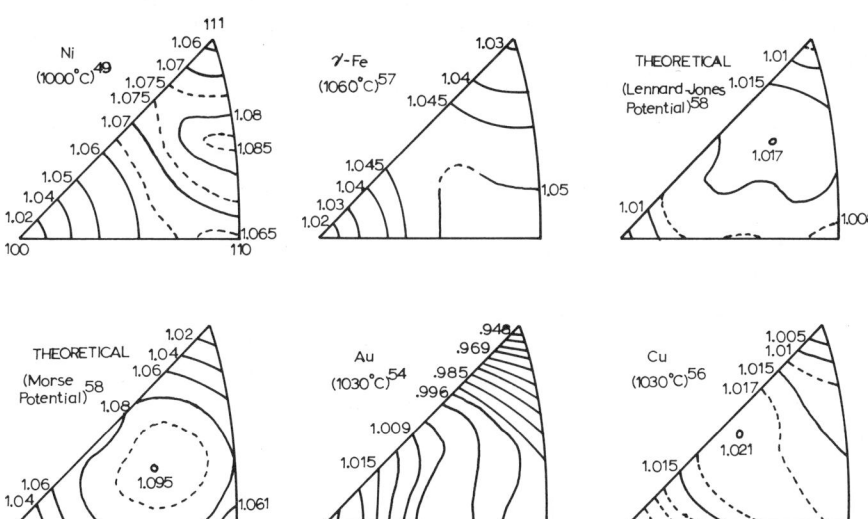

FIGURE 2.15 Experimental and theoretical γ-plots of fcc metals. Reference to the original date is given for each figure. Points plotted correspond to $F_{S(max)}/F_{S(min)}$; $F_{S(100)} > F_{S(111)}$.

various rotations about the axis of wires having the general features illustrated in Fig. 2.11(e). As a consequence, it is observed by comparison of the data in Fig. 2.15 that γ-plot determinations are not limited to specific interfacial intersections or crystal structures conducive to them. However, it should be apparent from Fig. 2.15, and from a consideration of similar data accumulated, that the extent of surface free energy anisotropy, $\gamma_{max}/\gamma_{min} \equiv F_{S(max)}/F_{S(min)}$, has not been more than 1.20 at high temperatures (generally near the melting point).

In addition to the data shown for copper in Fig. 2.15, Robertson and Shewmon[59] have obtained similar data from twin boundary-free surface intersections, and McLean and Mykura[60] have also used the twin boundary grooving technique to determine the temperature dependence of surface energy anisotropy in platinum. They found that $F_{S(max)}/F_{S(111)}$ decreased from 1.133 at 920°C to 1.078 at 1500°C.

A number of techniques have been employed in twin boundary groove measurements to determine the crystallography of the surface sections being considered. These include orientation determinations from the angles between nonconcurrent twin boundary traces in the specimen surface used by Mykura[49,61] (a modification of the stereographic methods outlined, for example, by Barrett and Massalski[62]), the use of etch pits whose geometric patterns can be correlated with specific crystallographic orientations, and x-ray techniques—for example, Laue back-reflection as employed in the work of Winterbottom and Gjostein.[54] In addition, modern research techniques such as the analysis of electron channeling patterns formed in the scanning electron microscope could lead to enhanced analysis of, for example, wire samples as in Fig. 2.11(e) using the techniques outlined in detail by McLean and Gale.[56]

It is apparent from the γ-plots obtained to date that cusps occur at {111} and {100} orientations, in agreement with theoretical predictions for cubic crystals. The measurements confirm the ideal γ-plot previously depicted in Fig. 1.2. Measurements also show that, for free surfaces (solid-vapor interface), $F_{S(100)} > F_{S(111)}$. The relative torque values obtained have, generally, been in the same range as those quoted in Section 2.2.3 (as associated, for example, with Fig. 2.9). Robertson and Shewmon[59] obtained $\gamma_{tb}/F_S \cong 0.010$ for pure copper, and extrapolated $\Sigma M/F_S \cong 0.1$ from their data (where ΣM is here considered a net effective torque because only a resultant or mean value is considered). Mykura,[49] on the other hand, obtained $\gamma_{tb}/F_S \cong 0.015$, and $\Sigma M/F_S \cong 0.2$ for nickel, while McLean and Mykura[57] obtained $\gamma_{tb}/F_S \cong 0.014$ and $\Sigma M/F_S \cong 0.1$ for γ-iron.

2.3.2 Twin Boundary-Grain Boundary Intersections

On the basis of the original equilibrium features of microstructures interpreted by Smith[8] and modified somewhat by Fisher,[63] Dunn[64] and Dunn and Lionetti[65] investigated the dependence of grain orientation by growing together single-crystal sheets of three grain orientations, affecting a situation essentially identical to that illustrated by Fig. 2.11(c), assuming the influence of boundary orientation (torque) to be negligible. Dunn et al.[66] extended this work to an investigation of the interfacial free energy of twin boundaries in silicon iron using an approximate method based on Herring's analysis [Eq. (2.4)], effectively replacing the surface free energies in Eq. (2.19) with the grain boundary free energies. X-ray methods[64] were used to determine the $\{112\}$ coherent twin planes in bcc Si-Fe tricrystal specimens, and the equilibrium angles at twin-grain boundary intersections were measured from the polished sample surfaces. A mean torque $\Sigma M/\gamma_{gb}$ (where γ_{gb} denotes the average grain boundary free energy) of 0.2 was determined from these measurements.

Fullman[67] modified the treatment of Dunn et al.[66] by preparing large-grained sheet structures, identical in concept to those shown in Figs. 2.3 and 2.11(c), for copper annealed for 40 hours at 950°C. Fullman based his analysis on the measurement of $\{111\}$ coherent twin bands with grain boundaries in the fcc copper, and assumed that torques associated with the junction equilibria were negligible because the interfacial energy balance occurred with respect to components parallel to the twin boundaries. Fullman ensured the measurement of true dihedral angles at twin-grain boundary intersections in the sheet samples by selecting primarily $\{110\}$ grain surface orientations and twin boundary intersections where the twin plane could be associated with the $\{111\}$ planes (at 90° to the $\{110\}$ surfaces), obtaining a distribution of values of γ_{tb}/γ_{gb}, with a mean value of approximately 0.035.

A similar optical metallographic analysis was performed on lead and lead-0.1 at.% silver alloy by Bolling and Winegard,[68] who obtained a value of $\gamma_{tb}/\gamma_{gb} = 0.05$ for lead and 0.08 for the dilute alloy. Bolling and Winegard interpreted the increase in the mean value of γ_{tb}/γ_{gb} with the small silver addition in lead to be the result of a lowering of the grain boundary free energy by silver segregation.

Inman and Khan[69] modified the technique of Fullman by employing transmission electron microscopy to measure the coherent twin boundary/grain boundary free energy ratio, γ_{tb}/γ_{gb}, in pure copper. In this technique, the true dihedral angles at random intersection systems could be determined as outlined in Section 2.2 [Eq. (2.8)] from thin foils prepared from annealed sheet specimens. In addition, it was determined by this technique that γ_{tb}/γ_{gb} did not change with the addition of 1 wt. of antimony to pure copper,[39] indicating that, in addition to antimony segregation to the grain boundaries,[41] antimony probably also segregated at the coherent twin boundaries.

Murr[70] demonstrated the significant accuracy attained by the use of electron transmission microscopy techniques compared with the optical metallographic approach of Fullman; these techniques have been utilized in measurements of γ_{tb}/γ_{gb} in a number of metals and alloys.[70-73] However, as was pointed out by Gjostein[74] and by Basterfield and Miller,[75] the results of electron microscope measurements of γ_{tb}/γ_{gb} reflected a variation of grain boundary free energy with crystallographic misorientation.[76]

Configurational Equilibrium at Twin-Grain Boundary Intersections in FCC Metals and Alloys. Based on Herring's generalized equation [Eq. (2.4)], and the equilibrium equations for a twin-grain boundary intersection system as described by

Murr,[70] the following equilibrium equations can be written for a twin band intersecting a grain boundary as depicted schematically in Fig. 2.16:

$$\gamma_{tb} + \gamma_{AB} \cos \Omega_1 + \gamma_{T_A B} \cos \Omega_2 + \Sigma M_I = 0$$
$$\gamma_{tb} + \gamma_{AB} \cos \Omega_4 + \gamma_{T_A B} \cos \Omega_3 + \Sigma M_{II} = 0 \quad (2.30)$$

where

$$\Sigma M_I = \frac{\partial \gamma_{AB}}{\partial \Omega} \sin \Omega_1 + \frac{\partial \gamma_{T_A B}}{\partial \Omega} \sin \Omega_2 \quad (2.31)$$

$$\Sigma M_{II} = \frac{\partial \gamma_{AB}}{\partial \Omega} \sin \Omega_4 + \frac{\partial \gamma_{T_A B}}{\partial \Omega} \sin \Omega_3 \quad (2.32)$$

and the true dihedral angles, Ω_i, are determined for a thin foil observed in the transmission electron microscope as outlined in Eq. (2.8). The associated torque components at each junction (illustrated by heavy vectors in Fig. 2.16) result from a right-handed convention in performing the cross-product in Eq. (2.4) (as in Fig. 2.2). The absence of a torque associated with γ_{tb} is a result of the fact that γ_{tb} is generally very small in comparison with γ_{gb}, and the fact that the twin boundary is at a cusp position (minimum free energy). Solving Eq. (2.30) simultaneously results in the more general forms:

$$\left. \begin{array}{l} \dfrac{\gamma_{tb}}{\gamma_{AB}} = C_{AB} - \dfrac{1}{\gamma_{AB}} \cdot \left(\dfrac{\Sigma M_I \cos \Omega_3 - \Sigma M_{II} \cos \Omega_2}{\cos \Omega_3 - \cos \Omega_2} \right) \\[2ex] \dfrac{\gamma_{tb}}{\gamma_{T_A B}} = C_{T_A B} - \dfrac{1}{\gamma_{T_A B}} \cdot \left(\dfrac{\Sigma M_{II} \cos \Omega_4 - \Sigma M_{II} \cos \Omega_3}{\cos \Omega_4 - \cos \Omega_1} \right) \end{array} \right\} \quad (2.33)$$

where

$$C_{AB} = \frac{\cos \Omega_2 \cos \Omega_4 - \cos \Omega_1 \cos \Omega_3}{\cos \Omega_3 - \cos \Omega_2} \quad (2.34)$$

and

$$C_{T_A B} = \frac{\cos \Omega_2 \cos \Omega_4 - \cos \Omega_1 \cos \Omega_2}{\cos \Omega_1 - \cos \Omega_4} \quad (2.35)$$

Equation (2.33) is similar in form to the mathematically consistent solutions given in Section 2.3.1 for annealing twin bands intersecting a free surface, and Fig. 2.16 could be considered to be phenomenologically similar to Fig. 2.12. Equation (2.33) contains two equations with five unknowns, and without reservation it must

INTERFACIAL EQUILIBRIA

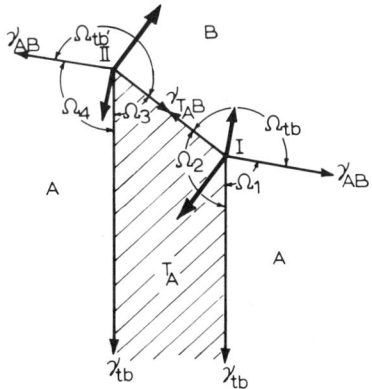

FIGURE 2.16   Equilibrated twin-grain boundary intersection.

be stated that no exact solution is possible. Alternatively, as has been done in previous studies (for example, Section 2.3.1), rational approximations can be made which might allow solutions to be achieved. The initial assumption in the work of Fullman and others that the torques be zero is tantamount to demanding that the grain boundary free energy be constant across the intersection system, and it must therefore follow that $\gamma_{AB} = \gamma_{T_AB} = \gamma_{gb}$. Substituting the previous condition that the torques are zero and the grain boundary free energy is constant for the twin-grain boundary configuration directly into Eq. (2.30) then results in a single expression:

$$\gamma_{tb}/\gamma_{gb} = \frac{1}{2} \sum_{i=1}^{4} (-\cos \Omega_i) \tag{2.36}$$

Since the intersection system of Fig. 2.16 represents an equilibrium configuration, it might be assumed that ΣM represents a resultant or net effective torque (as demonstrated schematically in Fig. 2.9) resolved parallel to the twin boundaries, having equal magnitude but opposite sense at I and II in Fig. 2.16. Equation (2.30) then becomes

$$(\gamma_{tb} + \Sigma M)/\gamma_{AB} = C_{AB}$$
$$(\gamma_{tb} - \Sigma M)/\gamma_{T_AB} = C_{T_AB} \tag{2.37}$$

where as a rough approximation in the absence of a direct proof it is assumed that

$$\Sigma M = \frac{\Sigma M_I \cos \Omega_3 - \Sigma M_{II} \cos \Omega_2}{\cos \Omega_3 - \cos \Omega_2} = -\frac{\Sigma M_I \cos \Omega_4 - \Sigma M_{II} \cos \Omega_1}{\cos \Omega_4 - \cos \Omega_1} \tag{2.38}$$

If it is assumed at this point that the grain boundary free energy does not change appreciably along the equilibrium configuration, then $\gamma_{AB} \cong \gamma_{T_A B} \equiv \gamma_{gb}$. Making this substitution in Eq. (2.37) and solving simultaneously then results in:

$$\gamma_{tb}/\gamma_{gb} = (C_{AB} + C_{T_A B})/2 \tag{2.39}$$

and

$$\Sigma M/\gamma_{gb} = (C_{AB} - C_{T_A B})/2 \tag{2.40}$$

Situations corresponding to low-torque and high-torque twin boundary configurations can be defined. In Fig. 2.17(a) a rotation of 180° about the twin direction coincident with <110> causes no change in the misorientation, Θ; that is, $\Theta_{AB} = \Theta_{T_A B}$. In Fig. 2.17(b), however, a 180° rotation about the twin direction produces a mirror reflection of the <110> direction in the matrix (grain A) to appear in the twin (T_A). The misorientation of <110> between A and B and T_A and B is now generally different; that is, $\Theta_{AB} \neq \Theta_{T_A B}$, except where <110> in grain B is parallel to the twin plane or trace direction in grain A. We now observe that, for $(\gamma_{AB}, \gamma_{T_A B}) = f(\Theta)$, the situations where $\Theta_{AB} = \Theta_{T_A B}$ will be conducive to $\gamma_{AB} \cong \gamma_{T_A B}$, while $\Theta_{AB} \neq \Theta_{T_A B}$ will imply $\gamma_{AB} \neq \gamma_{T_A B}$. These features have been demonstrated to apply with a confidence of approximately 80% or better for $(hkl)_A$ = (110), (100), and (112) grain surface orientations, and for similar situations where $(hkl)_A = (hkl)_B$; and misorientations were defined by the <110> directions in grains A and B (Fig. 2.17).[28,77] Horylev and Murr[28] have also shown that violations of the crystallographic conventions (Fig. 2.17) can generally be attributed to so-called anomalous cases where a marked variation in grain boundary plane inclination occurs between grains B and T_A and B and A in Fig. 2.18. In such situations, torque arises because of a variation in inclination—that is, $M = (\partial\gamma/\partial\theta) \sin \Omega$ or $(\partial\gamma/\partial\Omega) \sin \Omega$, where the variation in θ will be implicit in the variations of the dihedral angles. Again, these configurations could be readily identified by

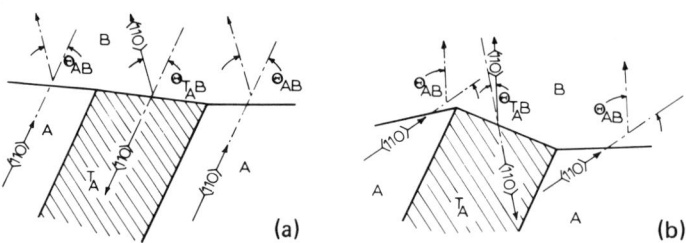

FIGURE 2.17 Crystallographic misorientation at twin-grain boundary intersections.[28] (a) Continuous misorientation (low-torque configuration). (b) Discontinuous misorientation (high-torque configuration).

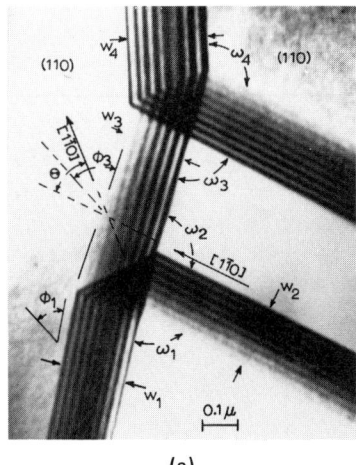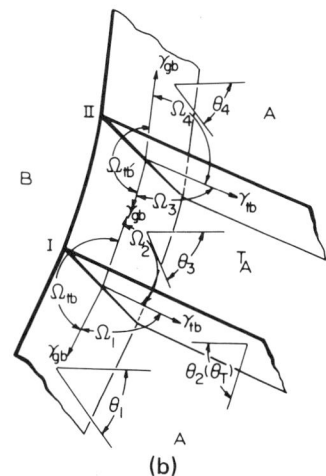

FIGURE 2.18 Twin–grain boundary intersection in a random thin section observed in the transmission electron microscope. (a) Bright-field electron transmission image showing the crystallographic and geometric parameters which characterize twin–grain boundary intersections. (b) Schematic section view of a twin–grain boundary intersection showing geometric parameters characterizing the interface planes.[28]

visual inspection of gross variations in the dihedral angles associated with measurable differences in the grain boundary projection widths.

Figure 2.19 illustrates several possible (idealized) equilibrium configurations at twin–grain boundary intersection systems, and Fig. 2.20 shows some examples of high-torque configurations. In Fig. 2.19 only the idealized surface (two-dimensional) views are presented, and the torques depicted could arise because of variations in Θ or θ. Figure 2.19(a)–(c) depicts the common types of twin–grain boundary intersections considered, and Fig. 2.19(d) and (e) depict so-called double junctions as modifications of the simple corner-twin systems shown in Fig. 2.19(f)–(h).

The differences apparent in the crystallographic-geometric distinctions which arise at low-torque and high-torque twin–grain boundary intersection configurations can be convincingly demonstrated by a comparison of Fig. 2.20(a) and Fig. 2.17(a). The differences are also apparent in the measurement of interfacial free energy ratios and torque ratios using Eqs. (2.39) and (2.40), respectively, as demonstrated in the histograms from the work of Horylev and Murr[28] reproduced in Fig. 2.21, from which mean values of γ_{tb}/γ_{gb} and $\Sigma M/\gamma_{gb}$ were observed to be 0.024 and 0.088, respectively. This configurational distinction for low-torque twin–grain boundary intersections has been employed in the measurement of mean values of γ_{tb}/γ_{gb} for a number of metals and alloys.[16,22,38,77-79] (Table 2.1).

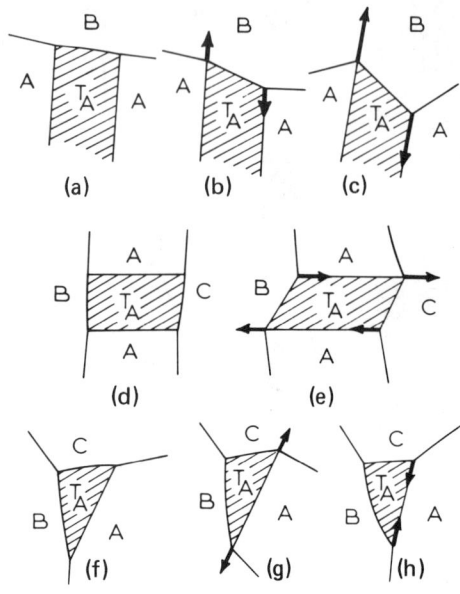

FIGURE 2.19 Equilibrium configurations at twin–grain boundary intersections.[28]

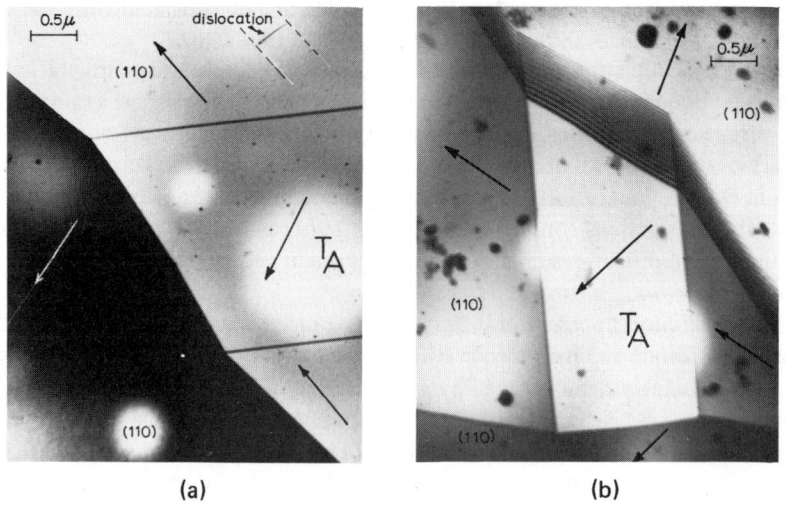

FIGURE 2.20 Electron micrographs of high-torque twin–grain boundary intersection configurations. (a) Twin–grain boundary intersection similar to Fig. 2.19(b) and (c). (b) Double junction system as depicted ideally in Fig. 2.19(e). Both observations were made in a 304 stainless-steel foil.

INTERFACIAL EQUILIBRIA

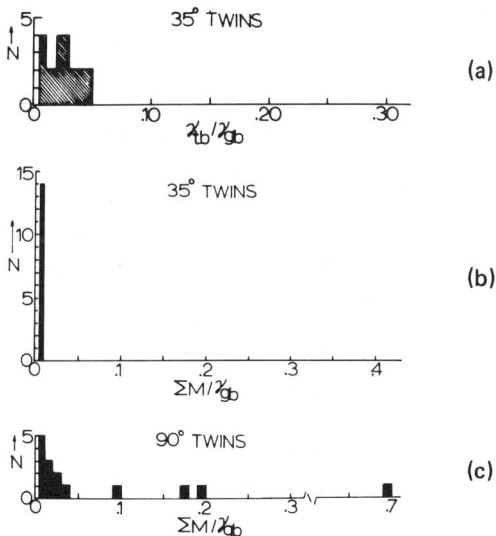

FIGURE 2.21 Histograms of relative interfacial free energy and relative torque ratios for (110) twin–grain boundary intersections in 304 stainless steel [$(hkl)_A = (hkl)_B = (110)$]. (a) Low-torque configurations–twin trace along <110>. (b) Low-torque configuration–twin trace along <110>. Corresponding torque values for (a). (c) High-torque configuration–twin trace along <112>. (From Horylev and Murr.[28])

It should be apparent from the development of a configurational distinction for twin–grain boundary intersections in this section that the torque concepts associated with twin band–free surface intersections are essentially identical to those for twin–grain boundary intersections. This feature is phenomenologically demonstrated by comparing Figs. 2.12 and 2.13 with Figs. 2.16, 2.19(b) and (c), and 2.20(a). It must be pointed out, in addition, that γ-plots for grain boundary plane orientations could also be constructed in a manner similar to that for free surface orientations.

Also, inasmuch as the work of Fullman[67] utilized primarily (110) orientations with the twin plane perpendicular to the surface (along <112> directions), values of γ_{tb}/γ_{gb} obtained would be expected to be erroneous, since such situations correspond to the high-torque configurations outlined above, and illustrated ideally in Fig. 2.18(b). Therefore, in addition to geometrical errors as previously noted,[39,70] optical metallographic measurements of γ_{tb}/γ_{gb} which seek to minimize this error by restricting measurements to situations where all interfaces are essentially normal to the specimen surfaces are phenomenologically incorrect, and errors are contributed by the occurrence of torques of appreciable magnitude.

It should be pointed out that, although only coherent twin intersections in (110), (100), and (112) grain orientations in fcc metals and alloys have so far been considered in any detail, the crystallographic concepts are applicable to any system.

Annealing Twin Equilibria; Origin of Annealing Twins in FCC Metals and Alloys.
The formation and behavior of annealing twins in fcc metals have been reviewed rather extensively by Burke and Turnbull,[80] Hall,[81] and Cahn.[82] The so-called growth accident mechanism of nucleation originally proposed by Carpenter and Tamura[83] was supported in the work of Burke[84] and of Bolling and Winegard,[68] and convincingly demonstrated in the thermionic electron emission microscope[16] observations of Grube and Rouze,[85] where it was shown[84,85] that annealing twins formed only in the corners (edges) of grain boundary triple junctions as the junction migrates during recovery and grain growth at elevated temperature. Consequently, the corner-twin equilibrium configurations illustrated schematically in Fig. 2.19(*f*)-(*h*) are formed initially, with a growth transition by grain boundary and twin boundary migration to the configurations of Fig. 2.19(*d*) and (*e*) and (*a*)-(*c*). Evidence has been obtained of two features conducive to twin boundary formation: (1) A small number of atoms in the migration junction edge become coincident with a $\{111\}$ plane in one of the grains composing it. This hexagonal stacking accident nucleates a twin and initially requires an energy equal to the stacking-fault free energy at the temperature at which the nucleation occurs. And (2) an abrupt change in crystallography in an existing grain may also act as an inducement for a twinning accident. The first of these features encompasses the coincidence site model of Kronberg and Wilson,[44] while both can be encompassed in the concept of the origin and growth of twins in the form of stacking faults or fault bundles at migrating grain boundaries during primary recrystallization as discussed by S. Dash and N. Brown [*Acta Met.*, **11**, 1067 (1963)] and subsequently extended to a general model of a two-dimensional nucleation process on the $\{111\}$ planes of growing fcc grain boundaries by H. Gleiter [*Acta Met.*, **17**, 1421 (1969)].

Fullman and Fisher[86] suggested that twin boundaries form at grain corners during annealing (grain growth) when the total free energy of the boundaries between the neighbor grains composing the particular corner in which nucleation will occur is higher than the total energy of the grain boundaries and the twin boundary after a coherent twin interface has formed. With reference to Fig. 2.22, it is observed that the free energy will be decreased by the formation of the twin boundary if[9]

$$(\gamma'_3 A'_2 + \gamma'_1 A'_3 + \gamma_{tb} A_T) < (\gamma_3 A_2 + \gamma_1 A_3) \tag{2.41}$$

where γ_i and A_i represent the interfacial free energies and associated boundary plane areas, respectively, with reference to Fig. 2.22(*a*). It is observed from Fig. 2.22(*a*) that, if the boundary plane inclinations remain essentially the same (no appreciable deviation of the interface plane occurs) during corner-twin formation, the areas A_2 and A_3 could be considered to be constant; that is, $A_2 \cong A'_2$, $A_3 \cong A'_3$. Consequently, for Eq. (2.41) to be satisfied necessitates $\gamma'_1, \gamma'_3 < \gamma_1, \gamma_3$. Considering the balance of interfacial energies implicit at the triple junction, it is observed that Eq. (2.41) is satisfied geometrically for all possibilities when

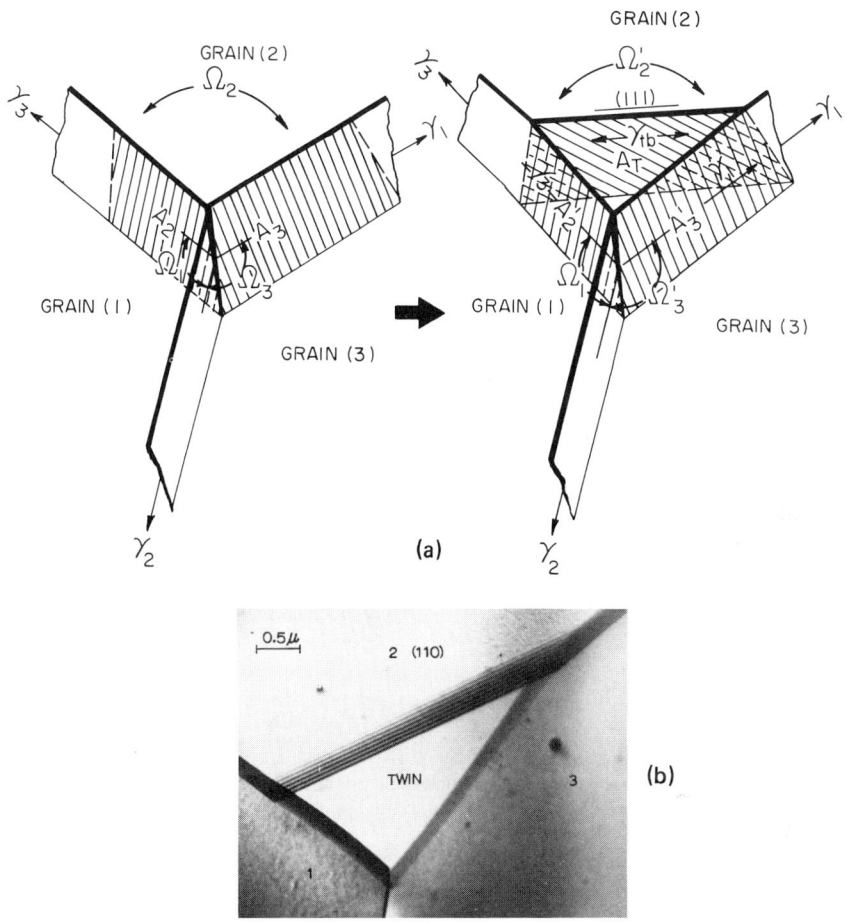

FIGURE 2.22 Coherent twin boundary formation in a grain corner (corner-twin formation). (a) Schematic representation in an idealized thin section—for example, a transmission electron microscope sample. (From Murr.[9]) (b) Electron micrograph of a corner twin in 304 stainless steel.

$\Omega_2' < \Omega_2$. Since for pseudoequilibrium the distribution of dihedral angles at a triple junction should occur at about 120°, Ω_2' would necessarily be less than 120° if γ_2 remains unchanged after corner-twin formation.

The latter feature, that $\Omega_2' < \Omega_2$ or $\Omega_T < 120°$ (where Ω_T is the mean dihedral angle in a twinned-grain corner), has been demonstrated by Fullman and Fisher[86] by optical metallographic measurements of the dihedral angles on the surface of annealed copper samples, and Murr[9] has shown this to be true for grain corner-twin observations in austenite in the transmission electron microscope. Similar results have been obtained for nickel and aluminum, as illustrated in Fig. 2.23.

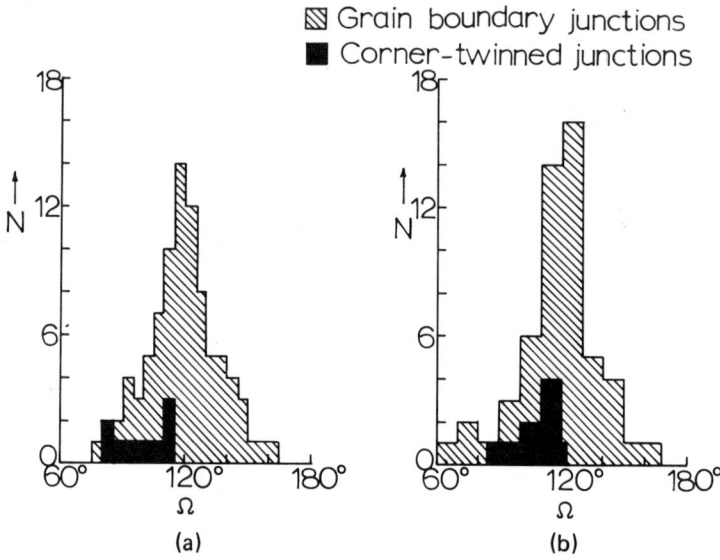

FIGURE 2.23 Distributions of dihedral angles at grain boundary triple junctions and corner corner-twin junctions measured by transmission electron microscopy. (a) Aluminum at 450°C. (b) Nickel at 1060°C.

Annealing twins become unstable when they terminate within a grain, and anneal out. In addition, if the free energy of a grain boundary at the end of a twin band—that is, where a twin band intersects a grain boundary—is sufficiently higher than that of the adjacent boundary, the twin may pull away from the grain boundary and anneal out.[86] Consequently, with reference to Fig. 2.16, this will occur when $\gamma_{T_A B} > \gamma_{AB}$, and one is immediately struck by the fact that grain boundary torque is probably an important feature of this process. This would be particularly true, for example, at twin-twin intersections as illustrated in Fig. 2.24. In the context of Fig. 2.16, $\gamma_{AB} = \gamma_{tb}$ in Fig. 2.24, while $\gamma_{T_A B}$ corresponds to a noncoherent boundary plane which could be assumed to have a value close to that of a random, high-angle grain boundary; that is, $\gamma_{T_A B} \cong \gamma_{gb}$ in Fig. 2.24. Since $\gamma_{tb}/\gamma_{gb} = 0.024$ for 304 stainless steel, we observe that $\gamma_{T_A B} \cong 40\gamma_{AB} (= \gamma_{tb})$ in Fig. 2.24. Note that, because such a noticeable deviation in the boundary asymmetry occurs at the junction, a rather large resultant torque would be associated with the dissociation process.

FIGURE 2.24 Twin dissociation (arrow) from another twin boundary. Electron micrograph of a 304 stainless-steel foil equilibrated at 985°C.

Fullman and Fisher[86] also considered the formation of noncoherent twin boundaries as a consequence of junction migration following grain corner-twin formation, and these features have been verified in the observations of Grube and Rouze.[85] In addition, Fullman and Fisher postulated that the abundance of annealing twins in fcc metals and alloys should decrease with an increasing ratio γ_{tb}/γ_{gb}. This feature has been demonstrated experimentally.[68] In addition, since a twin boundary and a stacking fault are generally considered to have a simple geometrical relationship in fcc materials, the stacking-fault free energy is approximately twice as large as the twin boundary free energy, as will be discussed in more detail in Chapter 3. Consequently, the frequency of annealing twins will decrease with an increase in twin boundary or stacking-fault free energy. Since annealing twin boundaries form at a grain corner or twin bands form by grain growth or repeated corner-twin formation at the same junction site, it would logically follow that, when the twin boundary or stacking-fault free energy become sufficiently large, the microstructure of a metal or alloy would exhibit only single twin boundaries or localized corner twins. These features are essentially observed for pure aluminum, which is considered to have one of the highest stacking-fault free energies of the pure fcc metals (and alloys).[79,87]

As Nielsen pointed out,[88] the proposal of Fullman and Fisher[86] regarding the energetics of twin boundary formation may not be a completely accurate description of the actual process that occurs when annealing twins originate. It is equally likely or even more likely that in many instances of three-dimensional grain growth an epitaxial geometric coalescence occurs where a growing grain meets another in such a way that the interface is at or near the coincidence boundary for a coherent twin—that is, near a $\{111\}$ plane. Nielsen has calculated such an encounter[89] in an fcc material to be roughly once per 16,000,[88] and experimental observations on the formation of coincidence boundaries as a result of twinning have been made in several zone-refined metals.[90-92] It is likely, in view of the observations by Grube and Rouze[85] and J. Dash and N. Brown [*Acta Met.*, **11**, 1067 (1963)], for example, that several mechanisms could nucleate twinning in a grain corner. However, the energetics of corner-twin equilibria implicit in the proposals of Fullman and Fisher appear to be experimentally well established. The structural features of annealing twins will be treated in more detail in Chapter 4.

2.4 MULTIPHASE EQUILIBRIA

Smith,[8] in his classic treatment of an interpretation of microstructures based on the simple concept that they (microstructures) result primarily from an attempt to balance the associated interfacial free energies (surface tensions), was the first to deal with the equilibrium of grains and phases. In the more than two decades since Smith's original paper, numerous applications of these concepts have been made in metal and alloy systems as well as ceramic and metal–ceramic systems.[93-95] The principles of multiphase equilibria established by Smith are applicable in all equilibrium situations primarily the result of high-temperature anneals where lattice coherency is not a contribution in the formation of a grain or phase boundary. As illustrated in Fig. 2.25, the simple Dupré equation[96] [Eq. (2.5)] governs the balance of interfacial free energies for three phases in equilibrium in the absence of interfacial torques:

$$\gamma_{23}/\sin \Omega_1 = \gamma_{13}/\sin \Omega_2 = \gamma_{12}/\sin \Omega_3 \tag{2.5a}$$

where γ_{12}, γ_{13}, and γ_{23} represent the phase boundary free energies, and Ω_1, Ω_2, and Ω_3 are the dihedral contact angles measured in phases 1, 2, and 3, respectively.

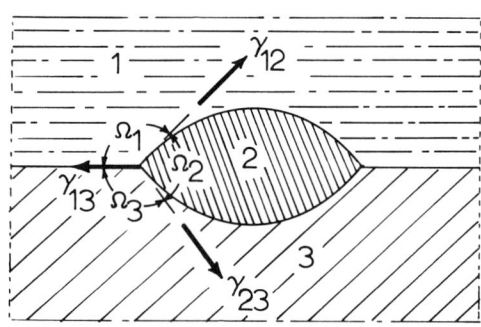

FIGURE 2.25   Interface equilibrium between three phases.

INTERFACIAL EQUILIBRIA

Phases 1, 2, and 3 may be solid, liquid, or vapor, or generally any variation thereof, or three immiscible liquid or solid systems in equilibrium.

2.4.1 Equilibration of a Particle (Crystal) in Contact with a Solid Substrate

Winterbottom[96] has developed a general solution for predicting the equilibrium shapes of small (crystalline) particles resting on a solid substrate as a special case of three-phase equilibrium well known for a liquid droplet on a rigid (solid) surface where gravitational influences are assumed to be negligible.[97,98] The essence of this concept, illustrated schematically in Fig. 2.26, is that the shape of a solid particle equilibrated on a substrate reflects the influence of the substrate as well as the surface energy anisotropy. The interfacial equilibrium is uniquely characterized by the dihedral contact angle, Ω_C, as a special case of the Dupré condition [Eq. (2.5a)] known as the Young[99] equation in the general form (Fig. 2.26):[100]

$$\gamma_{13} = \gamma_{23} + \gamma_{12} \cos \Omega_C \tag{2.42}$$

Treating the system as a solid particle (P) on a solid matrix or substrate (M), the balance equation is written,

$$F_{S(M)} = \gamma_{P/M} + F_{S(P)} \cos \Omega_C \tag{2.43}$$

where, assuming phase 1 in Fig. 2.26 to be vapor, $F_{S(M)}$ is the substrate or matrix surface (solid–vapor) free energy, $\gamma_{P/M}$ is the particle–substrate interfacial free energy, and $F_{S(P)}$ is the surface (solid–vapor) free energy of the particle.

Note in Fig. 2.26 that the generalized particle surface free energy corresponds to

$$K_W H = 2\gamma_{12} = 2F_{S(P)} \tag{2.44}$$

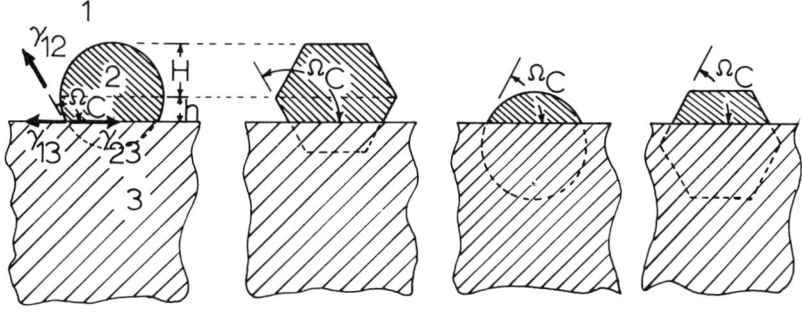

FIGURE 2.26 Equilibrium shapes of liquid droplets, supercooled-liquid (solid) particles, and crystals on a solid substrate.

where K_W is the Wulff constant and H corresponds to a radius vector of the perfect sphere (circle) for an ideal liquid, or λ for a Gibbs-Wulff construction (γ-plot) as defined by Eq. (1.17) (see Fig. 1.2). The particle-substrate interface is characterized by

$$K_W H = 2(\gamma_{23} - \gamma_{13}) = 2(\gamma_{P/M} - F_{S(M)}) \quad (2.45)$$

It can be observed from Fig. 2.26 that particle-substrate equilibria can be determined exclusively from particle shape (shape anisotropy), since, from Eqs. (2.44), (2.45), and (2.43),

$$-\cos \Omega_C = h/H \quad (2.46)$$

Consequently, as illustrated in the work of Pilliar and Nutting,[101] who used the electron microscope to examine the shapes of pure metal crystals equilibrated on the edges of thin ceramic crystal substrates, the ratios of interfacial free energies or the average particle-matrix (substrate) interfacial free energy can be determined by simply measuring the particle heights or shape anisotropy of equilibrated crystalline particles implicit in Fig. 2.26 and Eq. (2.46)–that is, H and h.

It should also be noted, as discussed by Winterbottom[96] and demonstrated in the earlier work of Sundquist,[102] that the particle shapes are related to a generalized Gibbs-Wulff construction (γ-plot), and that accurate measurements of particle shapes can be used to experimentally construct γ-plots. With the recent developments in scanning electron microscopy, it is possible to uniquely observe and characterize particles equilibrated on a substrate. These features are clearly demonstrated in the examples shown in Fig. 2.27. In a scanning electron microscope fitted with a nondispersive x-ray analyzer, it is even possible to accurately determine the particle or substrate identity.

Either the contact angle or the equilibrium particle shape can be employed in estimating the interaction energy or the energy of adhesion, E_{Ad}:

$$E_{Ad} = F_{S(P)} - (\gamma_{P/M} - F_{S(M)}) = F_{S(P)} (1 + \cos \Omega_C) \quad (2.47)$$

or

$$E_{Ad} = (H - h)(K_W/2) \quad (2.47a)$$

We note from Fig. 2.26 that, when $H = h$ ($\Omega_C = 180°$), the particle sits freely upon the substrate, and $E_{Ad} = 0$. However, under the conditions of complete wetting $\Omega_C = 0°$, and from Eq. (2.46) we obtain $h = -H$. Consequently

$$E_{Ad} = 2F_{S(P)} \quad (2.48)$$

Equation (2.47) therefore describes the various degrees of partial wetting illustrated

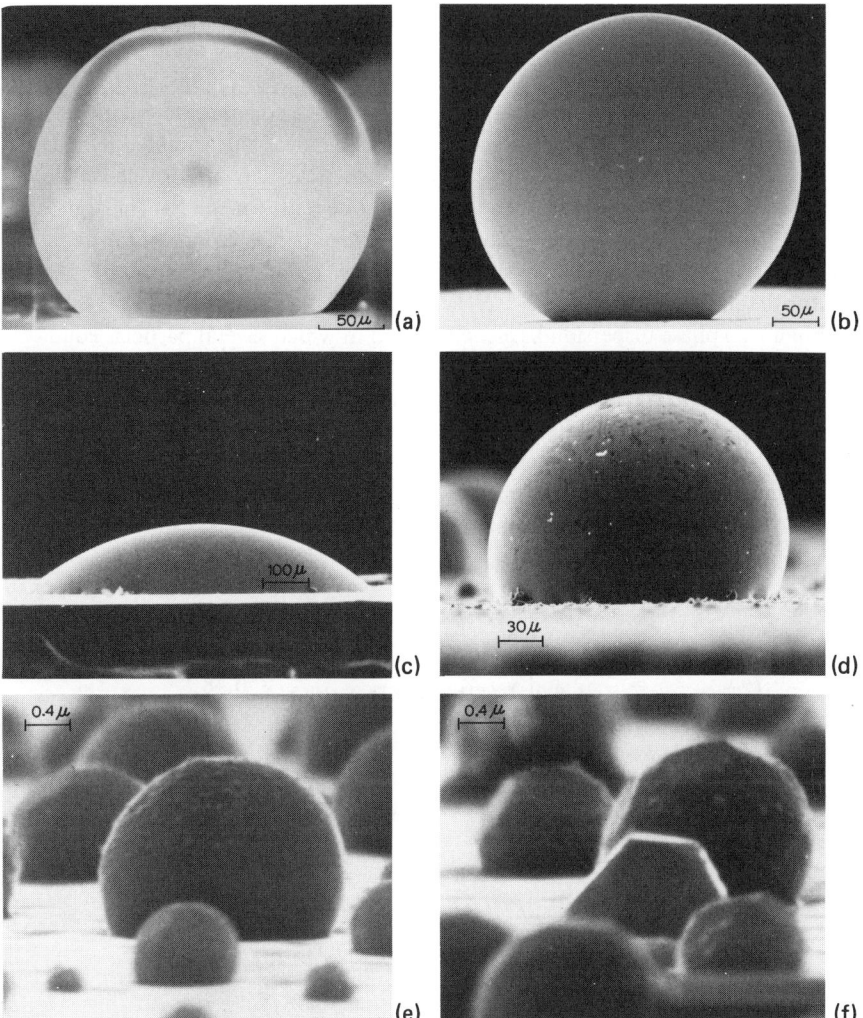

FIGURE 2.27 Particle-substrate equilibria observations in the scanning electron microscope. (a) Mercury droplet on glass. (b) Mercury droplet on aluminum. (c) Mercury droplet on palladium. (d) Stainless steel particle on alumina (Al_2O_3). (e) Nickel particle on thorianite (ThO_2). (f) 80/20 NiCr particles on thorianite (ThO_2). Particles in (a), (b), and (c) were equilibrated at 25°C. Particles in (d) were equilibrated from molten drops falling on a 1000°C substrate; (e) and (f) were equilibrated at 1200°C. All observations were made at 25°C.

ideally in Fig. 2.26. Equation (2.48), on the other hand, is an idealized case which cannot be measured in the sense of shape implicit in Fig. 2.26. This situation corresponds to the case where the "particle" spreads uniformly over the substrate.

Equilibrium of a Particle in a Matrix. The special case of a particle in a matrix is a logical modification of a particle on a matrix. In effect, we are dealing with a two-phase system, ideally considered to be a noncoherent dispersoid in a metal or alloy matrix. The particle interfacial energy, $\gamma_{P/M}$, is the noncoherent boundary free energy at the particle–matrix interface, and it can be ideally determined by measuring a matrix metal particle on a substrate of the particle material, or vice versa. Figure 2.27(e) and (f) illustrate the measurement of $\gamma_{P/M}$ for the Ni-ThO$_2$ and the NiCr-ThO$_2$ dispersion-hardened systems. In addition, observations of faceted dispersoid shapes within the matrix phase can allow for a determination of the γ-plot for the dispersoid.[103]

For a two-phase dispersion-hardened system in which several particles equilibrate as sintered or multigrain systems, it is possible to determine the ratio of particle–matrix interfacial free energy to particle–particle (particle grain boundary) interfacial free energy, $\gamma_{gb(P)}$. Figure 2.28 illustrates this feature. By choosing particle intersection systems where the boundary planes are essentially symmetrical, measurement of the dihedral angle [Ω_p in Fig. 2.28(a)] then allows one to obtain the interfacial free energy ratio from [Fig. 2.28(a)]:

$$\gamma_{gb(P)} = 2\gamma_{P/M} \cos(\Omega_p/2) \qquad (2.49)$$

The values obtained for the Ni-ThO$_2$ and NiCr (80/20)-ThO$_2$ systems are $\gamma_{gb(ThO_2)}/\gamma_{ThO_2/Ni} = 0.52$ and $\gamma_{gb(ThO_2)}/\gamma_{ThO_2/NiCr} = 0.48$.[105]

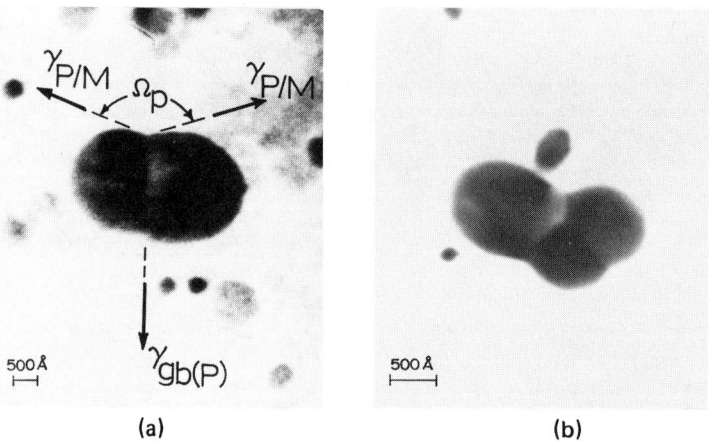

FIGURE 2.28 Electron transmission microscope observations of particle equilibria in dispersion-hardened systems. (a) ThO$_2$ particles in a NiCr matrix. (b) Equilibration of ThO$_2$ particles in a pure nickel matrix.[104]

Precipitate or Bubble at a Grain Boundary. The equilibrium shape of a precipitate or a (gas) bubble at a grain boundary can be treated as a special case of a particle on a matrix (or substrate), and, as Winterbottom pointed out,[96] the orientation dependence of the interfacial free energies and grain boundary free energies could be determined. The situation of a precipitate or a bubble on a grain boundary can be described generally by Fig. 2.25 and Eq. (2.5a). However, where the precipitate is a second phase in a metal or homogeneous (single-phase) alloy, and symmetrically disposed within the interface, the balance of interfacial free energies can be described by an equation having the essential form of Eq. (2.49), where in this instance $\gamma_{gb(P)} = \gamma_{gb}$, the average grain boundary free energy of the matrix. The implications of these features are illustrated in Fig. 2.29.

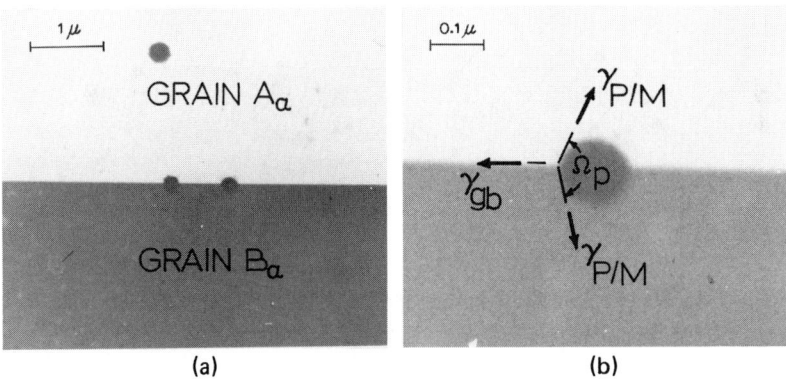

FIGURE 2.29 Carbide precipitates in annealed austenite observed in the electron microscope. (a) Low magnification view of several precipitates. (b) Magnified view of single precipitate showing interfacial free energy equilibrium.

2.4.2 Two- and Three-Phase Alloy Equilibria

As originally discussed by Smith,[8] three-phase equilibria in alloys can be treated phenomenologically as illustrated in Fig. 2.25 and described in terms of interfacial free energy balance in Eq. (2.5a). However, with reference to Fig. 2.25, either phases 1, 2, and 3 are all solid phases, or the system may be composed of a liquid or liquid phases equilibrating with solid phases or a single solid phase at temperature. Observations on such systems are normally made when all phases are solid and the equilibria established at temperature are "frozen in" with rapid cooling. As demonstrated in the original work of Smith, there is little error in assuming the mean value for a sufficiently large distribution of interphase angles to be the dihedral contact angle even for optical metallographic examinations of two-dimensional surface microstructures.[7] However, no accuracy can be assigned to individual equilibrium configurations unless the true dihedral angles can be measured as in the case of thin foil transmission electron microscopy, or where interfaces are equilibrated perpendicular to the section surfaces. It is worth noting, as Figure 2.30 illustrates schematically, that equilibrium interfacial free energy ratios for interphase interface

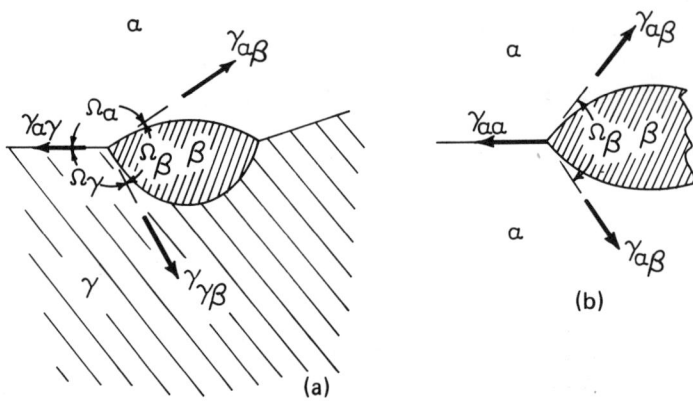

FIGURE 2.30 Equilibrium between three-phase and two-phase alloy systems. (a) General three-phase system. (b) Symmetric two-phase system; β represents the second phase.

equilibria are obtained as simple ratios of the dihedral angles according to the Dupré equation [Eq. (2.5a)] in the form

$$\gamma_{\alpha\gamma}/\gamma_{\gamma\beta} = \sin \Omega_\beta/\sin \Omega_\gamma; \quad \gamma_{\alpha\beta}/\gamma_{\gamma\beta} = \sin \Omega_\alpha/\sin \Omega_\gamma;$$
$$\gamma_{\alpha\beta}/\gamma_{\alpha\gamma} = \sin \Omega_\alpha/\sin \Omega_\beta \qquad (2.50)$$

with reference to Fig. 2.30(a). For a two-phase system [$\gamma = \alpha$ in Fig. 2.30(a)] where torques are negligible, the interphase boundaries will generally form symmetrical configurations where, with reference to Fig. 2.30(b),

$$\gamma_{\alpha\alpha}/\gamma_{\alpha\beta} = \gamma_{gb(\alpha)}/\gamma_{\alpha\beta} = 2 \cos (\Omega_\beta/2) \qquad (2.51)$$

It is apparent on inspection of Fig. 2.30(b) and Eq. (2.51) that, if the interphase boundary free energy is more than half that of the grain boundary free energy ($\gamma_{\alpha\alpha}$), the dihedral angle measured in the second phase (Ω_β) will be a positive (first-quadrant) angle; if $\gamma_{gb(\alpha)} = \gamma_{\alpha\beta}$, $\Omega_\beta = 120°$; if $\gamma_{\alpha\beta} > \gamma_{gb(\alpha)}$, $\Omega_\beta > 120°$. In addition, if $\gamma_{\alpha\beta}$ is less than one-half the grain boundary free energy, $\gamma_{gb(\alpha)}$, there is no value of Ω_β that can satisfy Eq. (2.51), and the second phase (β) will wet the grain boundary and spread indefinitely along the interface ($\alpha\alpha$). Figure 2.31 shows in idealized form several second-phase shapes for various values of Ω_β of small volumes of the second phase at a grain boundary triple junction, and some experimental examples of these features in alloy systems.

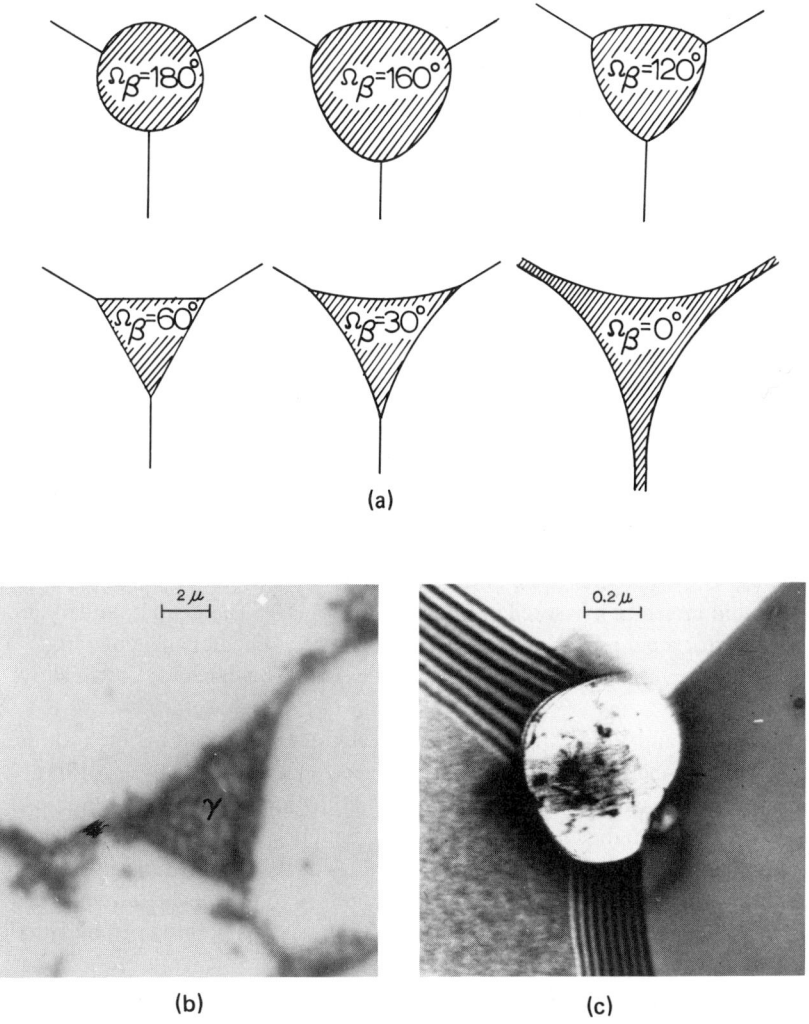

FIGURE 2.31 Second-phase equilibria at a grain boundary triple junction. (a) Idealized second-phase shapes as a function of the dihedral angle, Ω_β. (b) Optical metallographic observation of γ-phase in α titanium alloy. (c) Transmission electron micrograph of carbide in austenite (γ).

Equilibrium of Transformed Phases. Except for phase transformations or precipitation where a coherent interface forms, the simple balance of interfacial free energies governs the equilibrium configurations. However, even in the coherent cases, the interfacial free energy associated with the precipitate or coherent phase is paramount in determining the growth of a nucleus. Eutectic and martensitic phases possessing noncoherent interfaces are treated simply as interphase boundaries, as illustrated ideally in Fig. 2.30. Many such interphase interfaces possess some coherency, and in addition exhibit considerable variation with respect to matrix crystallography. Consequently, torque components are normally active in establishing equilibrium of a transformed phase, and assumptions made concerning the negligence of torques must be carefully considered.

Grain Boundary-Solid/Liquid Interface Intersection Equilibrium. A special equilibrium situation of interest is that of a surface in equilibrium with its own liquid which is intersected by a grain boundary. The equilibrium equation for a solid–liquid interface can be written ideally as*

$$\gamma_{LS} \equiv \gamma_{SL} = F_S - \gamma_{LV} \tag{2.52}$$

where F_S denotes the solid-vapor surface free energy as before, and γ_{LV} is the liquid-vapor surface free energy. The equilibrium of a grain boundary with the solid-liquid interface is essentially the same as Eq. (2.51) in form, namely:

$$\gamma_{\alpha\alpha}/\gamma_{SL} = \gamma_{gb}/\gamma_{SL} = 2\cos(\Omega_L/2) \tag{2.53}$$

where Ω_L is the dihedral angle at the intersection of the grain boundary with the solid-liquid interface measured in the liquid phase (L). The grain boundary groove angle, Ω_L, for a pure solid-melt system usually has a dihedral angle of zero.[106] Consequently, $\gamma_{SL} \cong 0.5\gamma_{gb}$. This is generally not to be expected for an alloy system.

Glicksman and Vold[107] have devised a technique for measuring the ratio of γ_{gb}/γ_{SL} in low-melting-point metals in the electron microscope by recording observations of partially melted foils. Their results indicated the mean value of γ_{gb}/γ_{SL} to be 1.6 in bismuth foils where γ_{gb} could be associated with a high-angle grain boundary ($\Theta > 15°$). The measured values of the dihedral angle were also observed to vary with misorientation, and therefore to reflect the influence of torques on the solid-liquid interfacial equilibrium, as was pointed out by Horylev and Murr.[108]

Miller and Chadwick[109] have considered the dihedral angle data in binary eutectic systems and estimated the expected value of γ_{gb}/γ_{SL} to be approximately 2.25. This is to be compared with values of γ_{gb}/γ_{SL} obtained by Hilliard and Cahn:[110] 2.76, 3.11, and 3.17 for gold, copper, and silver at 900°C, respectively.

---

*Woodruff[106] has argued that for most materials it is possible to write $\gamma_{SL} < F_S - \gamma_{LV}$ because a solid usually melts as soon as it reaches its melting point.

2.5 SUMMARY

This chapter has been concerned with a description of interface equilibrium treated as a condition of energy balance where the boundary plane does not vary significantly with crystallography, or of equilibrium configurations that require the resolution of a resultant or net effective torque to establish a vectorial balance at a designated junction point of crystallographically unique interface planes. In the case of metals and alloys, it has been shown that an examination of annealed, equilibrated microstructures—grains, phases, inclusions, etc.—by optical or electron metallography can be employed in the characterization of interfacial energetics through the measurement of (mean) relative interfacial free energy ratios or the determination of mean interfacial torque or net effective torque ratios.

It is the ratios of interfacial free energies that allow one to systematically evaluate the interfacial equilibria and to characterize structurally, geometrically, or crystallographically unique interfaces in metals and alloys according to their relative energies. Undoubtedly, the most effective way to summarize these distinctions for the range of boundary characteristics in metals and alloys discussed herein is to present a tabulation of the experimentally measured interfacial energy or torque ratios. These data are tabulated in Table 2.1 for single-phase systems. Table 2.2 summarizes most of the available data for multiphase systems, including solid interphase interfacial free energy ratios. *Tables 2.1-2.2, pp. 76-82.*

REFERENCES

1. J. W. Gibbs, "The Scientific Papers of J. Willard Gibbs," Vol. I, Dover Publications, New York (1961).
2. C. Herring, in "The Physics of Powder Metallurgy," W. E. Kingston (ed.), McGraw-Hill Book Co., New York, p. 143 (1951).
3. A. U. MacRae, E. R. Jones, J. T. McKinney, and M. B. Webb, *Phys. Rev.*, **151**, 476 (1966).
4. P. Wynblatt and N. A. Gjostein, *Surface Sci.*, **12**, 109 (1968).
5. R. W. Vook, S. Ouyang, and M. A. Otooni, *Surface Sci.*, **29**, 277 (1972).
6. C. S. Smith, *Met. Rev.*, **9**, No. 33 (1964).
7. D. Harker and E. R. Parker, *Trans. ASM*, **34**, 156 (1945).
8. C. S. Smith, *Trans. AIME*, **175**, 15 (1948).
9. L. E. Murr, *J. Appl. Phys.*, **39**, 5557 (1968).
10. R. L. Fullman, Ph.D. Dissertation, Yale University, New Haven, Connecticut (1951).
11. R. T. DeHoff and F. N. Rhines (eds.), "Quantitative Microscopy," McGraw-Hill Book Co., New York (1968).
12. E. E. Underwood, "Quantitative Stereology," Addison-Wesley Publishing Co., Reading, Massachusetts (1970).
13. G. Thomas, "Transmission Electron Microscopy of Metals," John Wiley & Sons, New York (1962).
14. P. B. Hirsch, A. Howie, R. B. Nicholson, D. W. Pashley, and M. J. Whelan, "Electron Microscopy of Thin Crystals," Butterworths, London (1966).
15. L. E. Murr, *Phys. Status Solidi*, **19**, 7 (1967).
16. L. E. Murr, "Electron Optical Applications in Materials Science," McGraw-Hill Book Co., New York (1970).

References continued, p. 83.

TABLE 2.1
Solid Interfacial Free Energy Ratios for Single-Phase Systems[a]

System (Metal or Alloy)	Method and Atmosphere	Temperature (°C)	γ_{gb}/F_S	γ_{tb}/F_S	γ_{tb}/γ_{gb}	γ_{TB}/γ_{gb}	$\Sigma M/F_S$	$\Sigma M/\gamma_{gb}$	Ref.
Aluminum	TEM Vac.	450			0.23				87
	OM Vac.	625			0.21				111
	OM —	—						0.15	26, 27
Chromium	Interfer Ar	1350	0.40						34
	Ar	1550	0.37						34
	Ar	1800	0.31						34
Cr–35 w/o Re	Interfer Ar	1350	0.30						34
	Ar	1800	0.30						34
Copper	Interfer H_2, Ar	850	0.36						112
	TEM(S) He	950	0.34						39
	Interfer H_2	1065	0.37						55
	H_2	1000		0.01					59
	OM Vac.	950			0.035	0.80			10
	TEM Vac.	700			0.034				78
	Vac.	800			0.034				113
	Vac.	875			0.034				73
	Vac.	900			0.034				78, 113

INTERFACIAL EQUILIBRIA

Cu–5 a/o Al	Vac.	700			0.10	
	TEM	700		0.032	0.01	78, 22, 77
Cu–11 a/o Al	Vac.	700		0.030		78
Cu–16 a/o Al	Vac.	700		0.018		78
Cu–0.26 a/o Sb	TEM(S)	950	0.30			41
Cu–0.57 a/o Sb	He	950	0.29			41
Cu–0.78 a/o Sb	He	950	0.36			41
Cu–5 w/o Zn	TEM	875		0.032		73
Cu–10 w/o Zn	Vac.	875		0.031		73
Cu–20 w/o Zn	Vac.	875		0.029		73
Cu–30 w/o Zn	Vac.	875		0.023		73
Gold	Interfer	1130	0.26			114
	Ar					
	Interfer	1035	0.28			115
	Vac.					
	Interfer	1027	0.28			114
	He					
	Interfer	850	0.25			116
	Vac.					
	TEM	1000		0.039		16
	Vac.					
Au–20 w/o Cu	Interfer	850	0.28			39, 116
Au–40 w/o Cu	Vac.	850	0.34			39, 116
Au–60 w/o Cu	Vac.	850	0.42			39, 116
Au–80 w/o Cu	Vac.	850	0.37			39, 116
Indium	TEM	110		0.055		72
	Vac.					

TABLE 2.1 (continued)

System (Metal or Alloy)	Method and Atmosphere	Temperature (°C)	γ_{gb}/F_S	γ_{tb}/F_S	γ_{tb}/γ_{gb}	γ_{TB}/γ_{gb}	$\Sigma M/F_S$	$\Sigma M/\gamma_{gb}$	Ref.
	TEM(S) Vac.	110	0.31						—
Iridium	TEM Vac.	1715			0.041				117
	TEM(S) He	1715	0.26						—
Iron									
α-phase	Interfer Vac.	880		0.12					39
δ-phase	Vac.	1100	0.40						118
	OPT. Ar	1400–1535	0.24						119
γ-phase	Ar	~1375	0.36						119
	Ar	~1000		0.014			0.10		57
Stainless steel (304)	TEM He	910			0.012				—
	He	980			0.016				120
	He	1060	0.38		0.024	0.253		0.10	22, 28
	He	1160	0.38		0.046			0.30	—
	He	1260	0.45						—
	He	1360	0.36						—
Fe–24% Al	Interfer Vac.	1150	0.27						121
Fe–3 w/o Si	TEM(S) H_2	~1375	0.32						122
Lead	OM Ar	220			0.05				68

INTERFACIAL EQUILIBRIA

Material	Method	Atmosphere	Temp				Ref
	TEM(S)	He	200	0.39			—
Pb–0.1 a/o Ag	OM	Ar	220			0.08	68
Molybdenum	Interfer	Vac.	1600	0.24			123
		Vac.	2000	0.23			123
		Vac.	2400	0.20			123
Mo–33 w/o Re	Interfer	Vac.	2200	0.25			34
	Interfer	Ar	2300	0.25		0.03	34
Nickel	Interfer	Ar	~1350	0.40			124
	Interfer	Vac.	1000		0.015		49
	TEM	He	1060	0.38		0.2	117
	Interfer	Vac.	900	0.26			125
	TEM	He	1200	0.39		0.050	—
Ni–20 w/o Cr	TEM	He	1060	0.35	0.009	0.046	38
	TEM(S)	He	1200	0.39		0.025	—
Ni~16 w/o Cr 8 w/o Fe (Inconel 600)	TEM	Vac.	1060	0.41		0.022	16, 71
Ni–2 v/o ThO$_2$	TEM	Vac.	1200			0.040	71 (corrected)

TABLE 2.1 *(continued)*

System (Metal or Alloy)	Method and Atmosphere	Temperature (°C)	γ_{gb}/F_S	γ_{tb}/F_S	γ_{tb}/γ_{gb}	γ_{TB}/γ_{gb}	$\Sigma M/F_S$	$\Sigma M/\gamma_{gb}$	Ref.
Ni-20 a/o Cu	Interfer Vac.	900	0.29						125
Ni-40 a/o Cu	Vac.	900	0.32						125
Ni-80 a/o Cu	Vac.	900	0.32						125
Niobium	OM Vac.	2250	0.36						126
Platinum	Interfer Vac.	1100	0.33	0.050			~0.07		127
	Vac.	1500	0.24	0.028					127
Silver	TEM Vac.	950	0.25		0.030	0.33			16
	Interfer Ar	900	~0.30						128
Tin	Interfer Vac.	223	0.24						129
Tungsten	Interfer Ar, He	3100	0.36						123
W–33 w/o Re	Interfer Ar, He	3100	0.36						123
Zinc	OM Vac.	350						~0.30	130

OM: Optical metallography (surface microscopy) — optical analysis of thermal grooves on wires.
Interfer: Interferometric measurement of thermal grooves on specimen surfaces.
TEM: Transmission electron microscopy — thin foil microscopy and electron shadowgraphy of thermal grooves on wires.

TEM(S): Transmission electron shadowgraphy of thermal grooves on wires.
Atmospheres: Vac. (vacuum); Ar (argon); He (helium); H_2 (hydrogen); etc.

γ_{gb}: Grain boundary free energy.
F_S: Solid–vapor surface free energy.
γ_{tb}: Coherent twin boundary free energy.
γ_{TB}: Noncoherent twin boundary free energy.
ΣM: Net effective or resultant (mean) torque.

[a] All ratio values given are experimentally determined values and represent *mean* values for the temperature or temperature range cited.

TABLE 2.2
Multiphase (Equilibrium) Interfacial Free Energy Ratios[a]

System (Composition in w/o)	Ratio Designation	Temperature (°C)	Method	Ratio	Ref.
Al–4 Cu (α) θ precipitates in	$\gamma_{\alpha\theta}/\gamma_{gb(\alpha)}$	200–475	TEM	0.91	131
Bismuth	γ_{gb}/γ_{SL}	271	TEM	1.6	107
Copper	γ_{gb}/γ_{SL}	900	OM	3.11	110
60 Cu–40 Zn	$\gamma_{\alpha\beta}/\gamma_{gb(\alpha)}$	600–700	OM	0.78	8
91 Cu–9 Al	$\gamma_{\alpha\beta}/\gamma_{gb(\alpha)}$	600	OM	0.71	8
86 Cu–14 Al	$\gamma_{\gamma\beta}/\gamma_{gb(\gamma)}$	600	OM	0.87	8
79 Cu–21 Sn	$\gamma_{\alpha\beta}/\gamma_{gb(\beta)}$	700	OM	0.87	8
84 Cu–12 Sn	$\gamma_{\alpha\beta}/\gamma_{gb(\alpha)}$	750	OM	0.74	8
Cu–liquid Pb	$\gamma_{Cu/Pb}/\gamma_{gb(Cu)}$	600–900	OM	0.58	8
79 Cu–17 Sn–4 Pb	$\gamma_{\alpha\beta}/\gamma_{gb(\alpha)}$	600	OM	0.74	8
Iron	$\gamma_{\alpha\gamma}/\gamma_{gb(\gamma)}$	910	EEM	0.95	95
	$\gamma_{\alpha\gamma}/\gamma_{gb(\alpha)}$	910	EEM	0.95	95
	$\gamma_{gb(\gamma)}/\gamma_{gb(\alpha)}$	910	EEM	1.00	95
	$\gamma_{\alpha\gamma}/\gamma_{gb(\alpha)}$	1000	OM	0.70	94
Fe–C	$\gamma_{\alpha\gamma}/\gamma_{gb(\gamma)}$	910	EEM	0.95	95
	$\gamma_{\alpha\gamma}/\gamma_{gb(\alpha)}$	910	EEM	0.95	95
Fe–Cu	$\gamma_{gb(Fe)}/\gamma_{gb(Cu)}$	1000	OM	0.61	94
Fe–30 Cu (liquid/solid)		1125	OM	0.31	8
Austenite/carbide precipitate (γFe–MC$_6$)	$\gamma_{gb(\gamma Fe)}/\gamma_{P/M}$	1200	TEM	0.82	—
Gold	γ_{gb}/γ_{SL}	830	OM	2.76	110
Ni/ThO$_2$	$\gamma_{gb(ThO_2)}/\gamma_{P/M}$	1200	TEM	0.52	—
80 Ni–20 Cr/ThO$_2$	$\gamma_{gb(ThO_2)}/\gamma_{P/M}$	1200	TEM	0.48	—
Silver	γ_{gb}/γ_{SL}	900	OM	3.17	110
90 Zn–6 Cu–4 Al	$\gamma_{\epsilon\eta}/\gamma_{gb(\epsilon)}$	375	OM	0.93	8
	$\gamma_{\epsilon\beta}/\gamma_{gb(\epsilon)}$	375	OM	0.74	8
	$\gamma_{\epsilon\beta}/\gamma_{gb(\beta)}$	375	OM	0.87	8

[a] OM: Optical metallography (surface microscopy).
TEM: Transmission electron microscopy (thin foil microscopy).
EEM: Electron emission microscopy (surface microscopy).

γ_{SL}: Solid–liquid interfacial free energy.

$\gamma_{gb(x)}$: Grain boundary free energy of phase (x).

$\gamma_{P/M}$: Particle–matrix interfacial free energy.

γ_{xy}: Interphase interfacial free energy between phase x and phase y.

17. C. S. Barrett and T. B. Massalski, "Structure of Metals," McGraw-Hill Book Co., New York, p. 30 (1966).
18. S. Tolansky, "Multiple-Beam Interference Microscopy of Metals," Academic Press, New York (1970).
19. L. E. Murr, *Phys. Status Solidi,* 25, 629 (1968).
20. C. Goux, *Mem. Sci. Rev. Met.,* 58, 661 (1961).
21. F. F. Lange, *Acta Met.,* 15, 311 (1967).
22. L. E. Murr, R. J. Horylev, and W. N. Lin, *Phil. Mag.,* 22, 515 (1970).
23. W. H. Hartt, G. H. Bishop, and G. Bruggeman, *J. Metals,* 20, 71A (1968).
24. J. Basterfield, W. A. Miller, and G. C. Weatherly, *Can. Met. Quart.,* 8, 131 (1970).
25. F. Weinberg, quoted in ref. 24, p. 139.
26. J. B. Hess, in "Metal Interfaces," ASM, Metals Park, Ohio, p. 134 (1952).
27. W. A. Miller and W. M. Williams, *Acta Met.,* 15, 1077 (1967).
28. R. J. Horylev and L. E. Murr, in "The Nature and Behavior of Grain Boundaries," H. Hu (ed.), Plenum Press, New York, p. 203 (1972).
29. B. Chalmers, R. King, and R. Shuttleworth, *Proc. Roy. Soc., Ser. A,* 193, 465 (1948).
30. W. W. Mullins, *J. Appl. Phys.,* 28, 333 (1957).
31. W. W. Mullins, *Trans. AIME,* 218, 354 (1960).
32. N. A. Gjostein, *Trans. AIME,* 221, 1039 (1961).
33. W. M. Robertson, *Trans. AIME,* 233, 1232 (1965).
34. B. C. Allen, *Trans. AIME,* 236, 915 (1966).
35. H. Mykura, *Proc. Phys. Soc., Ser. B,* 67, 281 (1954).
36. F. R. Tolmon and J. G. Wood., *J. Sci. Instrum.,* 33, 236 (1956).
37. H. Mykura, *J. Sci. Instrum.,* 40, 313 (1963).
38. L. E. Murr, R. J. Horylev, and G. I. Wong, *Surface Sci.,* 26, 184 (1971).
39. M. C. Inman and H. R. Tipler, *Met. Rev.,* 8, No. 30, 105 (1963).
40. A. P. Greenough, *Appl. Mater. Res.,* 4, 25 (1965).
41. M. C. Inman, D. McLean, and H. R. Tipler, *Proc. Roy. Soc., Ser. A,* 273, 538 (1963).
42. M. Drechsler and J. F. Nicholas, *J. Phys. Chem. Solids,* 28, 2609 (1967).
43. G. Friedel, "Leçons de Cristallographie," Berger-Levrault Publishers, Paris (1926).
44. M. L. Kronberg and F. H. Wilson, *Trans. AIME,* 185, 501 (1949).
45. D. G. Brandon, *Acta Met.,* 14, 1479 (1966).
46. P. H. Pumphrey and K. M. Bowkett, *Scripta Met.,* 5, 365 (1971).
47. C. Herring, in "Structure and Properties of Solid Surfaces," R. Gomer and C. S. Smith (eds.), University of Chicago Press, Chicago (1953).
48. H. Mykura, *Acta Met.,* 5, 346 (1957).
49. H. Mykura, *Acta Met.,* 9, 570 (1961).
50. F. R. Winslow and P. G. Shewmon, *Trans. AIME,* 227, 1078 (1963).
51. W. M. Robertson and P. G. Shewmon, *J. Chem. Phys.,* 39, 2330 (1963).
52. N. A. Gjostein, *Acta Met.,* 11, 957 (1963).
53. W. L. Winterbottom and N. A. Gjostein, *Acta Met.,* 14, 1033 (1966).
54. W. L. Winterbottom and N. A. Gjostein, *Acta Met.,* 14, 1041 (1966).
55. N. A. Gjostein and F. N. Rhines, *Acta Met.,* 7, 319 (1959).
56. M. McLean and B. Gale, *Phil. Mag.,* 20, 1033 (1969).
57. M. McLean and H. Mykura, *Acta Met.,* 12, 326 (1964).
58. J. F. Nicholas, *Australian J. Phys.,* 21, 21 (1968).
59. W. M. Robertson and P. G. Shewmon, *Trans. AIME,* 224, 804 (1962).
60. M. McLean and H. Mykura, *Surface Sci.,* 5, 466 (1966).
61. H. Mykura, *Bull. Inst. Metals,* 4, 102 (1958).
62. C. S. Barrett and T. B. Massalski, "Structure of Metals, 3rd ed., McGraw-Hill Book Co., New York (1966).
63. J. C. Fisher, *Trans. AIME,* 175, 906 (1948).
64. C. G. Dunn, *Trans. AIME,* 185, 72 (1949).

65. C. G. Dunn and F. Lionetti, *Trans. AIME,* **185,** 125 (1949).
66. C. G. Dunn, F. W. Daniels, and M. J. Bolton, *Trans. AIME,* **188,** 368 (1950).
67. R. L. Fullman, *J. Appl. Phys.,* **22,** 448 (1951).
68. G. F. Bolling and W. C. Winegard, *J. Inst. Metals,* **86,** 492 (1958).
69. M. C. Inman and A. R. Khan, *Phil. Mag.,* **6,** 937 (1961).
70. L. E. Murr, *Acta Met.,* **16,** 1127 (1968).
71. L. E. Murr, P. J. Smith, and C. M. Gilmore, *Phil. Mag.,* **17,** 89 (1968).
72. L. E. Murr, *phys. Status Solidi (a),* **1,** 487 (1970).
73. R. A. Queeney, *Scripta Met.,* **5,** 1031 (1971).
74. N. A. Gjostein, *Scripta Met.,* **3,** 1 (1969).
75. J. Basterfield and W. A. Miller, *Scripta Met.,* **3,** 343 (1969).
76. L. E. Murr, R. J. Horylev, and W. N. Lin, *Scripta Met.,* **3,** 347 (1969).
77. L. E. Murr, R. J. Horylev, and W. N. Lin, *Phil. Mag.,* **20,** 1245 (1969).
78. L. E. Murr, *Phys. Status Solidi (a),* **3,** 447 (1970).
79. L. E. Murr, *Scripta Met.,* **6,** 203 (1972).
80. J. E. Burke and D. Turnbull, and "Progress in Metal Physics," Pergamon Press, London, p. 282 (1952).
81. E. O. Hall, "Twinning," Butterworths, London (1954).
82. R. W. Cahn, *Advan. Phys.,* **3,** 363 (1954).
83. H. C. H. Carpenter and S. Tamura, *Proc. Roy. Soc., Ser. A,* **113,** 161 (1926).
84. J. E. Burke, *Trans. AIME,* **188,** 1324 (1950).
85. W. L. Grube and S. R. Rouze, *Can. Met. Quart.,* **2,** 31 (1963).
86. R. L. Fullman and J. C. Fisher, *J. Appl. Phys.,* **22,** 1350 (1951).
87. L. E. Murr, *Acta Met.,* **21,** 791 (1973).
88. J. P. Nielsen, *Acta Met.,* **15,** 1083 (1967).
89. J. P. Nielsen, *J. Metals,* **18** (1966).
90. K. T. Aust and J. W. Rutter, *Trans. AIME,* **218,** 1023 (1960).
91. K. T. Aust, *Trans. AIME,* **221,** 758 (1961).
92. G. Ferran, G. Cizeron, and K. T. Aust, *Mem. Sci. Rev. Met.,* **54,** 1067 (1967).
93. W. D. Kingery, *J. Amer. Ceram. Soc.,* **37,** 42 (1954).
94. L. H. VanVlack, *Trans. AIME,* **191,** 251 (1951).
95. N. A. Gjostein, H. A. Domian, H. I. Aaronson, and E. Eichen, *Acta Met.,* **14,** 1637 (1966) (1966).
96. W. L. Winterbottom, *Acta Met.,* **15,** 303 (1967).
97. A. W. Adamson, "Physical Chemistry of Surface," Wiley-Interscience, New York, p. 265 (1960).
98. J. J. Bikerman, "Physical Surfaces," Academic Press, New York (1970).
99. Th. Young, *Phil. Trans.,* **1,** 84 (1805).
100. J. E. McNutt and G. M. Andes, *J. Chem. Phys.,* **30,** 1300 (1959).
101. R. M. Pilliar and J. Nutting, *Phil. Mag.,* **16,** 181 (1967).
102. B. E. Sundquist, *Acta Met.,* **12,** 67 (1964).
103. R. J. Horylev and L. E. Murr, in "Proceedings Electron Microscopy Society of America," C. J. Arceneaux (ed.), Claitor's Publishing Division, Baton Rouge, Louisiana, p. 168 (1969).
104. L. E. Murr, *J. Mater. Sci.,* **9,** 1309 (1974).
105. L. E. Murr, to be published.
106. D. P. Woodruff, "The Solid-Liquid Interface," Cambridge University Press, London (1973).
107. M. E. Glicksman and C. L. Vold, *Acta Met.,* **17,** 1 (1969).
108. R. J. Horylev and L. E. Murr, *Scripta Met.,* **3,** 783 (1969).
109. W. A. Miller and G. A. Chadwick, *Acta Met.,* **15,** 607 (1967).
110. J. E. Hilliard and J. W. Cahn, *Acta Met.,* **6,** 772 (1958).

111. R. L. Fullman, General Electric Co. Research Report NO. RL-422 (1950).
112. G. Bailey and H. Watkins, *Proc. Phys. Soc., Ser. B*, **63**, 350 (1950).
113. L. E. Murr, *Scripta Met.*, **6**, 203 (1972).
114. F. H. Buttner, H. Udin, and J. Wulff, *Trans. AIME*, **197**, 313 (1953).
115. J. M. Blakely, *Trans. Faraday Soc.*, **57**, 1164 (1961).
116. J. E. Hilliard, M. Cohen, and B. L. Averbach, *Acta Met.*, **8**, 26 (1960).
117. L. E. Murr, O. T. Inal, and G. Wong, in "Electron Microscopy and Structure of Materials," G. Thomas (ed.), University of California Press, Berkeley, California, p. 417 (1972).
118. K. Mazanec and E. Kamenska, *Phys. Met. Metal. (USSR)*, **12**, 79 (1961).
119. A. T. Price, H. A. Holl, and A. P. Greenough, *Acta Met.*, **12**, 49 (1964).
120. L. E. Murr and R. J. Horylev, in "Proceedings of the Electron Microscopy Society of America," C. J. Arceneaux (ed.), Claitor's Publishing Division, Baton Rouge, Louisiana (1972).
121. R. W. Cahn and J. A. Coll, *Acta Met.*, **9**, 138 (1961).
122. B. Mills and G. M. Leak, *Acta Met.*, **16**, 303 (1968).
123. B. C. Allen, *Trans. AIME*, **245**, 2089 (1969).
124. E. R. Hayward and A. P. Greenough, *J. Inst. Metals*, **88**, 217 (1959).
125. T. M. Williams and P. Barrand, *J. Inst. Metals*, **93**, 447 (1965).
126. S. V. Radcliffe, *J. Less-Common Metals*, **3**, 360 (1961).
127. M. McLean and H. Mykura, *Acta Met.*, **13**, 1291 (1965).
128. E. R. Funk, H. Udin, and J. Wulff, *Trans. AIME*, **191**, 1206 (1951).
129. H. Mykura, *Acta Met.*, **3**, 436 (1955).
130. G. H. Bishop, W. H. Hartt, and G. A. Bruggeman, *Acta Met.*, **19**, 37 (1971).
131. H. B. Aaron and H. I. Aaronson, *Acta Met.*, **18**, 699 (1970).

PROBLEMS

(2.1) Examine the "grain" structure in a styrofoam coffee cup, and, using a hand lens (magnifier) if necessary, plot a distribution curve (for at least 100 measurements) of the number of sides (edges) per grain. Calculate the mean value of Δ, and compare it with the value of slightly more than 5 found for typical metals and alloys.

(2.2) Three interfaces meet at a common point in a pure metal. Two of the interfacial free energies resolved in the interfaces are equal, while the third has an interfacial free energy only one-third that of the other two. Sketch an equilibrium configuration, indicating the dihedral angles between the interfaces, assuming interfacial torques to be negligible.

(2.3) Thin sheets of pure copper and gold are heated separately in a vacuum or inert gas environment to form thermal grooves where the grain boundaries intersect the surface. Calculate the average groove angle formed at $1000°C$. Describe how you could experimentally determine whether your calculated value would be valid based on the equilibration of the grain boundaries. That is, discuss how you could prove the grooves were symmetrical.

(2.4) Calculate the true dihedral angles at the intersection of grain boundaries associated with grains 1, 2, and 3 in Fig. 2.6(a), and calculate the ratios of the associated grain boundary free energies.

(2.5) Calculate the mean value of the net effective torque (ΣM) associated with the grain boundary shown in Fig. 2.9 (note that the configuration is equilibrated at 1000°C, and assume the grain boundary plane to be perpendicular to the specimen surfaces). Compare this value with the energy of coherent (annealing) twin boundaries equilibrated at the same temperature.

(2.6) Find the surface-to-grain boundary free energy ratio in Fig. 2.11(f) and (g), and discuss any differences you calculate. What is a possible mechanistic reason for the difference in groove profile in these two cases? Support your argument by considering Eq. (2.15) and Eq. (2.17). From measurements of the groove profile in Fig. 2.11(g), determine whether the grooving was due to volume diffusion or surface diffusion, and support your argument with measured data.

(2.7) From measurements on Fig. 2.27(a), (b), (d), and (e), comment on the validity of Eq. (2.46). Explain any discrepancies after correcting for measurement error. If the energy of adhesion for mercury on palladium at 25°C is 720 ergs/cm^2 and for stainless steel on Al_2O_3 at 1000°C is 1440 ergs/cm^2 (see Table 5.2), calculate the corresponding values for the corresponding Wulff constants.

(2.8) Calculate the specimen thickness of the sample in Fig. 2.22(b) and, having also calculated the true dihedral angles, evaluate γ_{tb}/γ_{gb} from Eq. (2.41), assuming that $A_2 = A'_2, A_3 = A'_3$, and $\Omega_2 = \Omega_3 = 120°$. Compare this value with that reported experimentally in Table 2.1. Referring to Fig. 2.23, find the average change in the grain boundary energies in twinned grains of nickel and aluminum.

(2.9) Calculate the ratio of ThO_2-NiCr interfacial free energy to ThO_2 grain boundary free energy for the particle in Fig. 2.28(a), and compare it with the mean value of the same energy ratio measured from Fig. 2.28(b). What assumptions must be made in making these measurements? What is the ratio of the ThO_2 particle-matrix interfacial free energies in these systems? What can you conclude about ThO_2 particle stability in these two systems?

(2.10) Consult a text on electron microscopy (for example, L. E. Murr, "Electron Optical Applications in Materials Science," McGraw-Hill Book Co., New York, 1970). Referring to Fig. 2.31(c), find the ratio of the austenite grain boundary free energy to the carbide-austenite interfacial free energy by initially considering the grain boundary fringes to result from a <111> operating reflection at an imaging voltage of 200 kilovolts in the transmission electron microscope. Estimate the γ-phase/α-Ti interfacial free energy/α-Ti grain boundary free energy ratio in Fig. 2.31(a), assuming the angles measured to be the true dihedral angles.

3
INTERFACIAL FREE ENERGY

3.1 SURFACE ENERGY OF LIQUID METALS AND ALLOYS; THE CONCEPT OF SURFACE TENSION

The surface tension of any general liquid is, as described in Section 1.1, a measurable force existing through the surface, arising primarily from a combined effect of attractive forces between all atoms or molecules bringing them as close together as the repulsive forces arising from overlapping wave functions (or electron clouds) will allow. In a general liquid such as water, for example, the intermolecular forces can consist of overlapping electron repulsion and a number of intermolecular attractive forces commonly classified according to primary (chemical) bonds and secondary (physical) bonds which involve dispersion forces, Debye forces, and Keesom forces, all of which are polar or nonpolar combinations. Although these force contributions are important in connection with surface tensions of nonmetallic liquids, liquid metal and alloy surface tensions arise primarily by metallic interatomic force interactions.

One of the distinguishing features of liquids is that their surfaces rarely exhibit planar features common to solids. In addition, the atoms or molecules in a liquid surface (as shown in Fig. 3.1) can be considered to be acted upon by unsymmetrical forces, F_u (perpendicular to the surface), whose combined effect gives rise to an internal pressure acting equally in all directions, stabilized by a force, F_p, tangent to to the surface. In the absence of any measurable gravitational influence, the pressure due to the normal surfaces forces (F_u in Fig. 3.1) must be the same over the entire surface if the liquid is in equilibrium. This can occur only when the sum of the curvatures at the surface is identical at all points on the surface.

Young[1] and Laplace[2] originally recognized that the attractive forces between molecules in a liquid surface would create a pressure difference across a curved surface as illustrated in Fig. 3.1. Their derivation of the pressure in terms of the liquid surface tension forms the basis of all liquid surface tension and interfacial tension measurements. Laplace's fundamental equation governing the shape of all macroscopic menisci can be written (with reference to Fig. 3.1) as

$$\Delta p = \gamma_{LV}\left(\frac{1}{R_1} + \frac{1}{R_2}\right) \qquad (3.1)$$

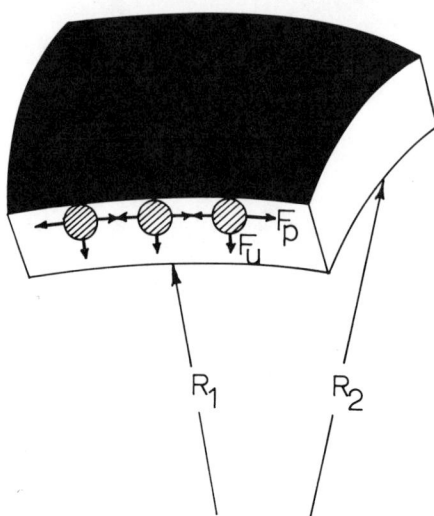

FIGURE 3.1 Curved surface element with principal radii R_1 and R_2.

where Δp is the pressure difference between any two sides of a surface element, and γ_{LV} is the surface (liquid-vapor) tension, identical to the surface free energy [Eq. (1.4)]. Although Eq. (3.1) is commonly referred to as the Laplace equation, it was arrived at somewhat simultaneously by Young and Laplace, and later by Gauss[3] and Poisson.[4]

Another of the obvious physical differences between the behavior of solids and liquids, and a distinguishing feature of specific surface free energy of solids and the surface tension of liquids, is that liquids will rise or be depressed in capillary tubes. Leonardo da Vinci[5] is believed to have been the first to observe and record the rise of a liquid in a capillary tube, from which the phenomenon came to be known as capillarity because the tube possessed a bore resembling a fine "hair" or *capillus* (Latin.) By the end of the seventeenth century, Hawksbee[6] performed the first systematic observations of the ascent of liquids in capillary tubes and between very closely spaced glass plates.

When a capillary tube is immersed in a liquid as shown in Fig. 3.2, the curved meniscus (M) can either rise above the liquid level or be depressed below it. This occurs because the capillary pressure under the horizontal surface is zero, whereas under the curved meniscus it can be either positive or negative depending on whether the meniscus is concave or convex. For a concave meniscus as in Fig. 3.2, the hydrostatic pressure, $g\rho h$, of the liquid in the capillary tube must equal the pressure difference between the liquid and vapor phases, or the excess pressure on the concave side over the convex side. Considering Eq. (3.1), we can therefore write

INTERFACIAL FREE ENERGY

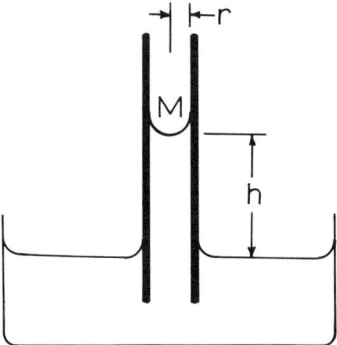

FIGURE 3.2   Rise of liquid in a capillary tube: M, meniscus.

$$\Delta p = g\rho_L h = \gamma_{LV}\left(\frac{1}{R_1} + \frac{1}{R_2}\right) \tag{3.2}$$

where g is the gravitational constant, ρ_L is the liquid density, and h is the capillary rise. If the capillary is a very narrow tube with a circular cross section, changes in hydrostatic pressure arising from gravitational effects are either zero or negligible, and the meniscus becomes part of a complete spherical drop: $R_1 = R_2$. When the liquid surface is tangential to the inner tube walls along the upper end of the meniscus, the contact angle, Ω_C, is zero, and the radius of curvature is equal to the internal tube radius, r. Equation (3.2) therefore assumes the form

$$g\rho h = 2\gamma_{LV}/r \tag{3.3}$$

If the capillary is very narrow, the contact angle may not be zero, since the solid material of the tube wall may influence the equilibrium conditions. If the contact angle, Ω_C, is not zero, Eq. (3.3) must be written

$$g\rho_L h = 2\gamma_{LV}\cos\Omega_C/r \tag{3.3a}$$

Conversely, when the capillary is not very narrow, the meniscus will not be a hemisphere, and some attempt must be made to account for the effect of shape on the surface tension. An approximation of the surface tension under these conditions can be obtained from[4,7]

$$\gamma_{LV} = \frac{g\rho_L r}{2}\left(h + \frac{r}{3} - \frac{0.129r^2}{h}\right) \tag{3.4}$$

when Ω_C is not equal to zero, and from

$$\gamma_{LV} = \frac{g\rho r}{2}\left(h + \frac{r}{3}\right) \qquad (3.4a)$$

when $\Omega_C = 0$.

Equation (3.3) may also be used to evaluate the vapor pressure inside a cavity of of radius r within a liquid, and to determine the vapor-pressure difference between a small droplet and a large flat surface of the same liquid. For a small gas bubble within a liquid[8],*

$$\gamma_{LV} = \frac{\rho_L rRT}{2M}\left(\ln \frac{P}{P_0}\right) \qquad (3.5)$$

where P is the vapor pressure of the small bubble of radius r, P_0 is the vapor pressure over a liquid surface of infinite radius, R is the gas constant, T is the temperature, and M is the molecular weight of the liquid having a density ρ_L. The stabilization of small air bubbles as a result of the lowering of the surface tension of water ($\sim$72 ergs/cm^2 at 20°C) by soaps, detergents, and flotation agents is an important consideration related to Eq. (3.5). In flotation, the interfacial tension (or interfacial free energy) between solid particles and the liquid phase is lowered in conjunction with the lowering of the liquid surface tension so that the particles are more readily wetted and floated after attachment to the included bubbles of air.

3.2 SURFACE TENSION MEASUREMENTS

As a method of measuring surface tension (surface free energy for liquids), capillary elevation can be used for metals and alloys in only a very limited number of cases because, as we have observed in Chapter 2 (see in particular Fig. 2.27), the contact angle, Ω_C, is not zero. Consequently, there results a dependence of capillary rise (h in Fig. 3.2) on the contact angle, and large errors can arise, particularly for liquid metal systems at high temperature.

In principle, all the techniques devised for surface tension measurements on organic liquids, for example,[9,10] can be employed in the measurement of the surface tensions of liquid metals and alloys.** However, as was pointed out above, shape changes occurring in the liquid surface as a result of equilibration associated with the container or support surface must normally be accounted for in order to obtain reliable surface tension values. In addition, the only measured values of γ_{LV} that have a strictly defined meaning are those values of the surface tension corresponding to equilibrium of the liquid with its saturated vapor.

The most satisfactory and accurate methods for measuring the surface tension of liquid metals, particularly at high temperature, are the sessile drop, the pendant

---

*Equation (3.5) is actually an approximation which neglects strain energy.

**See the article by E. D. Hondros in Chapter 8A of "Techniques of Metals Research," R. Bunshah (ed.), Vol. IV(A), John Wiley & Sons, New York (1970).

INTERFACIAL FREE ENERGY

drop, and the drop weight methods. A number of investigators have employed the maximum bubble pressure method or the capillary rise method for metals of moderate melting points; the so-called ring or plate removal method has also been used in the measurement of γ_{LV} for low-melting-point metals such as tin or lead. However, there are difficulties in accounting for certain parameters of the latter three methods, and only the sessile and pendant drop methods appear to be accurate over a wide range of temperatures for liquid metals and alloys. In addition, these methods are the simplest to apply.

3.2.1 The Sessile Drop Method

A sessile drop, as illustrated in Fig. 3.3, is a special case of capillary phenomena where the equilibrium between the capillary and hydrostatic pressure at some point, H, below the summit is expressed generally by a modification of Eq. (3.2):

$$\Delta p = \gamma_{LV}\left(\frac{1}{R_1} + \frac{1}{R_2}\right) = g\rho_L z + P_0 \tag{3.6}$$

where P_0 is the pressure at the apex of the meniscus, where $R_1 = R_2$. Bashforth and Adams[9] considered the radii of curvature at points Q and Q' to be R in the plane of Fig. 3.3(a), and $x/\sin\phi$ in a plane perpendicular to the plane of the figure (rotating about the vertical axis). At the apex of the meniscus [Fig. 3.3(a)], $R_1 = R_2 = b$, $z = 0$, and $P_0 = 2\gamma_{LV}/b$ in Eq. (3.6). Thus,

$$\gamma_{LV}\left(\frac{1}{R} + \frac{\sin\phi}{x}\right) = g\rho_L z + \frac{2\gamma_{LV}}{b} \tag{3.7}$$

Equation (3.7) can be rewritten as

$$2 + \beta\left(\frac{z}{b}\right) = \left(\frac{1}{R/b} + \frac{\sin\phi}{x/b}\right) \tag{3.8}$$

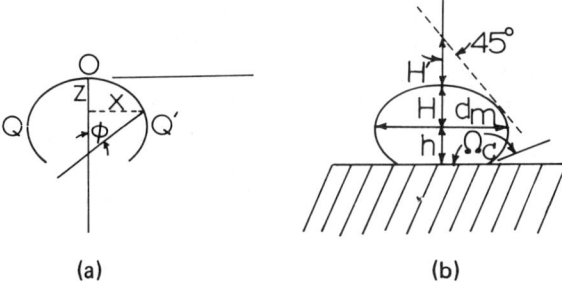

(a)   (b)

FIGURE 3.3 Sessile drop profile and identification of geometric parameters.

where

$$\beta = b^2 \rho_L g / \gamma_{LV} \tag{3.9}$$

Bashforth and Adams[9] prepared tables of β and x/z for $\phi = 90°$, and x/b for values of b and ϕ [Fig. 3.3(a)]. From a measurement of $d_m/2$ and H in Fig. 3.3(b), β and b can also be determined from tables,[9,10] and the surface tension calculated from

$$\gamma_{LV} = g\rho_L b^2 / \beta \tag{3.9a}$$

A convenient table was prepared by Koshevnik et al.[11] for the calculation of surface tension by the direct measurement of the drop profile in Fig. 3.3(b) for small drops:

$$\gamma_{LV} = D d_m^2 \rho_L g \tag{3.10}$$

where D is tabulated as a function of d_m/H as indicated in Table 3.1.[12] It is observed in Table 3.1 that, since D decreases with an increase in the ratio (d_m/H), sessile drops become more spherical as they decrease in size. On the other hand, an infinitely large sessile drop will have effectively no curvature at the summit of the drop, and it can be assumed that any portion of the circumference is straight. For these conditions the surface tension is given by

$$\gamma_{LV} = \rho_L g H^2 / 2 \tag{3.11}$$

Gibson[13] has shown that Eq. (3.11) is not applicable for water drops less than about 12 cm in diameter. However, Gibson found that Worthington's[14] modification of Eq. (3.11) in the form [with reference to Fig. 3.3(b)]

$$\gamma_{LV} = \frac{\rho_L g H^2}{2} \left(\frac{1.641(d_m/2)}{1.641(d_m/2) + H} \right) \tag{3.12}$$

was accurate for drops with diameters greater than 4 cm. Kemball[15] and Nicholas et al.[16] have also confirmed the accuracy of Eq. (3.12) for mercury drops having diameters greater than about 4 cm.

Some of the most accurate measurements of surface tension of metals and alloys were performed in the classic work of Humenik and Kingery,[17] who used the criteria of Bashforth and Adams [Eq. (3.9a)] and an alternative method of calculation suggested by Dorsey,[18*] based on Eq. (3.6). In this form, surface tension is calculated from a sessile drop as shown in Fig. 3.3(b) from the empirical relation

---

*The work of N. E. Dorsey was originally published in *J. Wash. Acad. Sci.*, **18**, 505 (1928). Reference 18 reviews this calculation. It should be noted that Eq. (3.12a) cannot be accurately applied when drop sizes are less than about 1 cm.

INTERFACIAL FREE ENERGY

TABLE 3.1
Sessile Drop Parameters for Surface Tension Calculations[a]

d_m/H	2.40	2.42	2.44	2.46	2.48	2.50
D	0.2373	0.2240	0.2122	0.2013	0.1914	0.1824
d_m/H	2.52	2.54	2.56	2.58	2.60	2.62
D	0.1740	0.1664	0.1592	0.1527	0.1466	0.1407
d_m/H	2.64	2.66	2.68	2.70	2.72	2.74
D	0.1334	0.1304	0.1257	0.1212	0.1171	0.1132
d_m/H	2.76	2.78	2.80	2.82	2.84	2.86
D	0.1095	0.1061	0.1027	0.0995	0.0966	0.0937
d_m/H	2.88	2.90	2.92	2.94	2.96	2.98
D	0.0911	0.0885	0.0861	0.0838	0.0816	0.0795
d_m/H	3.00	3.02	3.04	3.06	3.08	3.10
D	0.0775	0.0756	0.0737	0.0719	0.0702	0.0686
d_m/H	3.12	3.14	3.16	3.18	3.20	3.22
D	0.0670	0.0655	0.0641	0.0627	0.0613	0.0600
d_m/H	3.24	3.26	3.28	3.30	3.32	3.34
D	0.0588	0.0576	0.0564	0.0553	0.0542	0.0531
d_m/H	3.36	3.38	3.40	3.42	3.44	3.46
D	0.0521	0.0511	0.0502	0.0493	0.0483	0.0475
d_m/H	3.48	3.50	3.52	3.54	3.56	3.58
D	0.0467	0.0458	0.0450	0.0443	0.0435	0.0428
d_m/H	3.60	3.62	3.64	3.66	3.68	3.70
D	0.0421	0.0414	0.0408	0.0401	0.0395	0.0389
d_m/H	3.75	3.80	3.85	3.90	3.95	4.00
D	0.0374	0.0360	0.0347	0.0334	0.0322	0.0313

[a] Based on data in Bikerman.[12]

$$\gamma_{LV} = \frac{g\rho_L d_m^2}{4}\left(\frac{0.0520}{f} - 0.1227 + 0.0481f\right) \tag{3.12a}$$

where

$$f = (2H'/d_m) - 0.4142 \tag{3.12b}$$

Equation (3.12a) [as noted from Eq. (3.12b)] depends on measurements from the top of the drop to the intersection of the axis with a 45° tangent to the drop, and does not rely on interpolations or calculations based on tabulated parameters. Equation (3.12a) represents perhaps the most accurate method of measuring surface free energies of liquid metals and alloys, particularly when the sessile drop is small in diameter (<4 cm[18]).

Experimental Observations of Sessile Drops. To obtain a sessile drop of a metal or alloy it is necessary to heat a small mass of the material in a vacuum or purified inert atmosphere at a temperature equal to or greater than the melting point. The small metal mass is usually placed upon a polished ceramic or refractory oxide support. In the arrangement of Humenik and Kingery,[17] a high-frequency generator heated a tubular enclosure through which the temperature was measured with an optical pyrometer at one end, while the sessile drop image was recorded by self-illumination from the other. In many experiments, the sessile drop is illuminated from one side, and its shadow is photographed as illustrated in Fig. 3.4. Figure 3.4(c) shows an illuminated shadowgraph of a sessile drop of water at room temperature which is normally used as a check on the accuracy of the measurement technique or as a means to calibrate the experimental profiles.

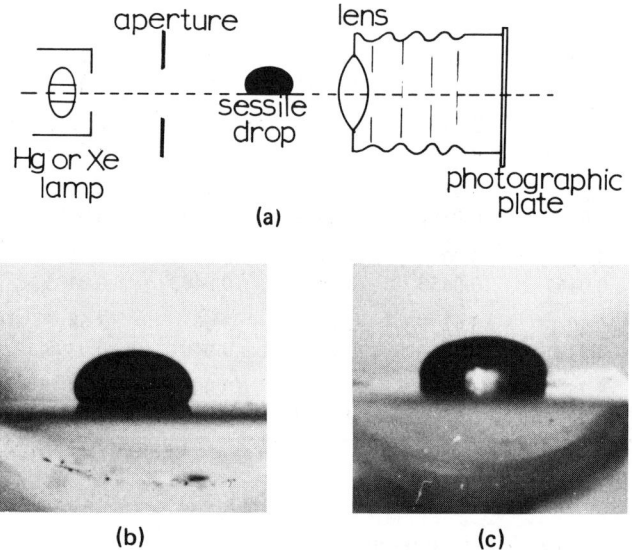

FIGURE 3.4 Recording the shape of a sessile drop. (a) Photographing an illuminated sessile drop. (b) Photograph of a mercury sessile drop (25°C). (c) Photograph of a water sessile drop (25°C). [(b) and (c) are courtesy of U. M. Ahmad.]

Since the substrate material can have an important influence on the shape of a sessile drop and on the measured value of γ, care must be taken in selecting both a substrate and an atmosphere. For experiments involving systematic variations in γ with, say, temperature, use of the same experimental conditions can help in eliminating random errors. It should also be cautioned that the measurements are

INTERFACIAL FREE ENERGY 95

accurate only when the drop is symmetrical. For this reason, several different views of the drop shape should be recorded for comparison.

3.2.2 The Pendant Drop Method

A pendant drop, as illustrated in Fig. 3.5, can also be considered as a special case of capillary phenomena. Gravitation, which flattens a sessile drop, elongates a pendant drop. The balance of capillary and hydrostatic pressures (or surface tension and gravitational forces) is again expressed by Eq. (3.2), where $g\rho_L h$ in Eq. (3.2) refers to the capillary pressure at h cm above the bottom of the drop.

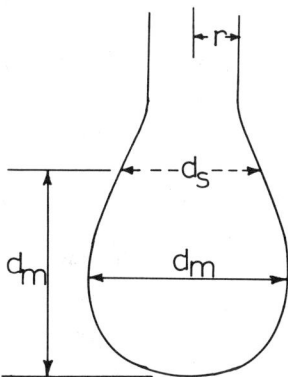

FIGURE 3.5 Pendant drop profile and associated geometric parameters.

Although pendant drop menisci are completely described by the equations of Bashforth and Adams,[9] Andreas et al.[19] showed that the surface tension of a liquid may be derived from the dimensions of simple construction lines applied to a pendant drop (as shown in Fig. 3.5) in the form

$$\gamma_{LV} = Jg\rho_L d_m^2 \qquad (3.13)$$

where g is the gravitational constant, ρ_L is the liquid density, d_m is the maximum drop diameter, and J is the drop shape factor. Andreas et al.[19] showed that J maintained a fixed relationship to the ratio d_s/d_m (Fig. 3.5). Values of J are tabulated as a function of d_s/d_m in Table 3.2.[20]

Experimental Observations of Pendant Drops. Allen[22] measured the surface tensions (liquid surface free energies) of a large number of pure metals and claims that the pendant drop method and drop weight method (to be discussed below) are perhaps the most accurate. One significant feature of the pendant drop is that it

TABLE 3.2
Pendant Drop Parameters for Surface Tension Calculations[a]

ds/dm	0.30	0.31	0.32	0.33	0.34	0.35	0.36
J	7.0984	6.5400	6.0400	5.5908	5.1861	4.8203	4.4887
ds/dm	0.37	0.38	0.39	0.40	0.41	0.42	0.43
J	4.1877	3.9138	3.6643	3.4357	3.2258	3.0326	2.8548
ds/dm	0.44	0.45	0.46	0.47	0.48	0.49	0.50
J	2.6911	2.5401	2.4003	2.2709	2.1507	2.0391	1.9352
ds/dm	0.51	0.52	0.53	0.54	0.55	0.56	0.57
J	1.8384	1.7481	1.6637	1.5848	1.5109	1.4416	1.3766
ds/dm	0.58	0.59	0.60	0.61	0.62	0.63	0.64
J	1.3155	1.2581	1.2040	1.1531	1.1050	1.0597	1.0168
ds/dm	0.65	0.66	0.67	0.68	0.69	0.70	0.71
J	0.9764	0.9380	0.9018	0.8674	0.8347	0.8038	0.7744
ds/dm	0.72	0.73	0.74	0.75	0.76	0.77	0.78
J	0.7464	0.7198	0.6945	0.6704	0.6474	0.6255	0.6046
ds/dm	0.79	0.80	0.81	0.82	0.83	0.84	0.85
J	0.5846	0.5655	0.5473	0.5298	0.5131	0.4970	0.4817
ds/dm	0.86	0.87	0.88	0.89	0.90	0.91	0.92
J	0.4669	0.4527	0.4391	0.4260	0.4134	0.4012	0.3895
ds/dm	0.93	0.94	0.95	0.96	0.97	0.98	0.99
J	0.3780	0.3671	0.3564	0.3460	0.3359	0.3259	0.3160

[a] Based on tables of Fordham[20] and Stauffer.[21]

can be formed in a high vacuum or inert atmosphere by electron bombardment or laser excitation of a pure rod of the test material, thereby eliminating possible contact contamination or reaction products as in the case of sessile drop formation. Figure 3.6 illustrates a niobium rod of diameter $2r$ with a molten pendant drop heated by electron bombardment. Pendant drop images at high temperatures can be photographed directly, or illuminated in a manner identical to that shown in Fig. 3.4(a).

3.2.3 Drop Weight Method

Were the drop formed in Fig. 3.6 to detach itself from the rod, it could be reasoned that the weight of such a drop would just balance the surface tension for the condition of detachment. The relationship between the drop weight (the detachment force) and the surface tension is given by

$$\gamma_{LV} = m_0 g / 2\pi r f \qquad (3.14)$$

where m_0 is the weight of a static drop hanging from a wetted vertical rod of radius r (as in Fig. 3.6), and f is the fraction of "ideal" drop separating from the rod—that

INTERFACIAL FREE ENERGY 97

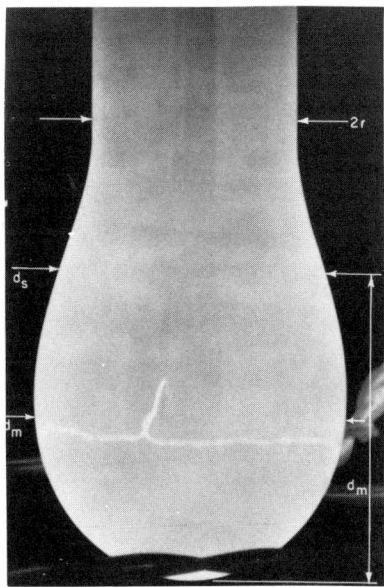

FIGURE 3.6 Niobium pendant drop at the melting point (~2415°C); r = 0.25 cm. (Courtesy of Allen.[22])

is, a correction factor for the true surface tension. The fraction, f, is a single-valued function of $r/(m_0/\rho_L)^{1/3}$, where ρ_L is the liquid density, with values usually in the range 0.6 to 0.8.[23] Since (m_0/ρ_L) represents the detached drop volume, V, Harkins and Brown[23] were able to correct experimental values of drop weights using tables relating V/r^3 and a correction factor, F, in the form

$$\gamma_{LV} = (m_0 g/r) \cdot F \tag{3.15}$$

To find F, the value of V/r^3 was derived experimentally, and the corresponding value of F was found from tables. Lando and Oakley[24] obtained extensive tables relating values of $r/V^{1/3}$ or $r/(m_0/\rho_L)^{1/3}$ and F. Table 3.3 lists a range of values of the coefficient F along with corresponding values of $r/(m_0/\rho_L)^{1/3}$.

The surface tension values obtained from drop weight experiments utilizing the corrections of Table 3.3 in Eq. (3.15) are accurate to at least ±0.1 erg/cm².[10] Drop weight values of surface tension are normally assumed to be accurate to 0.1% or better.[23]

Determination of Drop Weight. The drop weight, m_0, is found by weighing one or more drops collected in a weighing container. The container should be a tall

TABLE 3.3
Drop Weight Correction Factors[a]

$r/(m_o/\rho_L)^{1/3}$	0.30	0.31	0.32	0.33	0.34	0.35	0.36
F	0.2166	0.2183	0.2201	0.2218	0.2235	0.2251	0.2267
$r/(m_o/\rho_L)^{1/3}$	0.37	0.38	0.39	0.40	0.41	0.42	0.43
F	0.2283	0.2299	0.2314	0.2328	0.2343	0.2357	0.2371
$r/(m_o/\rho_L)^{1/3}$	0.44	0.45	0.46	0.47	0.48	0.49	0.50
F	0.2384	0.2397	0.2410	0.2423	0.2435	0.2447	0.2458
$r/(m_o/\rho_L)^{1/3}$	0.51	0.52	0.53	0.54	0.55	0.56	0.57
F	0.2469	0.2480	0.2490	0.2501	0.2510	0.2520	0.2529
$r/(m_o/\rho_L)^{1/3}$	0.58	0.59	0.60	0.61	0.62	0.63	0.64
F	0.2538	0.2546	0.2554	0.2562	0.2570	0.2577	0.2584
$r/(m_o/\rho_L)^{1/3}$	0.65	0.66	0.67	0.68	0.69	0.70	0.71
F	0.2590	0.2596	0.2602	0.2608	0.2613	0.2618	0.2622
$r/(m_o/\rho_L)^{1/3}$	0.72	0.73	0.74	0.75	0.76	0.77	0.78
F	0.2626	0.2630	0.2634	0.2637	0.2640	0.2642	0.2644
$r/(m_o/\rho_L)^{1/3}$	0.79	0.80	0.81	0.82	0.83	0.84	0.85
F	0.2646	0.2648	0.2649	0.2650	0.2650	0.2650	0.2650
$r/(m_o/\rho_L)^{1/3}$	0.86	0.87	0.88	0.89	0.90	0.91	0.92
F	0.2650	0.2649	0.2648	0.2646	0.2644	0.2642	0.2640
$r/(m_o/\rho_L)^{1/3}$	0.93	0.94	0.95	0.96	0.97	0.98	0.99
F	0.2637	0.2634	0.2630	0.2626	0.2622	0.2618	0.2613
$r/(m_o/\rho_L)^{1/3}$	1.00	1.01	1.02	1.03	1.04	1.05	1.06
F	0.2608	0.2602	0.2597	0.2590	0.2584	0.2577	0.2570
$r/(m_o/\rho_L)^{1/3}$	1.07	1.08	1.09	1.10	1.11	1.12	1.13
F	0.2563	0.2555	0.2547	0.2538	0.2529	0.2520	0.2511
$r/(m_o/\rho_L)^{1/3}$	1.14	1.15	1.16	1.17	1.18	1.19	1.20
F	0.2501	0.2491	0.2480	0.2470	0.2459	0.2447	0.2435

[a] From data of Lando and Oakley.[24]

receptacle placed below the sample to catch the drops and avoid possible splashing or related loss of material. If the container rests upon a balance arrangement, drop weights can be recorded automatically. If a number of drops are collected, the average weight for a single drop can be measured.

3.2.4 Maximum Bubble Pressure Method

If a gas bubble is pressed through a capillary immersed in a liquid metal or alloy, the maximum pressure required can be related to the surface tension of the liquid. In addition, if a liquid metal or alloy of different density is pressed through the liquid metal or alloy in question, the interfacial energy can be determined in a

INTERFACIAL FREE ENERGY

similar manner, provided the two separate liquids are immiscible. The theory of this method was developed by Cantor,[25] Verschaffelt,[26] and Schrodinger,[27] resulting in an equation of the form[12,28]

$$\gamma_I = \frac{1}{2} r P_M \left(1 - \frac{2}{3} \frac{g(\rho_{L1} - \rho_{L2})r}{P_M} - \frac{1}{6} \frac{g^2(\rho_{L1} - \rho_{L2})^2 r^2}{P_M^2} \right) \quad (3.16)$$

where r is the capillary radius (Fig. 3.7), P_M is the maximum bubble pressure, g is the gravitational constant, and ρ_{L1} and ρ_{L2} are the corresponding liquid (L) densities. As shown in Fig. 3.7, the maximum bubble pressure, P_M, is obtained by considering the liquid pressure head.

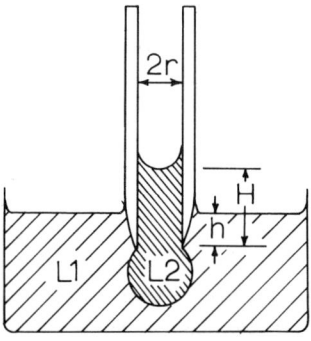

FIGURE 3.7 Maximum bubble pressure measurement of liquid interfacial free energies.

If the liquid, $L2$, in the capillary of Fig. 3.7 is replaced by an inert gas such as helium of density ρ_V (where $\rho_V \ll \rho_{L1}$), the surface tension of the liquid metal or alloy can be obtained from a modified form of Eq. (3.16):

$$\gamma_{LV} = \frac{1}{2} r P_M \left(1 - \frac{2}{3} \frac{g \rho_{L1} r}{P_M} - \frac{1}{6} \frac{g^2 \rho_{L1}^2 r^2}{P_M^2} \right) \quad (3.17)$$

The maximum pressure of gas bubble formation when Eq. (3.17) is used can be determined manometrically, keeping in mind that accurate pressure measurements must involve a hydrostatic correction owing to the distance of the orifice below the surface of the liquid metal or alloy (h in Fig. 3.7).

It should be apparent from Fig. 3.7 that, although the maximum bubble method affords the only technique of directly measuring liquid interfacial free energies, it is limited in accuracy because of the necessity of using a capillary tube of different composition than either liquid involved, and the probability of contamination or interaction induced by any of the components in the system. The method of maximum bubble pressure and other capillary methods such as capillary rise are generally not suitable for high-temperature systems. Several reasonably accurate measurements have, however, been made for mercury,[29] sodium,[30,31] tin,[32] bismuth,[33] lead,[34] and related low-melting metal or alloy systems. Measurements should be made as rapidly as possible to avoid contamination by diffusion, particularly when gases such as hydrogen are used to form the bubbles. It is a recommended practice to use mercury or water as a means of testing the measuring devices and the accuracy of the technique, since these liquids are readily available and the surface tensions are well established.

3.2.5 Surface Tensions at the Melting Point and Temperature Coefficients of Liquid Metals and Alloys

The variation of surface tension of liquid metals and alloys with temperature is ideally described by the thermodynamic considerations outlined in Chapter 1. For pure metals (single-component systems or metallic elements) the temperature coefficient, $d\gamma_{LV}/dT$, is given by Eq. (1.26). Since vacancies are not a factor at temperatures above the melting point, the variation in surface tension for binary systems will be given by Eq. (1.37a).

A considerable number of measurements of surface tension of liquid metals and binary alloys have been made over the past four or five decades, but many are of questionable accuracy, considering the experimental techniques and the high coincidence of contamination attendant to earlier measurements. A thorough evaluation of the measured values of surface tension of liquid metals and alloys has been made. Table 3.4 lists the values or mean values where possible of the measured surface tensions at the melting point for metals and alloys, and the associated linear temperature coefficients. Note that, although temperature coefficients refer to the liquid state, it is not unreasonable in many cases to apply these values in the solid state at temperatures close to the melting point.

If the surface free energies of metals at their melting points are plotted against the liquid-state temperature coefficients, it is generally observed that the magnitude of $d\gamma_{LV}/dT$ increases with increasing γ_{LV}. This is illustrated in Fig. 3.8 utilizing the data in Table 3.4. In Fig. 3.8 the block of data points consisting of Re, Os, W, Mo, Ir, etc., may be in error, since these values are estimates calculated by Allen.[50] Values of $d\gamma_{LV}/dT$ would appear to be generally too low for these metals; this apparently occurs because of the inability to measure the critical temperatures or liquid- and vapor-phase densities at the critical point with any accuracy [see Eq. (3.28)].

It is observed from Table 3.4 that the temperature coefficients are generally negative as necessitated by thermodynamics. However, there are several cases, specifically for zinc and cadmium, where positive values are observed. It must be concluded, in the absence of any compelling rationalizations, that positive temperature coefficients simply reflect the occurrence of surface reactions.

TABLE 3.4
Surface Free Energy of Liquid Metals and Alloys[a]

Metal or Alloy	Surface Energy, γ_{LV} (ergs/cm²)[b]	Temperature (°C)	$d\gamma_{LV}/dT$ (ergs/cm² °C)	Ref.
Aluminum	866	600 (m.p.)	−0.50	35–40
Al–5 w/o Cu	467	700	−	28
Al–33 w/o Cu	452	700	−	28
Al–11 a/o Mg	811	605 (m.p.)	−0.35	28, 41
Al–27 a/o Mg	693	515 (m.p.)	−0.33	28, 41
Al–40 a/o Mg	631	435 (m.p.)	−0.31	28, 41
Al–70 a/o Mg	570	436 (m.p.)	−0.15	28, 41
Al–92 a/o Mg	571	600 (m.p.)	−0.30	28, 41
Al–12 a/o Zn	900	433 (m.p.)	−0.50	28, 41
Al–22 a/o Zn	730	479 (m.p.)	0.00	28, 41
Al–40 a/o Zn	818	560 (m.p.)	−0.15	28, 41
Al–79 a/o Zn	890	622 (m.p.)	−0.30	28, 41
Antimony	380	640	−0.07	28, 34, 42, 43
Sb–41 w/o Cd	381	460	−0.15	28
Sb–46 w/o Cd	388	460	−0.17	28
Sb–59 w/o Cd	398	460	−0.10	28
Sb–70 w/o Cd	425	460	−0.22	28
Sb–24 w/o Cu	400	1100	−0.01	28
Sb–60 w/o Cu	570	1100	+0.06	28
Sb–85 w/o Cu	790	1100	+0.07	28
Sb–20 w/o Pb	395	590	−0.08	28
Sb–61 w/o Pb	405	510	−0.10	28
Sb–81 w/o Pb	427	450	−0.11	28
Sb–90 w/o Pb	430	450	0.05	28
Sb–30 w/o Zn	499	−	−	28
Sb–65 w/o Zn	392	−	−	28
Barium	224	720	−	44
Beryllium	1100	1500	−	45
Bismuth	375	~350	−0.07	42, 43, 46
Bi–30 w/o Pb	400	260	−	28
Bi–50 w/o Pb	409	260	−	28
Bi–35 w/o Sn	440	260	−	28
Bi–50 w/o Sn	461	260	−	28
Bi–2 a/o Te	362	350	−0.07	28
Calcium	337	850	−	44
Cesium	60	29 (m.p.)	−	46
Cadmium	590	321 (m.p.)	+(m.p. −415°C) −(>415°C)	47
Cd–91 a/o Na	170	300	−0.14	48
Cd–95 a/o Na	160	300	−0.09	48
Cd–98 a/o Na	163	300	−0.07	48

TABLE 3.4 *(continued)*

Metal or Alloy	Surface Energy, γ_{LV} (ergs/cm²)[b]	Temperature (°C)	$d\gamma_{LV}/dT$ (ergs/cm² °C)	Ref.
Cd–30 w/o Sb	425	460	–0.22	28
Cd–41 w/o Sb	398	460	–0.10	28
Cd–54 w/o Sb	388	460	–0.17	28
Cd–59 w/o Sb	381	460	–0.15	28
Chromium	1590	1950	–	49
Cobalt	1880	1495 (m.p.)	–0.70	50, 51
Co–80 w/o Fe–C	1700	1425	–	52
Columbium	1900	2473 (m.p.)	–0.24	50
Copper	1300	1083 (m.p.)	–0.45	50, 53, 54
Cu–67 w/o Al	452	700	–	28
Cu–95 w/o Al	467	700	–	28
Cu–20 w/o Au	1200	1300	–	55
Cu–50 w/o Au	1150	1300	–	55
Cu–80 w/o Au	1110	1300	–	55
Cu–77 w/o Fe	1324	1550	–	56
Cu–83 w/o Fe	1377	1550	–	56
Cu–88 w/o Fe	1460	1550	–	56
Cu–93 w/o Fe	1535	1550	–	56
Cu–96 w/o Fe	1620	1550	–	56
Cu–20 w/o Ni	1125	1300	–0.45	57
Cu–40 w/o Ni	1310	1300	–0.55	57
Cu–60 w/o Ni	1460	1300	–0.60	57
Cu–80 w/o Ni	1700	1300	–0.60	57
Cu–44 w/o Ag	1035	1000	–0.10	55
Cu–72 w/o Ag	965	1000	–0.10	55
Cu–15 w/o Sb	790	1100	+0.07	28
Cu–40 w/o Sb	570	1100	+0.06	28
Cu–76 w/o Sb	400	1100	–0.01	28
Cu–30 w/o Sn	820	1100	+0.08	28
Cu–50 w/o Sn	640	1100	+0.06	28
Cu–75 w/o Sn	545	1100	–0.01	28
Gallium	720	30.3 (m.p.)	–	46, 58
Germanium	610	960	–	46
Gold	1140	1063 (m.p.)	–0.52	52
Au–31 w/o Ag	1029	1108	–	59
Au–55 w/o Ag	982	1108	–	59
Au–67 w/o Ag	945	1108	–	59
Au–79 w/o Ag	917	1108	–	59
Au–20 w/o Cu	1110	1300	–	55
Au–50 w/o Cu	1150	1300	–	55
Au–80 w/o Cu	1200	1300	–	55
Hafnium	1630	2230 (m.p.)	–0.21	50

TABLE 3.4 *(continued)*

Metal or Alloy	Surface Energy, γ_{LV} (ergs/cm²)[b]	Temperature (°C)	$d\gamma_{LV}/dT$ (ergs/cm² °C)	Ref.
Indium	560	156 (m.p.)	−0.09	47
In–93 a/o Na	165	400	−0.09	48
In–95 a/o Na	175	400	−0.08	48
In–99 a/o Na	190	400	−0.08	48
Iridium	2250	2454 (m.p.)	−0.31	50
Iron	1880	1535 (m.p.)	−0.43	46, 50, 51
Fe–C	1980	1350	−	52
Fe–C–20 w/o Co	1700	1425	−	52
Fe–C–19 w/o Ni	1800	1350	−	52
Fe–4 w/o Cu	1620	1550	−	56
Fe–7 w/o Cu	1535	1550	−	56
Fe–12 w/o Cu	1460	1550	−	56
Fe–17 w/o Cu	1377	1550	−	56
Fe–23 w/o Cu	1324	1550	−	56
Fe–25 w/o Ni	1480	m.p.	−	60
Fe–50 w/o Ni	1510	m.p.	−	60
Fe–75 w/o Ni	1550	m.p.	−	60
Fe–7 w/o P	1130	1400	+0.83	28
Fe–0.3 w/o Si	1400	1550	−	61
Fe–1 w/o Si	1425	1550	−	61
Fe–3 w/o Si	1440	1550	−	61
Fe–9 w/o Si	1425	1550	−	61
Fe–1 w/o Sn	1328	1550	−	56
Fe–7 w/o Sn	987	1550	−	56
Fe–19 w/o Sn	813	1550	−	56
Fe–43 w/o Sn	550	1550	−	56
Fe–48 w/o Sn	532	1550	−	56
Fe–18 w/o Cr, 8 w/o Ni	1172	1475 (m.p.)	−2.48	c
Lead	450	327 (m.p.)	−0.08	28, 46
Pb–50 w/o Bi	409	260	−	28
Pb–70 w/o Bi	400	260	−	28
Pb–10 w/o Sb	430	450	−0.05	28
Pb–19 w/o Sb	427	450	−0.11	28
Pb–39 w/o Sb	405	510	−0.10	28
Pb–80 w/o Sb	395	590	−0.08	28
Lithium	398	180 (m.p.)	−0.14	31
Magnesium	540	651 (m.p.)	−0.32	28, 41, 46
Manganese	1060	1245 (m.p.)	−	50
Mg–8 a/o Al	571	600 (m.p.)	−0.30	28, 41
Mg–30 a/o Al	570	436 (m.p.)	−0.15	28, 41
Mg–60 a/o Al	631	435 (m.p.)	−0.31	28, 41
Mg–73 a/o Al	693	515 (m.p.)	−0.33	28, 41
Mg–89 a/o Al	811	605 (m.p.)	−0.35	28, 41
Mg–10 a/o Zn	576	565 (m.p.)	−0.30	28, 41
Mg–20 a/o Zn	557	460 (m.p.)	−0.10	28, 41

TABLE 3.4 *(continued)*

Metal or Alloy	Surface Energy, γ_{LV} (ergs/cm^2)[b]	Temperature (°C)	$d\gamma_{LV}/dT$ (ergs/cm^2 °C)	Ref.
Mg–30 a/o Zn	478	340 (m.p.)	–0.10	28, 41
Mg–50 a/o Zn	438	525 (m.p.)	–0.30	28, 41
Mg–67 a/o Zn	–	590 (m.p.)	–0.33	28, 41
Mg–83 a/o Zn	640	500 (m.p.)	–0.10	28, 41
Mercury	475	20	–0.22	28, 62
Molybdenum	2250	2620 (m.p.)	–0.30	46, 50
Nickel	1780	1455 (m.p.)	–1.20	50, 51, c
Ni–20 w/o Cu	1700	1300	–0.60	57
Ni–40 w/o Cu	1460	1300	–0.60	57
Ni–60 w/o Cu	1310	1300	–0.55	57
Ni–80 w/o Cu	1125	1300	–0.45	57
Ni–25 w/o Fe	1550	m.p.	–	60
Ni–50 w/o Fe	1510	m.p.	–	60
Ni–75 w/o Fe	1480	m.p.	–	60
Ni–81 w/o Fe–C	1800	1350	–	52
Niobium	1900	2473 (m.p.)	–0.24	50
Osmium	2500	3000 (m.p.)	–0.33	50
Palladium	1500	1547 (m.p.)	–0.22	46, 50
Platinum	1800	1773 (m.p.)	–0.17	50
Potassium	101	64 (m.p.)	–0.11	31
Rhenium	2700	3200 (m.p.)	–0.34	50
Rhodium	2000	1966 (m.p.)	–0.30	50
Rubidium	76	39 (m.p.)	–	46
Ruthenium	2250	2250 (m.p.)	–0.31	50
Selenium	106	220	–	46
Silicon	730	1410 (m.p.)	–0.10	46, 60
Si–91 w/o Fe	1425	1550	–	61
Si–97 w/o Fe	1440	1550	–	61
Si–99 w/o Fe	1425	1550	–	61
Si–99.7 w/o Fe	1400	1500	–	61
Silver	895	1000	–0.30	46, 59, 60, 63, 64
Ag–28 w/o Cu	965	1000	–0.10	55
Ag–56 w/o Cu	1035	1000	–0.10	55
Ag–21 w/o Au	917	1108	–	59
Ag–33 w/o Au	945	1108	–	59
Ag–45 w/o Au	982	1108	–	59
Ag–69 w/o Au	1029	1108	–	59
Sodium	191	98 (m.p.)	–0.10	31
Na–2 a/o Cd	163	300	–0.07	48
Na–5 a/o Cd	160	300	–0.09	48

INTERFACIAL FREE ENERGY

TABLE 3.4 *(continued)*

Metal or Alloy	Surface Energy, γ_{LV} (ergs/cm²)[b]	Temperature (°C)	$d\gamma_{LV}/dT$ (ergs/cm² °C)	Ref.
Na–9 a/o Cd	170	300	–0.14	48
Na–1 a/o In	190	400	–0.08	48
Na–5 a/o In	175	400	–0.08	48
Na–7 a/o In	165	400	–0.09	48
Tantalum	2150	3020 (m.p.)	–0.25	50
Technetium	2100	2220 (m.p.)	–	50
Tellurium	178	460	–	46
Te–98 a/o Bi	362	350	0.07	28
Thallium	440	318	–	46
Tin	550	232 (m.p.)	–0.08	28, 46
Sn–50 w/o Bi	461	260	–	28
Sn–65 w/o Bi	440	260	–	28
Sn–25 w/o Cu	545	1100	–0.01	28
Sn–50 w/o Cu	640	1100	+0.06	28
Sn–70 w/o Cu	820	1100	+0.08	28
Sn–52 w/o Fe	532	1550	–	56
Sn–57 w/o Fe	550	1550	–	56
Sn–81 w/o Fe	813	1550	–	56
Sn–93 w/o Fe	987	1550	–	56
Sn–99 w/o Fe	1328	1550	–	56
Sn–6 a/o Pb	529	250	–0.06	28, 41, 43, 65
Sn–25 a/o Pb	492	200	–0.05	28, 41, 43, 65
Sn–36 a/o Pb	477	250	–0.06	28, 41, 43, 65
Sn–63 a/o Pb	469	280	–0.09	28, 41, 43, 65
Sn–84 a/o Pb	450	350	–0.07	28, 41, 43, 65
Sn–96 a/o Pb	450	350	–0.09	28, 41, 43, 65
Sn–25 a/o Zn	580	254 (m.p.)	–0.08	28, 41
Sn–50 a/o Zn	584	234 (m.p.)	–0.08	28, 41
Sn–75 a/o Zn	620	370 (m.p.)	–0.09	28, 41
Sn–90 a/o Zn	690	388 (m.p.)	–0.11	28, 41
Titanium	1650	1730 (m.p.)	–0.26	50
Thorium	978	1750 (m.p.)	–0.14	66
Tungsten	2400	3410 (m.p.)	–0.29	50, 67
Uranium	1550	1132 (m.p.)	–0.14	68
Vanadium	1950	1710 (m.p.)	–0.31	50
Zinc	770	420 (m.p.)	+0.2 (–0.17)	28, 46, 69, 70

TABLE 3.4 *(continued)*

Metal or Alloy	Surface Energy, γ_{LV} (ergs/cm²)[b]	Temperature (°C)	$d\gamma_{LV}/dT$ (ergs/cm² °C)	Ref.
Zn-21 a/o Al	890	622 (m.p.)	−0.30	28, 41
Zn-60 a/o Al	818	560 (m.p.)	−0.15	28, 41
Zn-78 a/o Al	730	479 (m.p.)	0.00	28, 41
Zn-88 a/o Al	900	433 (m.p.)	−0.50	28, 41
Zn-35 w/o Sb	392	–	–	28
Zn-70 w/o Sb	499	–	–	28
Zn-17 a/o Mg	640	500 (m.p.)	−0.10	28, 41
Zn-33 a/o Mg	–	590 (m.p.)	−0.33	28, 41
Zn-50 a/o Mg	438	535 (m.p.)	−0.30	28, 41
Zn-70 a/o Mg	478	340 (m.p.)	−0.10	28, 41
Zn-80 a/o Mg	557	460 (m.p.)	−0.10	28, 41
Zn-90 a/o Mg	576	565 (m.p.)	−0.30	28, 41
Zn-10 a/o Sn	690	388 (m.p.)	−0.11	28, 41
Zn-25 a/o Sn	620	370 (m.p.)	−0.09	28, 41
Zn-50 a/o Sn	584	234 (m.p.)	−0.08	28, 41
Zn-75 a/o Sn	580	254 (m.p.)	−0.08	28, 41
Zirconium	1480	1857 (m.p.)	−0.20	50

[a] Values are generally experimentally determined. Where more than one value has been determined by either the sessile drop, pendant drop, drop weight, or maximum bubble pressure method, the value entered is a mean value. In many cases, the best value has already been evaluated in previous work. Semiconductor elements are also included in this tabulation.

[b] Values of surface free energy (surface tension) are given in units of ergs/cm². This unit is identical to the more international unit, millijoules per square meter (mJ/m²).

[c] U. M. Ahmad, Ph.D. Dissertation, New Mexico Tech., Socorro (1975).

Surface Tension and Critical Temperature. It is well known that surface tensions of pure liquids become zero at the critical point, forming the thermodynamic basis for such processes as critical-point drying, etc. In essence, this means that at some critical temperature (at an attendant pressure) both the surface tension, γ_{LV}, and the interfacial free energy, γ_I, between two dissimilar liquids will be zero. The physical significance of this phenomenon is that at or near the critical temperature, T_c, the surface molecules are so weakly bound that surface definition, as indicated for example in Fig. 3.1, is lost, and the density of the vapor and that of the liquid become identical. Van der Waals[71] proposed the following relationship to account for the response of surface or interfacial free energy with temperature and critical temperature:

$$\gamma = \gamma_0(1 - T/T_c)^n \tag{3.18}$$

where γ_0 is independent of temperature, and $n = 1.23$.

Eötvös,[72] assuming that bodies in corresponding states are similar with respect to the forces between their corresponding parts and their energies, developed an equation of the form

INTERFACIAL FREE ENERGY

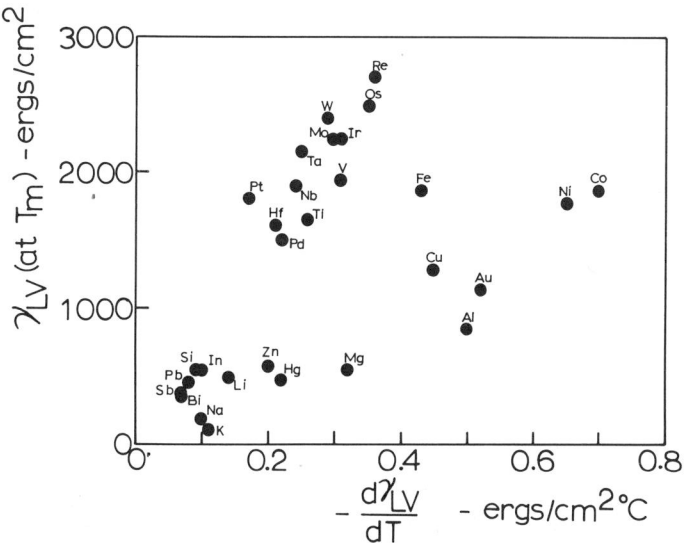

FIGURE 3.8   Surface tension of some metals at their melting points versus temperature coefficient.

$$d[\gamma(MV)^{2/3}]/dT = K_E \tag{3.19}$$

which on integration within the condition that $\gamma_{LV} = 0$ when $T = T_c$ yields

$$\gamma(MV)^{2/3} = K_E(T_c - T) \tag{3.20}$$

where M is the molecular weight, V is the specific volume, and K_E is a constant usually called the Eötvös constant. By comparing Eq. (3.18) with Eq. (3.20), the Eötvös constant is found to be

$$K_E = \frac{\gamma_0(1 - T/T_c)^n}{(T_c - T)(MV)^{2/3}} = \frac{\gamma_0(1 - T/T_c)^{n-1}}{T_c(MV)^{2/3}} \tag{3.21}$$

It is observed from Eq. (3.21) that the Eötvös constant relates the surface entropy to the action of intermolecular forces through γ_0, and to the mean distance of intermolecular separation through the value of $(MV)^{2/3}$. Furthermore, γ_0 represents the surface free energy at absolute zero ($T = 0°K$); therefore Eq. (3.18) is an expression of the surface free energy over the full phase range of a substance, particularly metals and alloys which are solid well above $T = 0°K$. One might therefore argue that, if the Eötvös equation is applicable, it will not be unreasonable to consider solid surface free energies and surface free energies of liquids (surface tensions) to be similar or even linear in a temperature range above and below the melting point. In this regard, a simple linear relation between γ_{LV} and T can be written in the form[12]

$$\gamma_{LV} = \gamma_m \frac{T_c - T}{T_c - T_m} \qquad (3.22)$$

or

$$\frac{d\gamma_{LV}}{dT} = \frac{\gamma_m}{T_c - T_m} \qquad (3.23)$$

which indicates that ideally $d\gamma_{LV}/dT$ would be constant in the liquid state (for temperatures above T_m, the melting temperature). We observed in Table 3.4 that this seems to be a reasonable approximation except for cadmium, zinc, and certain of their alloys. In principle, T_c can be calculated from Eq. (3.23), and estimates of T_c have in fact been made for metals.[73]

Katayama[74] and Guggenheim[75] considered the surface energy with respect to the critical temperature as

$$\gamma [M/(\rho_L - \rho_V)]^{2/3} = K_E T_c (1 - T/T_c) \qquad (3.24)$$

and

$$\gamma = \gamma_0 (1 - T/T_c)^{11/9} \qquad (3.25)$$

where ρ_L and ρ_V are the corresponding liquid- and vapor-phase densities. If these equations are combined, it is possible to eliminate the term $(1 - T/T_c)$:

$$\gamma/(\rho_L - \rho_V)^{11/3} = \text{constant} \qquad (3.26)$$

This equation was derived empirically by Macleod[76] and rewritten by Sugden[77] in the form

$$P = \frac{M\gamma^{1/4}}{(\rho_L - \rho_V)} \qquad (3.27)$$

where P is defined as the parachor, and M is the molecular weight. The parachor appears to be related to the number of valence electrons of atoms composing molecules in which only dispersion forces and polarizability operate, and a relationship has been shown between parachors and molecular properties for many organic liquids.[10] However, since liquid metals or alloys are not governed by the same intermolecular force considerations as organic liquids, it is doubtful that

INTERFACIAL FREE ENERGY   109

similar relationships will be observed, although there is no evidence to clarify this contention. There is evidence to suggest that surface tensions of metals are influenced by electronic structure as shown in Fig. 3.9, where it appears that high surface tensions are associated with prominent electron bonding, decreasing with a prominence of s-electron bonding.

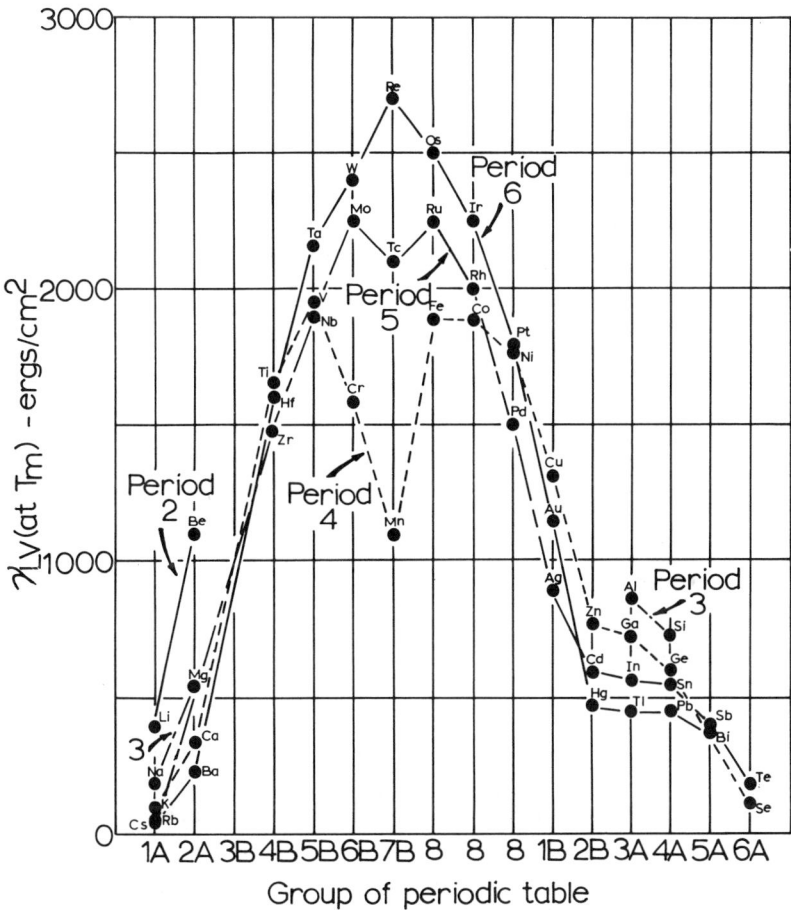

FIGURE 3.9   Periodic arrangement of surface tension of liquid metals (and semiconductors) at their melting points.

Since it is well known that thermal properties of matter are influenced by electronic structure, and keeping in mind the implications of the temperature-surface tension relationship (particularly the melting temperature, T_m), as expressed in Eqs. (3.22) and (3.23), some relationship of γ_{LV} at the melting point and T_m might be expected. This feature is illustrated in Fig. 3.10, which represents the data for pure metals listed in Table 3.4.

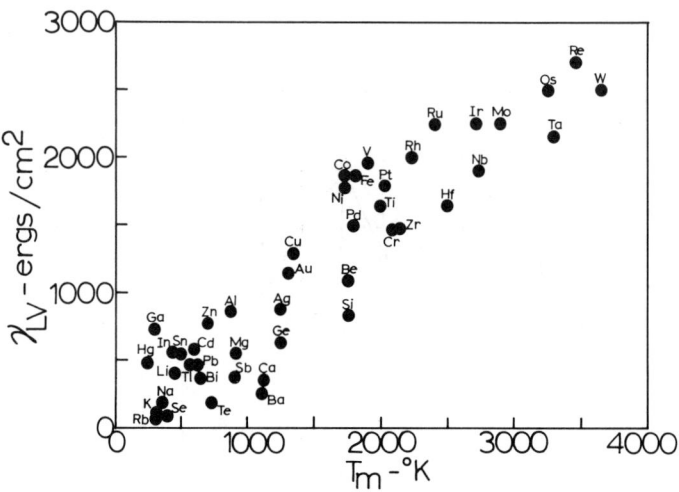

FIGURE 3.10 Surface tension of liquid metals at their melting point versus T_m (including semiconductor elements).

Except in certain cases as cited previously, Eq. (1.26) is applicable to pure metals in the liquid state. In addition, the expectations of sign change for the temperature coefficient of surface energy as discussed in Chapter 1 for alloy or multicomponent systems have also been observed in a number of instances. Several examples of these features are given in Fig. 3.11. We observe in Fig. 3.11 that, for alloy systems where $d\gamma_{LV}/dT$ is positive, the system must in most instances be suspect of surface adsorption of one or more components, or some propensity for surface ordering.[69] For example, when the cohesive energies of the elements are plotted according to increasing atomic number, minima occur for zinc, cadmium, and mercury in addition to the so-called inert gases. These observations suggest that electronic structure may play some role in the surface energetics of liquid metals and alloys as evidenced in Fig. 3.9. Closed valence shells appear to exhibit extra stability as a result of strong interatomic interaction and an associated lower surface entropy in liquids such as zinc, cadmium, and possibly mercury at temperatures very close to T_m for mercury. A change of sign associated with $d\gamma_{LV}/dT$ may be attributed to an electronic transformation of the type $ns^2 \rightarrow ns^1 np^1$, which for Cd [69] would involve $5s^2 \rightarrow 5s^1 5p^1$.

If, as Eq. (3.24) suggests, the surface tension, γ_{LV}, drops to zero at the critical

INTERFACIAL FREE ENERGY

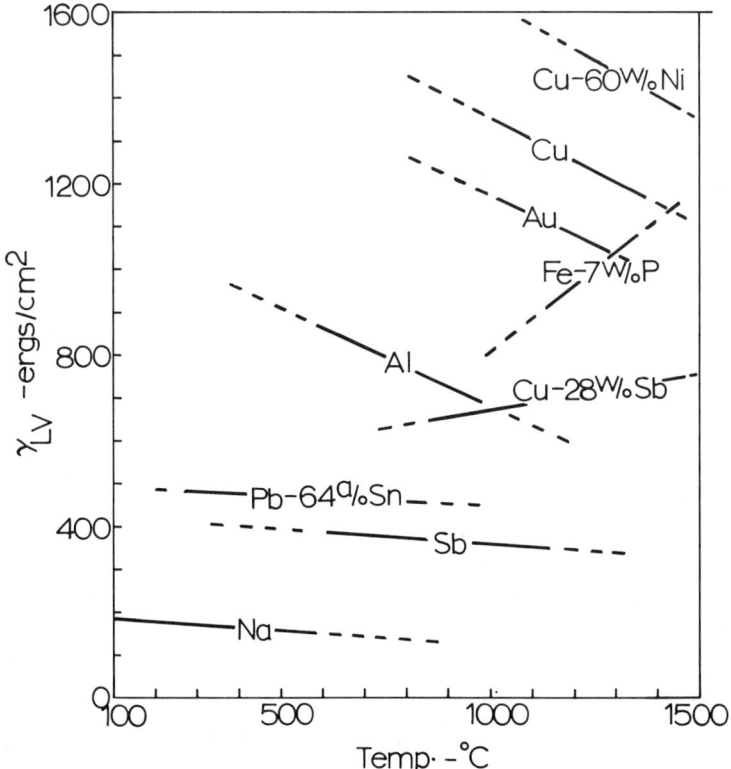

FIGURE 3.11 Variation of surface tension (γ_{LV}) for metals and alloys with temperature above the melting point.

temperature, it would be expected that the temperature coefficient will be negative at sufficiently elevated temperatures. In the case of linear temperature coefficients, particularly for pure metals, $d\gamma_{LV}/dT$ can be derived from Eq. (3.24) in the form

$$\frac{d\gamma_{LV}}{dT} = \frac{K_E}{[M/(\rho_L - \rho_V)]^{2/3}} \left[\frac{2(T_c - T)}{3(\rho_L - \rho_V)} \frac{d(\rho_L - \rho_V)}{dT} - 1 \right] \quad (3.28)$$

from which values of the temperature coefficient can be estimated when values of T_c are known or estimated.*

---

*Note that estimations will depend on the accuracy of T_c values and values of ρ_L and ρ_V. An approximate estimate of T_c can be obtained from $T_c \simeq 1.55 T_V$, where T_V is the boiling point or vaporization temperature [R. Hartmann and R. Schneider, *Z. Anorg. Chem.*, **180**, 275 (1929)].

Surface Tension and Heat of Vaporization. The change in thermal energy of a system whose surface undergoes an expansion per unit area is defined as the latent heat of surface formation by

$$\Delta H_s = -T(d\gamma/dT) \tag{3.29}$$

Since the surface free energy, γ_{LV}, is the work required to displace molecules from the bulk to form unit area of surface, the total energy associated with this process will be given by $U_s = \gamma_{LV} - \Delta H_s$, or

$$U_s = \gamma_{LV} - T(d\gamma_{LV}/dT) \tag{3.30}$$

It is apparent from Eq. (3.30) that, at $T = 0°K$, the total surface energy and the surface free energy (surface tension) will be equal, assuming the system to be liquid at $0°K$. Obviously this is physically impossible for metals and alloys or any other matter. For this condition to be valid, we must allow that a system possess a surface tension, or surface free energy in both the liquid and solid state, which is approximately the same entity.

Skapski[78] has argued that the total molar surface free energy, U_0 (in kilocalories per mole), of a liquid with molar surface area A_L is related to the latent heat of vaporization, ΔH_V, at absolute zero by

$$U_0 = A_L \left(\gamma_{LV} - T \frac{d\gamma_{LV}}{dT} \right) = \frac{Z - Z_s}{Z} \Delta H_V \tag{3.31}$$

where Z is the number of atoms in the bulk of the liquid system, and Z_s represents the number of surface atoms (assuming equal bond energies). Oriani[79] has shown that, for close-packed liquids such as aluminum and gold, $Z - Z_s/Z = 4/12 = 0.25$, and for bcc liquids such as iron, tungsten, and vanadium, the value is 0.363. Allen[50] has shown that a plot of U_0 against ΔH_V values extrapolated from Stull and Sinke[80] follows the trend predicted by Eq. (3.31) for close-packed pure liquid metals, and follows the trend in Fig. 3.10 which indicates that high surface free energies are associated with high bond strengths. We might expect, on examining the implications of this feature, that surface free energies of solid, close-packed metals or alloys would be expected to increase with increasing mechanical strength, since it depends to a measurable extent on bond strength. This is observed experimentally when the solid surface free energies of metals and alloys are plotted against Young's modulus at constant temperature.*

3.2.6 Surface Tension Isotherms for Alloy Solutions

As discussed in Section 1.3.2 and noted specifically in Eqs. (1.38) through (1.40), variations of component concentrations in binary or multicomponent alloy systems

*See Fig. 3.19.

INTERFACIAL FREE ENERGY

are expected to influence the interfacial free energy, as illustrated generally in Fig. 1.9. With liquid surfaces, certain components in the system can act as surface-active agents which are selectively absorbed as a concentration profile at the liquid-vapor interface. Figure 3.12 illustrates several examples of variations of surface tensions with composition in binary systems representative of the idealized responses depicted in Fig. 1.9.

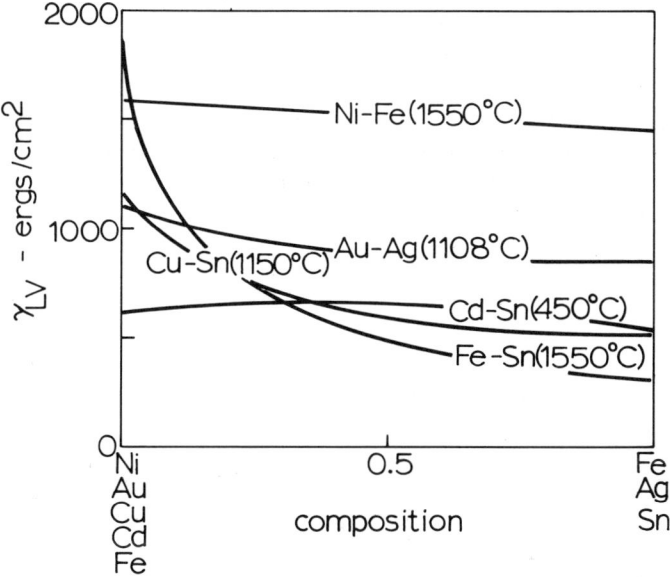

FIGURE 3.12 Surface tension isotherms for several binary solutions.[17,28,56,59]

Semenchenko[28] discussed the integration of the Gibbs adsorption isotherm [Eq. (1.38)] by considering the concept of activity coefficients and the volume relationships between the components. The essential features involve the identity of the liquid-vapor interface, or more specifically the ability to differentiate the surface from the bulk, and the thickness of the surface "layer." Generally the treatment involves a statistical consideration of the surface as a monolayer where the only inhomogeneity in the binary system is considered to occur at its topmost "atomic layer." If the liquid-vapor interface ("surface layer") and the bulk are considered to behave as ideal solutions,[81,82] the surface tension isotherm can be written as[59]

$$e^{-\gamma_{LV} A_0/kT} = x_1 e^{-\gamma_{LV_1} A_0/kT} + x_2 e^{-\gamma_{LV_2} A_0/kT} \tag{3.32}$$

where γ_{LV} is the solution surface tension, γ_{LV_1} and γ_{LV_2} are the surface tensions of the pure liquid components, x_1 and x_2 are the corresponding mole fractions of components 1 and 2, A_0 is the common value of the atomic area of components 1 and 2, k is Boltzmann's constant, and T is the absolute temperature. Obviously, the justification of Eq. (3.32) or a similar expression depends on a number of considerations including electronic compatibility of the constituent ions, their atomic radii, and their phase equilibria. Reasonable representations have been obtained for a number of binary systems[83,84] including the Ni-Fe [17] and Au-Ag [59] systems represented in Fig. 3.12.

As discussed in Section 1.3.2, multicomponent systems can exhibit adsorption-desorption phenomena; this applies equally to solid and liquid systems. There are few experimental data for liquid ternary systems, except for the work of Kaufman and Whalen[52] on liquid ternary Fe-C-Co and Fe-C-Ni systems where the variations of surface tension with cobalt and nickel compositions are as illustrated in the data reproduced in Fig. 3.13.

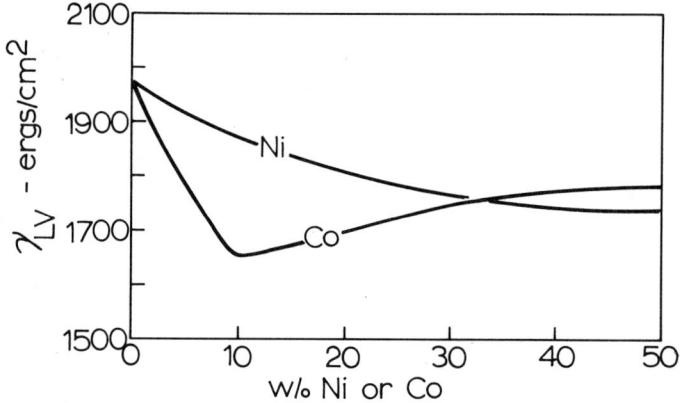

FIGURE 3.13 Surface tensions of ternary Fe-1.6 w/o C-Ni and Fe-1.5 w/o C-Co alloys (1425°C). (After Kaufman and Whalen.[52])

The data in Fig. 3.13 are characterized by the Gibbs adsorption isotherm for ternary systems [Eq. (1.48)] in the form

$$d\gamma_{LV} = -\Gamma_C \, d\mu_C - \Gamma_X \, d\mu_X \tag{3.33}$$

where X refers to component 3—nickel or cobalt in Fig. 3.13, for example. Since the form of Eq. (3.33) is predicated on the definition of the liquid alloy surface such that $\Gamma_1 \equiv \Gamma_{Fe} = 0$, then $\Gamma_2 = \Gamma_{2(1)}$ and $\Gamma_3 = \Gamma_{3(1)}$ [see Eq. (1.40)], where

INTERFACIAL FREE ENERGY

$2 \equiv$ carbon and $3 \equiv$ nickel or cobalt with reference to Fig. 3.13. Independent solutions of Eq. (3.33) are obtained in the form

$$\Gamma_C = \frac{\alpha_C + \beta_{CX}\alpha_X}{1 - \beta_{CX}\beta_{XC}} \qquad (3.34)$$

and

$$\Gamma_X = \frac{\alpha_X + \beta_{XC}\alpha_C}{1 - \beta_{CX}\beta_{XC}} \qquad (3.35)$$

where

$$\alpha_C = -(\partial \gamma_{LV}/\partial \mu_C)_{n_X, n_{Fe}} \qquad (3.36)$$
$$\alpha_X = -(\partial \gamma_{LV}/\partial \mu_X)_{n_C, n_{Fe}} \qquad (3.37)$$

and

$$\beta_{CX} = (\partial n_C/\partial n_X)_{\mu_C, \mu_{Fe}} \qquad (3.38)$$
$$\beta_{XC} = (\partial n_X/\partial n_C)_{\mu_X, \mu_{Fe}} \qquad (3.39)$$

n_j being the mole concentration of the jth component [Eq. (1.28)]. The mathematical details involved in evaluating Eqs. (3.34) through (3.39) have been described by Whalen et al.,[85] and it is observed that for liquid Fe-C-X ternary alloys, where X is a substitutional element, $\alpha_C \ll \alpha_X$, Eqs. (3.34) and (3.35) may be rewritten approximately as

$$\Gamma_C \equiv \frac{\beta_{CX}\alpha_X}{1 - \beta_{CX}\beta_{XC}} \qquad (3.34a)$$

$$\Gamma_X \equiv \frac{\alpha_X}{1 - \beta_{CX}\beta_{XC}} \qquad (3.35a)$$

As was pointed out by Whalen et al.,[85] the surface behavior exhibited by these systems can be predicted simply from the sign of β_{CX}, since, for negative values of α, a negative value of β_{CX} denotes a positive surface excess of one component (adsorption) and a negative surface excess or deficiency (desorption) of the other. A positive value of β_{CX}, on the other hand, implies that both solutes occur as surface excesses.

Since, as Fig. 3.13 shows, additions of nickel and cobalt result in a lowering of the surface tension, γ_{LV}, for Fe-C alloys, α_X in Eq. (3.37) must be negative. Although detailed activity measurements essential for evaluating β_{CX} and β_{XC} have not been made for the Fe–C–Co or Fe–C–Ni systems, the assumption that these systems are qualitatively similar to the Fe–C–Si alloys[85] would indicate a positive adsorption of cobalt and nickel and a desorption of carbon at the liquid surface. The essence of this feature is that cobalt or nickel atoms replace carbon at the liquid–vapor interface.

3.3 SURFACE ENERGY OF SOLID METALS AND ALLOYS

The measurement of solid surface free energy has been reviewed by Inman and Tipler,[86] Greenough,[87] Bikerman,[88] and Hondros,[89] to name a few. Bikerman[88] erroneously concluded that techniques devised to measure solid surface free energy are generally in error as a result of faulty logic or mistakes in mathematical manipulation. It is indeed generally agreed, as indicated in Chapter 1, that the concept of solid surface free energy must be carefully considered, particularly for an extension of this concept to solids at low temperatures where $dF_A/dA \neq 0$ in Eq. (1.4). However, at high temperatures near the melting point where diffusion is rapid and grain growth occurs in polycrystalline metals and alloys, surface equilibrium will be restored during slow deformation. As a consequence, it would appear that the only experimentally feasible method of measuring approximate solid surface free energies would involve one in which these conditions are met. The only presently known method of this kind involves the zero-creep technique where a polycrystalline wire or thin foil is elongated at high temperature, and the grain boundaries intersecting the solid–vapor interface (the free surfaces) sustain equilibrium by the movement of atoms at the grain boundary interface.

3.3.1 Zero-Creep Techniques for the Measurement of Solid–Vapor Surface Energies

The zero-creep method of determining the surface energy of solid metals and alloys was first used by Sawai and Nishida[90] and Tamman and Boehme,[91] and modified by Udin et al.,[92] Udin,[93] Pranatis and Pound,[94] and Hondros.[95] The technique is predicated on the fact that very thin wires or foils contract when heated to temperatures close to the melting point because, for a large surface-to-volume ratio, the surface tension (or shrinkage force) will exceed the static stress associated with the weight of the wire or foil sample.

Zero Creep of Polycrystalline Thin Foils. As we discussed in Chapter 2 (Section 2.2), grain boundaries in polycrystalline foils annealed under conditions to attain complete pseudoequilibrium* will intersect both free surfaces of a thin sheet at right angles as illustrated in Fig. 2.3(a). Hondros[95] considered a simple network of grain boundaries equilibrated in a thin sheet as illustrated schematically in Fig. 3.14 in which the grains were assumed to approximate a grid of squares of side S, with

---

*Pseudoequilibrium is distinguished from true equilibrium which occurs only when the grain boundaries anneal out and the solid becomes a single crystal.

INTERFACIAL FREE ENERGY

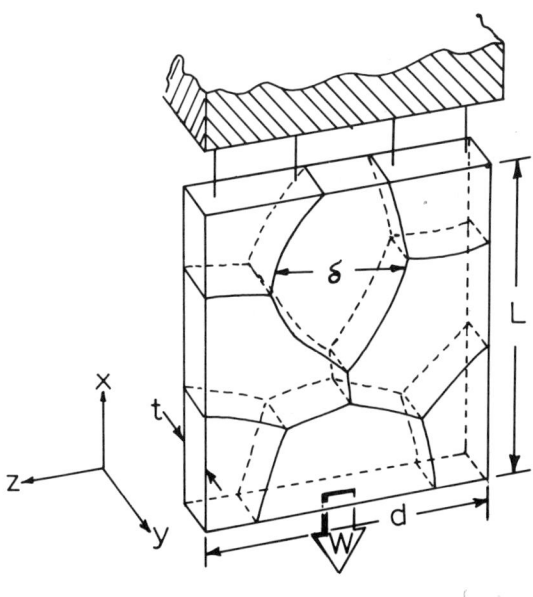

FIGURE 3.14   Equilibrium creep of thin foil loaded in tension by a weight W.

one side of each cellular grain parallel to the foil axis along the x-direction in Fig. 3.14. The principal stresses for such an equilibrium configuration (Fig. 3.14) are

$$\sigma_x = \frac{Wg}{dt} - \frac{2F_S}{t} - \frac{\gamma_{gb}}{\delta}$$

$$\sigma_y = -\frac{2F_S}{t} - \frac{\gamma_{gb}}{\delta}$$

$$\sigma_z = -\frac{2\gamma_{gb}}{\delta} - \frac{2F_S}{d} \tag{3.40}$$

where W is the total weight acting in grams (including the weight due to the foil itself), g is the gravitational constant, d is the foil width, t is the thickness, F_S is the surface free energy (solid-vapor interfacial free energy), and γ_{gb} is the grain boundary free energy.

The initial change in length of such thin foil specimens will be governed mainly by a redistribution of dislocations,[96] offsetting at inclined grain boundaries,[97] and orientation changes which reduce the specific grain boundary energies[98] as discussed in Chapter 2. The linear extension or contraction of the specimen is generally assumed to be controlled by a Nabarro-Herring[98,99] mechanism where material transport is affected by a flow of vacancies from the grain boundaries to the external surfaces (solid–vapor interface) or vice versa. This so-called "vacancy creep" is particularly effective at high temperature where a metal or alloy foil specimen can be considered as an isotropic viscous medium for which the axial strain rate and the principal stresses are related by

$$\dot{\epsilon}_x = \xi[\sigma_x - \nu(\sigma_y + \sigma_z)] \equiv dL/dt \tag{3.41}$$

where ξ is a constant referring to the viscosity, and ν is Poisson's ratio for the condition of equilibrium (at temperature).

If W_0 represents the load required to just balance the surface tension or contractile forces, the condition for zero creep or zero strain rate in the x-direction in Fig. 3.14 ($\epsilon_x = 0$) is

$$W_0 g = F_S d - \gamma_{gb}\nu(td/\delta) \tag{3.42}$$

Hondros[95] assumed that thin foil creep specimens did not change volume, and let $\nu = 0.5$; however, it is well known that ν for most metals and alloys will vary from roughly 0.25 to 0.45. But since the average grain size of thin metal or alloy foils is ~0.5 mm, the specimen dimensions t and d may be manipulated so as to make the grain boundary term negligible when compared with the surface free energy term in Eq. (3.42). Large grain size is not, however, a necessary condition for the use of Eq. (3.42), since a second equation relating γ_{gb} and F_S can be obtained by accurate observations of the groove angle at the intersection of the grain boundary with the free surfaces, as discussed in Chapter 2. Indeed, if the mean dihedral angle, Ω_S, is obtained from the test specimen, substitution of Eq. (2.13) for γ_{gb} in Eq. (3.42) results in a single expression for the average surface free energy:

$$F_S = \frac{W_0 g}{d[1 - 2\nu(t/\delta)\cos(\Omega_S/2)]} \tag{3.42a}$$

Determination of Surface Free Energies from Thin Foils. Although it is possible to monitor the creep of flat, polycrystalline foils freely suspended and loaded uniformly at one end, an ingenious modification by Hondros[95] permits the experiment to be performed as a cylindrical unit from which a number of data points representing strain rate versus load can be obtained simultaneously. In this arrangement, illustrated in Fig. 3.15(a), a specimen foil is shaped into a tube, or a tubular specimen of the desired material is utilized at the outset and loaded at intervals along its length so as to constitute five or more different specimen elements. Cylindrical flat-ended weights with a central hole for evaluation and of the same material are sintered or spot-welded in a sequence such that the lower elements might contract, while the upper, more heavily stressed elements elongate. The flat tops of the

INTERFACIAL FREE ENERGY

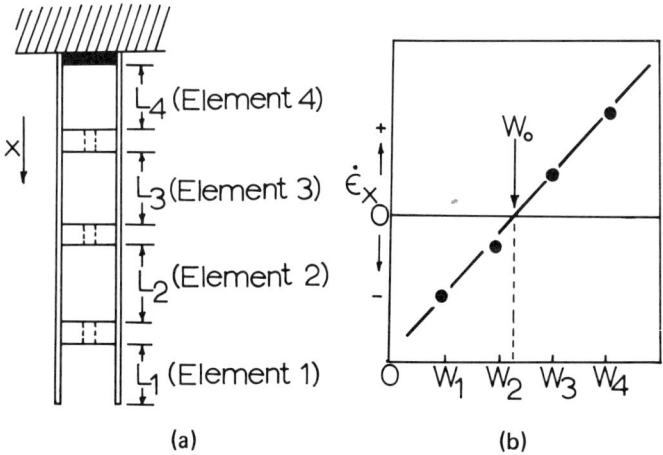

FIGURE 3.15 Cylindrical thin foil specimen assembly arrangement (a) and residual data tabulation (b) for determining average solid surface free energies. Strain rates are determined over the course of $\sim 10^2$ hours, depending on the temperature.

weights are used as fiducial marks, and the entire assembly is recorded by x-ray radiography at various times at temperature directly within the furnace arrangement. The effective weight corresponding to the strain rate for each element is then plotted as illustrated in Fig. 3.15(b), from which the zero-creep weight, W_0, is obtained as shown.

A source of consistent error when dealing with foils involves lateral constraint adjacent to the loads, as Fisher and Dunn pointed out.[100] Hondros[95] found, however, that the necessary correction for this effect was opposite to that required for corrections due to the presence of the grain boundaries in calculating the surface free energy [the second term in Eq. (3.42)], and sufficiently small that it could be neglected in the same sense that the grain boundary term could be neglected if the specimen shape was parameterized as suggested previously. In practice the foil thickness is in the range 10^{-3} cm and the length is chosen to be $>6d$,[89] or in the case of cylindrical samples $>12\pi r$ (where r is the mean radius).

After an initial settling-down period of 10 to 20 hours, the specimens are usually allowed to creep for periods ranging from 50 to several hundred hours, depending on the temperature. Specimens, normally enclosed in a larger cylinder of the same material to maintain vapor equilibrium at the specimen surface, are annealed in a high-vacuum furnace or in an inert or reducing atmosphere such as hydrogen to avoid oxidation, surface reactions, or embrittlement by gas diffusion to the grain boundaries. Purified helium at a positive pressure of 1 to 3 atmospheres is normally utilized in zero-creep experiments on metals and alloys.

Zero Creep of Polycrystalline Fine Wires. In some cases wire samples are preferred to foils because of the simpler stress system, and because the grain boundaries in fine wires are more easily described in the geometry of the system, with the result that the grain boundary free energy contribution can be more accurately determined. Under the conditions of zero creep, the grain boundaries align themselves normal to the wire axis, forming a so-called "bamboo" structure (Fig. 3.16). If we consider a virtual work argument, a fine wire of radius r carrying

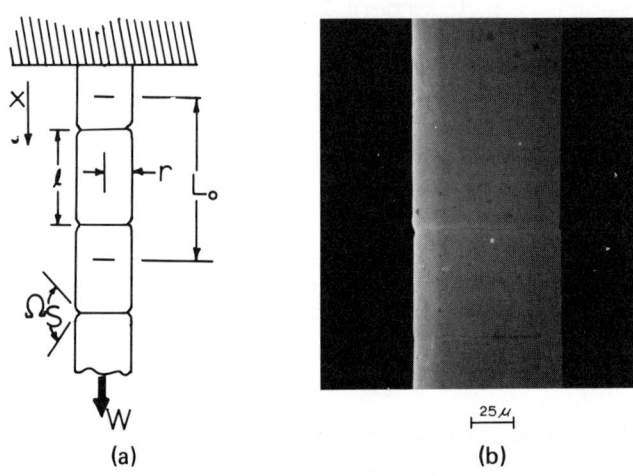

FIGURE 3.16 Bamboo structure of fine wires crept at high temperature. (a) Schematic view showing gauge length, L_0. (b) Scanning electron micrograph of stainless-steel wire crept in purified helium at 1060°C.

a balance load, W_0, at the zero-creep condition can be assumed to undergo a small extension, δx, accompanied by a concomitant change in grain boundary area, δA:

$$W_0 g \delta x = \pi r F_S \delta x - \gamma_{gb} \delta A \qquad (3.43)$$

If within a distance x between gauge marks the average grain length is measured to be ℓ, the average grain boundary area becomes $\pi r^2 (x/\ell)$, and Eq. (3.43) becomes

$$W_0 g = \pi r [F_S - \gamma_{gb}(r/\ell)] \qquad (3.44)$$

which is similar in form to Eq. (3.42).

Since, as noted previously in Chapter 2, fine wires are ideally suited to measuring the dihedral angle Ω_S, using the transmission electron microscope [as illustrated in Fig. 2.11(f)], Eq. (3.14) can be substituted for γ_{gb} in Eq. (3.44) to obtain

INTERFACIAL FREE ENERGY

$$F_S = \frac{W_0 g}{\pi r [1 - 2(r/\ell) \cos(\Omega_S/2)]} \quad (3.44a)$$

Jones[101] recently examined the grain boundary term in the zero-creep condition and showed that this term becomes positive when grains are sufficiently longer in the direction of the applied stress (the x-direction in our convention) and may dominate the zero-creep expression.

Determination of Surface Free Energies from Fine Wires. Fine wires having diameters $<10^{-2}$ cm are prepared with various weights spot-welded onto their free ends and hung in a cylindrical enclosure of the same material after being scribed circumferentially with a razor edge to form fiducial grooves, or cross-pieces of the same material are spot-welded at $\sim$2 cm or more intervals to act as gauge marks. The wires are annealed at temperature in a vacuum or gas atmosphere furnace for an initial period of 10 to 20 hours, and the gauge length is measured. The anneal is then continued for a long period of time (50 to 200 hours) during or after which the strain rate is determined from

$$\dot{\epsilon}_x \equiv d\epsilon_x/dt = (L - L_0)/tL_0 \quad (3.45)$$

where L_0 is the initial gauge length [Fig. 3.16(a)] and L is the gauge length after creep anneal at temperature for a time t.

The strain rates for each wire are then plotted against the effective weights determined by cutting the wire at the mid-point of the gauge length and weighing the wire and weight below this point. The zero-creep weight, W_0, is then determined as shown typically in Fig. 3.17(a), identical to Fig. 3.15(b). The average grain length, ℓ, is determined by examination of the wire samples in an optical metallograph or in the scanning electron microscope as shown in Fig. 3.16(b), while the mean value of Ω_S is determined from electron shadowgraphs as shown in Fig. 2.11(f). Mean values of these experimental parameters are then substituted into Eq. (3.44a) to obtain a mean value for F_S.

It is generally observed experimentally that $\ell \cong 2r$; that is, the mean grain length approximately equals the mean wire diameter. Many experiments have also shown that $\gamma_{gb} \cong F_S/3$ (see Table 2.1).

Self-Diffusion Coefficients of Solid Metals and Alloys. Nabarro[99] proposed that creep can occur by the diffusional exchange of vacancies between the grain boundaries and the free surface, setting up an effective diffusion current in each grain of a solid metal or alloy. Both Nabarro[99] and Herring[98] developed expressions for the creep rate which when applied to the creep of fine wires can be used to obtain an expression for the self-diffusion coefficient, D, in the form

$$D = 2\dot{\epsilon}\ell r RT/\beta \Omega \sigma$$

where $\dot{\epsilon}$ is the strain rate, ℓ is the mean grain length, r is the wire radius, R is the gas

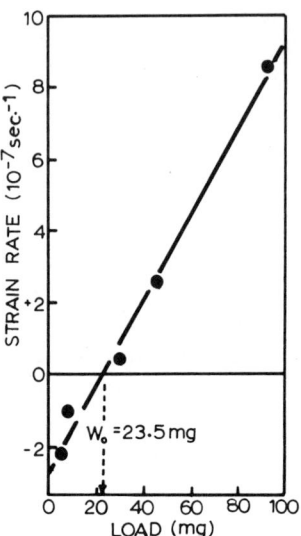

FIGURE 3.17a Strain-rate load graph for Chromel-A, and 80/20 NiCr alloy wires annealed 30 hours in purified helium at 1060°C (3-mil diameter). (After Murr et al.[102])

constant, T is the temperature, Ω is the volume/gm atom, σ is the resultant stress on the wire, and β is a constant which has a value of ~ 12 when $\ell \geqslant 2r$.

Experimental data obtained in zero-creep studies provide support for the form and quantitative accuracy of Eq. (3.46), and therefore the validity of the Nabarro-Herring argument[98,99] to Fig. 3.15(b).

3.3.2 Surface Free Energies and Temperature Coefficients of Solid Metals and Alloys

Measurements of surface energies for solid metals and alloys have been less extensive than surface tension measurements for liquids. Because of the critical conditions necessary to carry out zero-creep experiments, and the precision required in measuring variations in surface energy with temperatures near the melting point in the solid state, there have been few instances of experimentally determined temperature coefficients of surface free energy. Just as the variations of liquid surface tensions are ideally described by the thermodynamic considerations outlined in Chapter 1, the temperature variations of solid surface free energies are ideally described by Eq. (1.26) for pure metals (where vacancies are not a major consideration) and

$$\frac{dF_S}{dT} = - [S_S - \Gamma_1 S_1 - \sum_j (\Gamma_j S_j)] - j \sum_j \left[\left(\Gamma_j - \frac{x_j}{x_1} \Gamma_1 \right) \frac{d\mu_j}{dT} \right] \quad (3.47)$$

INTERFACIAL FREE ENERGY

using the notation of Chapter 1 for multicomponent systems.

Table 3.5 summarizes the known measurements of surface free energies and associated temperature coefficients determined for the most part by zero-creep techniques. In most cases the measurements represent mean values. Nearly all measurements listed were performed in purified, inert atmospheres, primarily argon and helium. It can be noted from Table 3.5 that the average value of the temperature coefficient of surface energy for pure metals is $\sim$$-0.45$ erg/cm^2 °C, and this value might be considered in estimating surface energies at temperatures below T_m for pure solid metals.

Figure 3.17(b) illustrates some examples of the variation of the surface free energy of solid metals and alloys with temperature. These examples are representative of a pure metal (nickel), a dilute binary alloy (Fe-Si), and a complex (multicomponent) alloy system (304 stainless steel).

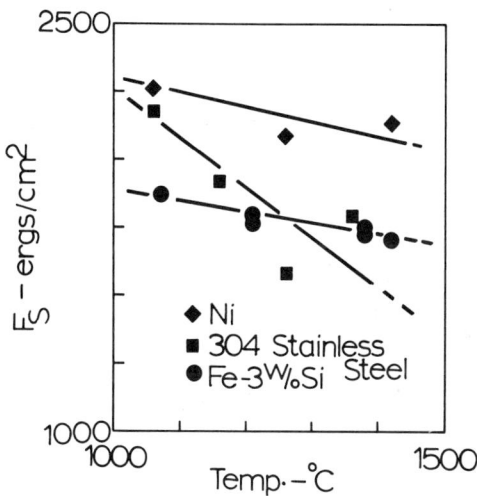

FIGURE 3.17b   Variation of surface free energy (F_S) for metals and alloys with temperature below the melting point. [Data from Hondros,[89] Murr,[104] and Murr (unpublished data).]

Surface Energy at the Melting Point; Calculation of F_S from the Corresponding Value of γ_{LV}. Allen and Kingery[51] have shown that the variations of surface energy with temperature of pure metals in the liquid state are offset abruptly at the melting point from the variation in the solid state. This abrupt change in energy is given approximately by[124]

$$\Delta \gamma_m = \Delta H_f / A_0 \tag{3.48}$$

TABLE 3.5
Surface Free Energy of Solid Metals and Alloys

Metal or Alloy	Surface Energy, F_S (ergs/cm²)[a]	Temperature (°C)	dF_S/dT (ergs/cm² °C)	Ref.
Aluminum	980[b]	450	−0.40	103, 104
Al–84 a/o Cu	1720	800	−	−
Beryllium	~2000[b]	700	−	105
Bismuth	~550	250	−	86, 106, Murr (unpublished data)
Chromium	2300	1400	−	107, 108
Cr–80 w/o Ni	2160	1060	−	102
Cr–35 w/o Re	~2500	1800	−	107
Cobalt	1970	1354	−	109
Copper	1780	925	−0.50	86, 92, 94, 110, 111
Cu–16 a/o Al	1720	800	−	Murr (unpublished data)
Cu–20 a/o Au	1160	850	−	86, 112
Cu–40 a/o Au	930	850	−	86, 112
Cu–60 a/o Au	910	850	−	86, 112
Cu–80 a/o Au	1140	850	−	86, 112
Cu–0.26 a/o Sb	980	950	−	111
Cu–0.57 a/o Sb	950	950	−	111
Cu–0.78 a/o Sb	900	950	−	111
Cu–60 w/o Ni	1800	950	−	Murr (unpublished data)
Cu–30 w/o Zn	1750	850	−	Murr (unpublished data)
Gold	1400	1000	−0.43	95, 110, 113
Au–20 a/o Cu	1140	850	−	86, 112
Au–40 a/o Cu	910	850	−	86, 112
Au–60 a/o Cu	930	850	−	86, 112
Au–80 a/o Cu	1160	850	−	86, 112
Iron (δ phase)	1950	1450	−0.9	86, 95
Iron (γ phase)	2100	1350	−3.0	86, 95, 114
δ-Fe–0.05 w/o P	800	1450	−	95
δ-Fe–0.09 w/o P	580	1450	−	95
δ-Fe–0.36 w/o P	400	1450	−	95
δ-Fe–0.15 w/o P	1690	1350	−	95

INTERFACIAL FREE ENERGY

TABLE 3.5 *(continued)*

Metal or Alloy	Surface Energy, F_S (ergs/cm²)[a]	Temperature (°C)	dF_S/dT (ergs/cm² °C)	Ref.
δ-Fe-0.27 w/o P	1670	1350	–	95
δ-Fe-0.36 w/o P	1605	1350	–	95
Fe-3 w/o Si	1850	1100	–0.36	89, 115
Fe-Cr-Ni (18-8 Stainless steel (304)	2190	1060	–1.76	116
Molybdenum	~2400	2350	–0.2	107, 108
Mo-33 w/o Re	~1900	2300	–0.2	107
Nickel	2280	1060	–0.55	117, 118, Murr (unpublished data)
Ni-20 w/o Cr	2160	1060	–	102
Niobium	2100	2250	–	119
Platinum	2200	1300	–0.60	110, 120
Silver	1100	950	–0.47	86, 121
Tin	685	223	–	122
Tungsten	2800[c]	2000	–	108, 123

[a] Equivalent to the international unit millijoules per square meter (mJ/m²).
[b] Estimated value obtained by a method not involving zero-creep techniques.
[c] Mean value based on zero-creep measurements and other reliable techniques.

where ΔH_f is the latent heat of fusion and A_0 is the area per atom in the metal surface. The solid surface free energy at the melting point is then approximated by

$$(F_S)_m = \Delta \gamma_m + (\gamma_{LV})_m \tag{3.49}$$

where $(\gamma_{LV})_m$ is the liquid surface energy (surface tension) at the melting point (T_m).

Figure 3.18 shows the surface energy response of the noble metals at the melting point obtained by plotting the experimental data in Tables 3.4 and 3.5. Note particularly in Fig. 3.18 that the measurable values of $\Delta \gamma_m$—namely, $\Delta \gamma_m(\text{Cu}) = 380$ ergs/cm², $\Delta \gamma_m(\text{Au}) = 245$ ergs/cm², and $\Delta \gamma_m(\text{Ag}) = 195$ ergs/cm² —are essentially coincident with those that can be calculated from Eq. (3.48) with values of ΔH_f.[51,125]

Skapski[126] has devised a more rigorous equation for calculating the solid surface energy from the corresponding liquid surface tension at the melting point in the form

$$(F_S)_m = \frac{Z - Z_S}{Z}\left(\frac{\Delta H_f}{A_S}\right) + \left(\frac{\rho_S}{\rho_L}\right)^{2/3}(\gamma_{LV})_m \tag{3.50}$$

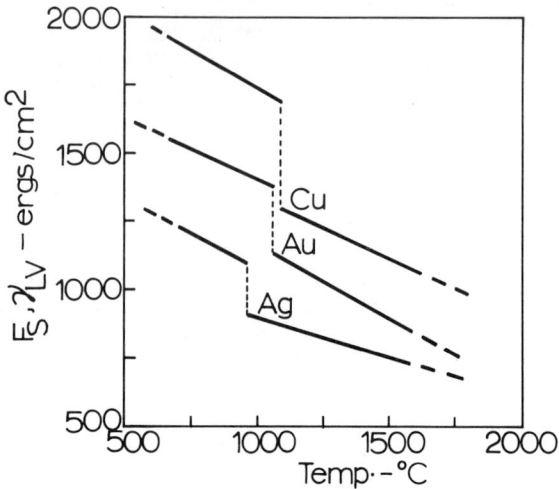

FIGURE 3.18 Temperature dependence of surface free energy and the solid/liquid-phase transition.

where Z and Z_S refer to the liquid bulk and surface atom coordination numbers, ΔH_f is the latent heat of fusion, A_S is the molar solid surface area, and ρ_S and ρ_L are the solid and liquid densities at temperature, respectively. As noted in Eq. (3.31), Oriani[79] has shown that $Z - Z_S/Z \cong 0.25$ for fcc metals and ~0.36 for bcc metals. If a value of $Z - Z_S/Z \cong 0.3$ is assumed for cubic metals where $A_S = 1.12 N^{1/3} (M/\rho_S)^{2/3}$ (N is Avogadro's number, and M is the molecular weight) and values of ΔH_f[125] are substituted into Eq. (3.50) assuming a 2.5% volume increase on fusion,[51] it is observed that $(F_S)_m \cong 1.2(\gamma_{LV})_m$. This value is essentially identical to the relationship derived experimentally from the data of Fig. 3.18 and Tables 3.4 and 3.5. It is possible, therefore, on considering an approximate expression of $dF_S/dT \cong -0.45$ to apply generally to cubic metals, to estimate the solid surface free energies from

$$F_S \simeq 1.2(\gamma_{LV})_m + 0.45(T_m - T) \tag{3.51}$$

where T_m is the melting temperature (°C), and T is the temperature (below the melting point) at which the equilibrium solid surface free energy is desired. Equation (3.51) can be shown to be approximately valid for pure fcc and bcc metals, and it would not appear unreasonable to apply this relationship in studies of most solid metals and alloys at high temperatures. This argument is supported to some degree by the data presented in Fig. 3.19, which also illustrates the dependence of mechanical strength on surface free energy for a number of (primarily cubic) metals and alloys.

INTERFACIAL FREE ENERGY 127

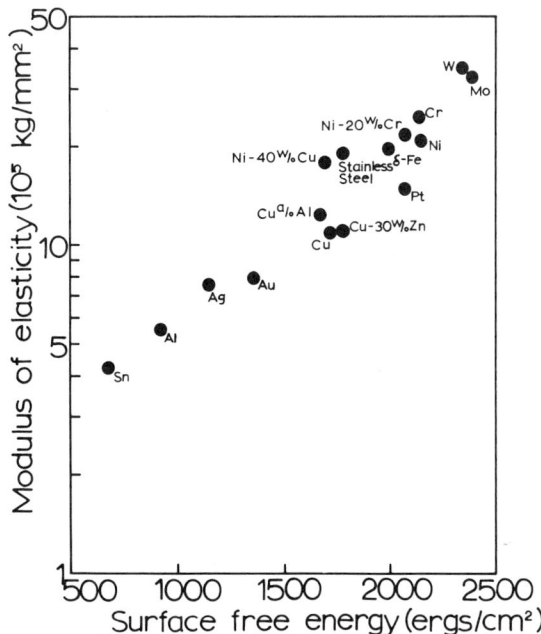

FIGURE 3.19 Surface free energy of solid metals and alloys versus modulus of elasticity in tension (Young's modulus) at $0.9T_m$ (T_m = melting temperature in °C). (Data based on Tables 3.4 and 3.5 and "Metals Handbook."[125])

3.3.3 Surface Energy Isotherms for Solid Alloys

The influence of various concentrations of components on surface energies of solid alloy systems will be governed by Eqs. (1.38) through (1.40) and Eq. (3.47), and the response will, in many cases, follow that illustrated for liquid metal alloys in Section 3.2.6. Figure 3.20 illustrates the variation of surface free energy for the Cu-Au system using the data in Table 3.5. The variation of surface free energy in Fig. 3.20 is similar to the liquid systems illustrated in Fig. 3.12 and both are representative of the idealized responses depicted in Fig. 1.9.

As indicated in Chapter 1, it is necessary, for systems of high concentrations of "solute" as depicted in Fig. 3.20, to make some assumptions about the relative surface areas of the solute and solvent atoms in order to evaluate the Γ_j's individually. Because of the complications involved, few meaningful data exist for concentrated binary multicomponent solid alloy systems.

Some data are available for dilute alloy systems, and several systems such as Cu-Sb and Fe-P have been investigated in considerable detail. Figure 3.21 illustrates the variation in surface energy with solute concentration in several dilute alloy systems.

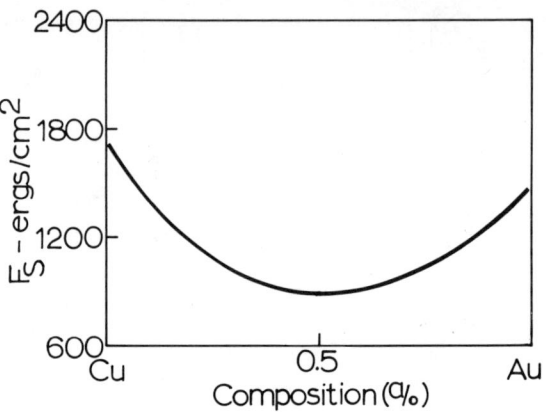

FIGURE 3.20 . Surface energy isotherm for the solid Cu–Au system (850°C).

By application of the Gibbs adsorption theorem to the data of Fig. 3.21, one can arrive at a quantitative description of solute segregation to the surfaces (solid-vapor interface). If one considers Eq. (1.38), and then applies the relation $d\mu_2 = RT\, d \ln x_2$, the general equation simplifies to

$$\left(\frac{dF_S}{d \ln x_2}\right)_{T,P} = -RT\left(\Gamma_2 - \frac{x_2}{x_1}\Gamma_1\right) \tag{3.52}$$

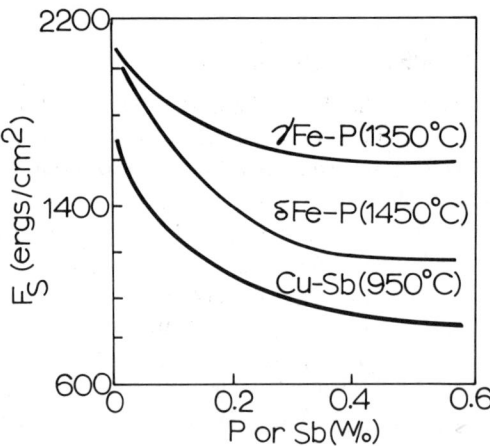

FIGURE 3.21 Influence of bulk phosphorus or antimony on the surface energy of iron and copper, respectively. (After Hondros[95] and Inman et al.[111])

INTERFACIAL FREE ENERGY

For a dilute system containing one solute species of molar concentration x_2, the response is described approximately by Eq. (1.44) in the form

$$\Gamma_2 = \frac{1}{RT}\left(\frac{dF_S}{d \ln x_2}\right) \tag{3.53}$$

Inman et al.[111] have shown by activation analysis of surface composition in Cu–Sb alloys (Fig. 3.21) that the average surface excess of Sb in equivalent number of monatomic layers is approximately 0.37, which, when applied to Eq. (3.52), results in a value of 0.6×10^{-9} mole/cm^2 for the right-hand parentheses. Hondros[95], assuming Eq. (3.53) to apply in the case of dilute Fe–P alloys (Fig. 3.21), obtained a surface excess of 2.3×10^{-9} mole/cm^2 which can be visualized as roughly one to two phosphorus atoms for each iron atom in the surface—that is, approximately a monolayer. The experimental results are therefore in qualitative agreement with Gibb's theory.

Solid Surface Adsorption and Positive Temperature Coefficients of Surface Energy. Whereas an equilibrium segregate at the surface of a solid metal or alloy will generally lower the surface free energy with increasing temperature, an adsorbed layer or layers of surface-active solute will normally reverse this trend and result in a positive temperature coefficient of surface free energy. This effect is also possible in liquid systems as discussed previously, but the liquid surface is not so well defined as in solid systems.

The first term on the right-hand side of Eq. (3.47) can be considered in terms of the excess surface entropy compared with the entropy of the same quantity of material in the bulk phase and can be approximated by the entropy of melting, ΔS_m. The second term, as discussed by Jones and Leak,[127] accounts for the effects due to variation in temperature of solute segregation or chemisorption, and approximately related to the heat of adsorption, ΔH_a. Rewriting Eq. (3.47) in this approximate notation we have

$$dF_S/dT \cong -\Delta S_m + (\Gamma \Delta H_a/T) \tag{3.47a}$$

where Γ represents the quantity of adsorbate per unit area of surface. It is obvious from Eq. (3.47a) that, for strongly adsorbing surfaces or for the accumulation of surface contamination, dF_S/dT would tend to be positive.

If the entropy of melting is taken to be 0.5 erg/cm^2 °C, the average temperature coefficient for pure metals from Table 3.5, and the heat of impurity adsorption is taken to be approximately 100 kcal/mole,[112] we obtain $\Gamma \cong 10^{-9}$ mole/cm^2 at $T = 1500°$ to 2000°C and $dF_S/dT \cong +1$ erg/cm^2 °C. This value of Γ, as observed from our earlier discussion, corresponds to roughly one-half a monolayer of chemisorbed interstitial impurities at the surface, and satisfactorily accounts for temperature coefficients of +0.71, +0.65, and +0.55 erg/cm^2 °C observed for chromium, molybdenum, and tungsten, respectively, in the work of Allen.[108] In the case of silicon iron investigated by Jones and Leak,[127] where dF_S/dT was observed to be +2.4 ergs/cm² °C, at temperatures near the melting point in the solid state, Γ corresponds to roughly a monolayer of surface-active oxygen and sulfur known to be supplied from the sample bulk.

Measurement of Solid Surface Segregation. Several methods are available for detecting surface segregation including low-energy electron diffraction techniques (LEED)[128] and radiotracer methods where a radioactive solute is alloyed with a metal which, after some equilibrium heat treatment, is systematically reacted in an appropriate acid to remove a detectable thickness of the surface.[28] Although LEED is capable of monolayer resolution, it is sometimes difficult to obtain a quantitative analysis of surface segregation. There have been only a few reported studies of physical adsorption,[129] but investigations of chemisorption are much more numerous.[130] In a study of hydrogen adsorption on the (100) surface of tungsten, for example, coverage exceeding half a monolayer of hydrogen atoms caused a detectable change in the LEED pattern recognized as a splitting of spots into four new spots which gradually moved apart and disappeared at saturation.[131]

Besides yielding diffraction data, the experimental arrangement in a LEED system allows other supplementary measurements. The most important of these from the standpoint of surface analysis is the measurement of Auger peaks in the energy spectrum of the inelastically scattered and secondary electrons which emerge from the specimen surface.[132] Auger spectroscopy affords a simple and powerful method for the study of surface segregation. Allen[108] has employed Auger electron spectroscopy to explore the possibility of surface-active contaminants from molybdenum specimens and found sulfur to be the agent, as evidenced by a strong characteristic peak in the Auger spectrum. Bonzel and Aaron[133] have also discussed the measurement of equilibrium surface segregation using Auger electron spectroscopy for the equilibrium surface segregation of silver in a Cu–14.3 w/o Al alloy.

3.4 INTERFACIAL FREE ENERGIES IN SOLID METALS AND ALLOYS

3.4.1 Measurement of Grain Boundary Free Energy

The direct measurement of average grain boundary free energies is a consequence of zero-creep measurements outlined in Section 3.3.1, and indicated previously in Chapter 2 by Eq. (2.13). Techniques of measuring grain boundary free energies from equilibrium configurations of interfaces have been treated somewhat extensively in Section 2.2, and the ratios of interfacial energies involving the equilibration of grain boundaries with other interfaces determined experimentally have been presented in Tables 2.1 and 2.2. If the average values of surface free energies for metals and alloys tabulated in Table 3.5 are inserted into the ratios of γ_{gb}/F_S presented in Table 2.1, average values of γ_{gb} are obtained at temperature as illustrated in Table 3.6.

TABLE 3.6
Grain Boundary Free Energy in Metals and Alloys

Metal or Alloy	Grain Boundary Energy, γ_{gb} (ergs/cm^2)	Temperature (°C)	$d\gamma_{gb}/dT$ (ergs/cm^2 °C)	Ref.
Aluminum	324	450	−0.12	103, 104
Chromium	920	1400	−	107, 108
Cr–35 w/o Re	750	1800	−	107
Cobalt	650	1354	−	109
Copper	625	925	−0.10	86, 92, 94, 110, 111, Murr (unpublished data)
Cu–16 a/o Al	550	800	−	107
Cu–0.26 a/o Sb	290	950	−	111
Cu–0.57 a/o Sb	280	950	−	111
Cu–0.78 a/o Sb	320	950	−	111
Cu–60 w/o Ni	550	950	−	Murr (unpublished data)
Cu–30 w/o Zn	595	850	−	Murr (unpublished data)
Gold	378	1000	−0.10	95, 110, 111, 113, Murr (unpublished data)
Au–20 a/o Cu	320	850	−	86, 112
Au–40 a/o Cu	310	850	−	86, 112
Au–60 a/o Cu	390	850	−	86, 112
Au–60 a/o Cu	430	850	−	86, 112
Iron (δ phase)	468	1450	−0.25	86, 95
Iron (γ phase)	756	1350	−1.0	86, 95
δ-Fe–0.05 w/o P	200	1450	−	95
δ-Fe–0.09 w/o P	145	1450	−	95
δ-Fe–0.36 w/o P	100	1450	−	95
γ-Fe–0.15 w/o P	620	1350	−	95
γ-Fe–0.27 w/o P	600	1350	−	95
γ-Fe–0.36 w/o P	570	1350	−	95
Fe–3 w/o Si	617	1100	−0.07	89, 115
Fe–Cr–Ni (304) Stainless steel	835	1060	−0.49	116
Molybdenum	575	2350	−	107, 108
Mo–33 w/o Re	475	2300	−	107

TABLE 3.6 *(continued)*

Metal or Alloy	Grain Boundary Energy, γ_{gb} (ergs/cm²)	Temperature (°C)	$d\gamma_{gb}/dT$ (ergs/cm² °C)	Ref.
Nickel	866	1060	−0.2	117, 118, Murr (unpublished data)
Ni–20 w/o Cr	756	1060	−	102
Niobium	756	2250	−	119
Platinum	660	1300	−0.18	110, 120
Silver	375	950	−0.10	86, 121
Tin	164	223	−	122
Tungsten	1080	2000	−	108, 123
Zinc	340	300	−	103

It must be cautioned that grain boundary free energies obtained by zero-creep techniques or by the true equilibrium geometry at grain boundary–interfacial intersections represent average values of high-angle grain boundary free energies and are not specific to any one grain boundary, nor are such values absolute. From an engineering viewpoint, these values are therefore meaningful, since they represent a general feature of solid metals and alloys.

Åstrom[103] has devised a precise calorimetry technique for the determination of grain boundary energies which is based upon the measurement of the total grain boundary energy obtained by measuring the energy released during grain growth in granules of metals a few millimeters in diameter. Åstrom found that for aluminum, silver, and zinc the total grain boundary energy, E_{gb}, as given by the following modification of Eq. (1.2),

$$E_{gb} = \gamma_{gb} - T(d\gamma_{gb}/dT) \tag{3.54}$$

was 625, 790, and 1740 ergs/cm², respectively. Using an approximation for the temperature coefficient (entropy), in terms of the heats of fusion, ΔH_f, Åstrom then derived values of γ_{gb} = 340, 420, and 340 ergs/cm² for aluminum, silver, and zinc, respectively, at a temperature of approximately 300°C.

While Åstrom's calorimetric method of measuring grain boundary energy is a very attractive one because of its direct approach, it has not been extensively utilized by other investigators, primarily because of the precision required in obtaining accurate calorimetric results.

3.4.2 Effects of Temperature and Composition on Grain Boundary Free Energy

The temperature coefficients for grain boundary free energy have been measured for only small numbers of metals and alloys as indicated in Table 3.6; the experi-

INTERFACIAL FREE ENERGY

mentally determined dependence of grain boundary free energy with temperature for nickel and stainless steel is illustrated in Fig. 3.22. The general temperature dependence of grain boundary free energy is ideally described by Eq. (3.47), which, for the case of pure nickel in Fig. 3.22, reduces to a simple statement for grain boundary entropy when vacancy contributions are neglected. It can be noted, on comparing the temperature coefficients for grain boundaries in Table 3.6 with those for the surface free energy in Table 3.5, that $d\gamma_{gb}/dT$ values are also negative, and proportionately smaller than values of dF_S/dT.

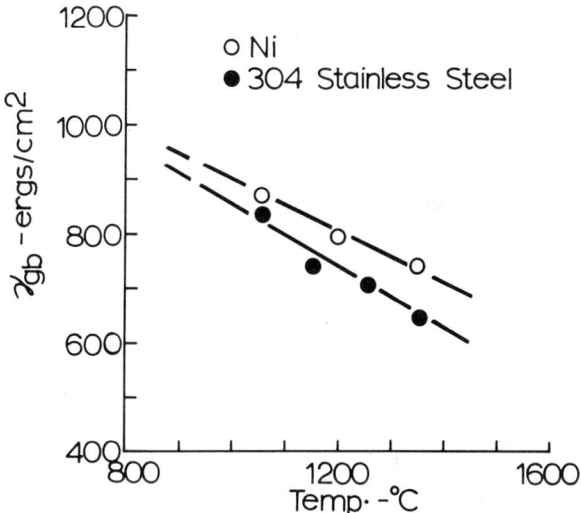

FIGURE 3.22   Variation of grain boundary free energy (γ_{gb}) for nickel and stainless steel with temperature below the melting point. (Data for stainless steel from Murr et al.[116])

In solid alloys, McLean[134] has shown by a statistical thermodynamic argument that segregation is related to temperature by an expression of the form

$$X_{gb} = \alpha x e^{Q/RT}/(1 + \alpha x e^{Q/RT}) \tag{3.55}$$

where X_{gb} is the grain boundary concentration, x is the bulk concentration, α is a constant, R is the gas constant, Q is the excess lattice energy due to the presence of the solute, and T is the absolute temperature.

Grain Boundary Free Energy Isotherms.   Typical of the influence of bulk concentrations of solute in dilute alloys is the effect on grain boundary free energy as depicted in Fig. 3.23. The reduction in grain boundary free energy is consistent

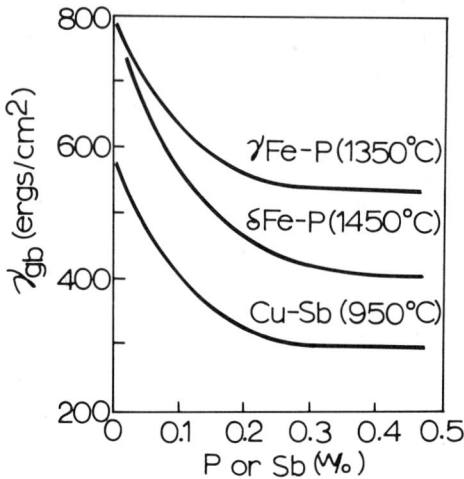

FIGURE 3.23   Influence of bulk phosphorus or antimony on the grain boundary free energy of iron and copper respectively. (After Hondros[95] and Inman et al.[111])

with segregation described by the Gibbs adsorption isotherm [Eq. (1.38)] essentially of the form given by Eq. (1.42), the Cahn-Hilliard criterion.

It must be pointed out that for any given grain boundary structure (to be described in Chapter 4) there may be some atomic structure which minimizes the system's interfacial free energy, having a grain boundary free energy $\gamma_{gb}(I)$. As Bitler[135] has argued, the assumptions implicit in the Cahn-Hilliard criterion [Eq. (1.42)] would be reasonable if the solvent or matrix atomic structure were fixed and solute atoms substituted. However, a grain boundary structure may exist in the same system with a grain boundary free energy, $\gamma_{gb}(II)$, which differs from $\gamma_{gb}(I)$ as illustrated schematically in Fig. 3.24. For the situation depicted, structure II would be the equilibrium structure for compositions in excess of X_1, where X is the mole fraction of solute and X_e is the solubility limit of the solute. Figure 3.24 suggests that grain boundary structure changes occur without bulk changes, and this phenomenon has in fact been suggested previously by Hart.[136] The essence of Fig. 3.24 is that the adsorption and segregation of solute in excess of those indicated by the Cahn-Hilliard criterion for dilute alloy systems could occur at certain grain boundary structures which would permit solute interactions over distances in excess of the interfacial plane (a monolayer). For example, the solute adsorption per unit length of a straight dislocation is infinite,[137,138] consequently the solute excess can be very large for certain elastic field arrangements. It is therefore not unlikely that solute concentrations in the region of a grain boundary can occur as an equilibrium

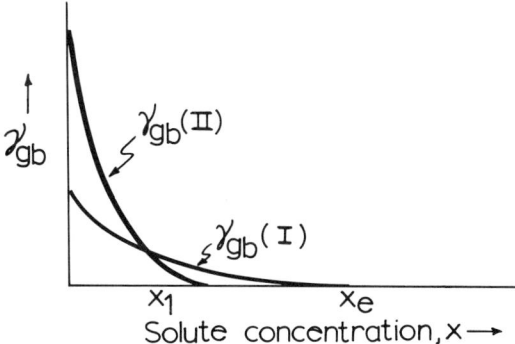

FIGURE 3.24 The dependence of grain boundary free energy on composition for two distinct grain boundary structures. (After Bitler.[135])

segregation which for all practical purposes is in qualitative agreement with the Gibbs theorem, and where precipitates will not form.

We shall deal with some of these concepts in greater detail in Chapters 4 and 5.

Measurement of Grain Boundary Segregation. The methods described in Section 3.3.3 for the measurement of adsorption and segregation to solid surfaces are generally applicable to qualitative or quantitative measurements of segregation to grain boundaries. Radio tracer measurements can be applied to studies of the surfaces created by intergranular fracture; LEED and Auger spectroscopy are also applicable where grain boundary surfaces are exposed by intergranular fracture within the vacuum chamber of the LEED apparatus as illustrated typically by the work of Stein et al.[139],* Figure 3.25 illustrates the essential features of an Auger spectroscopic analysis of grain boundary segregation inducing brittle, intergranular fracture in a W–Re alloy, a topic to be discussed in more detail in Chapter 5.

By utilizing the electron microprobe or the nondispersive x-ray spectrometer attachment of a scanning electron microscope, it is also possible to qualitatively examine the concentration of a solute or the systematic depletion of a component at a grain boundary by scanning across the grain boundary interface on the polished surface of a metal or alloy sample of interest. The depletion of chromium at the grain boundaries by carbide formation in NiCr alloys observed by electron probe microanalysis in the work of Fleetwood[140] is perhaps typical of this capability which becomes increasingly quantitative as the analytical precision of the technique is refined. In this regard ion spectroscopy techniques recently developed may prove

*The reader is also referred to G. A. Somorjai, *Surface Sci.*, **34**, 156 (1973), and R. E. Weber, *R/D Magazine*, **23**, 22 (October 1972).

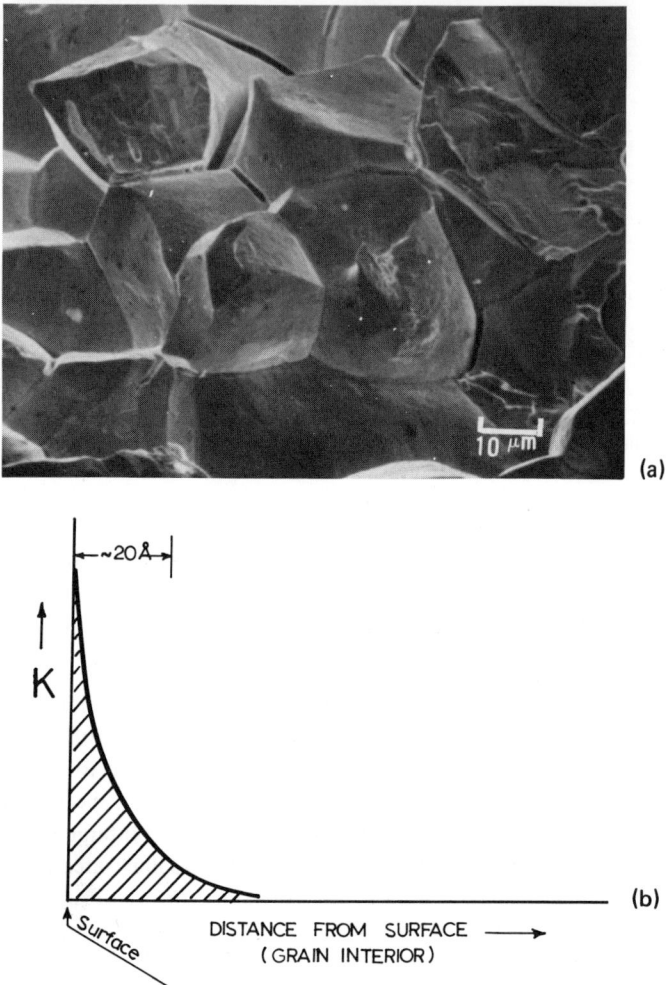

FIGURE 3.25   Auger spectroscopic analysis of potassium impurity segregation to grain boundaries in tungsten. (a) Scanning electron micrograph showing the exposed grain surfaces after brittle, intergranular fracture in vacuum. (b) Normalized Auger spectra for the surface in (a) showing potassium excess at the fracture surface (grain boundary). [(a) is courtesy of R. Simpson.]

to be extremely sensitive to grain boundary segregation or small compositional changes at or near the interface.[141]*

*Direct evidence of boron segregation to grain boundaries has been obtained by ion mass spectrometry as recently discussed by J. M. Walsh and B. H. Kear [*Met. Trans.*, **6**, 226 (1975)].

INTERFACIAL FREE ENERGY

Finally, it should be pointed out that, with refinements in the technique of atom-probe field-ion microscopy as devised by Müller et al.,[142] it is possible in principle to observe and identify individual solute atoms within a grain boundary or their disposition and density relative to the interface. Figure 3.26 illustrates this feature schematically with respect to impurity atoms in the vicinity of a grain boundary in iridium. Not only is it possible in the case of single-particle resolution

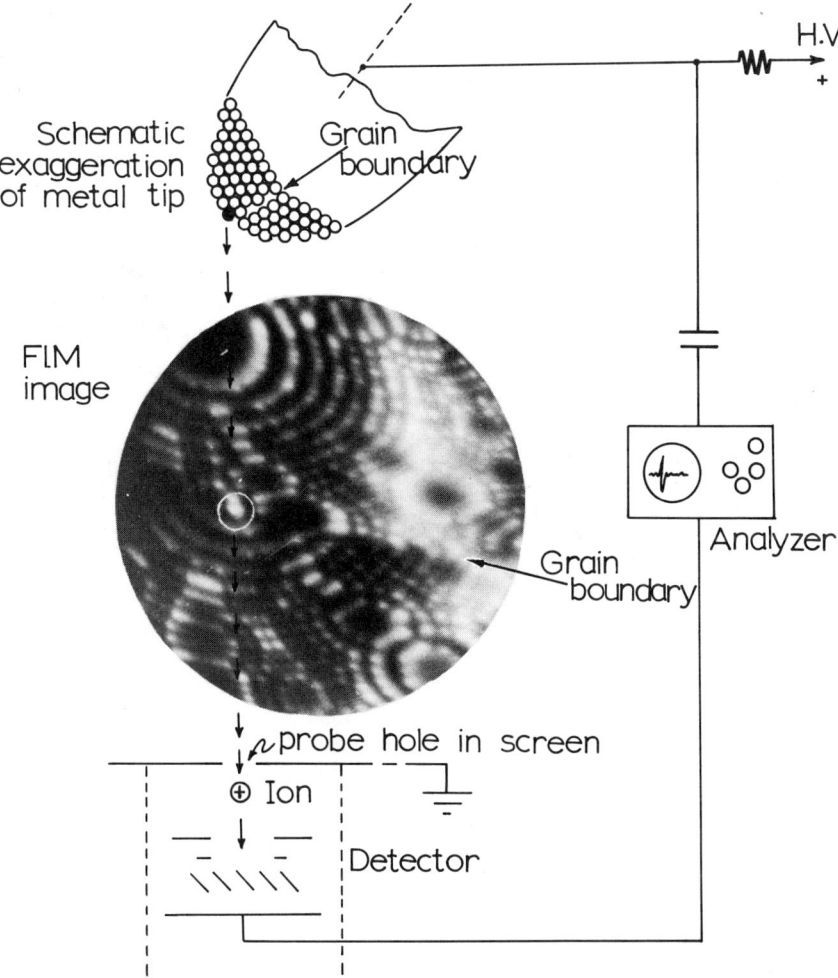

FIGURE 3.26 Conceptual view of the analysis of single solute atoms in the vicinity of a grain boundary observed in the field-ion microscope. The metal tip is adjusted so that the solute atom image of interest falls within a hole in the screen of the field-ion microscope. The atom is then field-evaporated by a short high-voltage pulse, and the ion flight is detected as an analyzable pulse which can be calibrated for values of atomic number, Z.

as illustrated in Fig. 3.26 to make an exact quantitative measurement of solute excess at a grain boundary, but it is additionally possible to determine the distribution of solute relative to the interface plane, and to quantitatively confirm the notion of long-range equilibrium segregation. Experiments of this type have, at present, not been performed.

3.4.3 Energetics of Twin Boundaries in Solid Metals and Alloys

Measurements of twin boundary free energy are indirect and depend on a knowledge of the average surface or grain boundary free energy, and its substitution into values of twin boundary/surface or grain boundary free energy ratios [(γ_{tb}/F_S) or $(\gamma_{tb}/\gamma_{gb})$] determined by equilibrium conditions described in Sections 2.3.1 and 2.3.2 and tabulated in Table 2.1. If the average values for surface and grain boundary free energy as tabulated in Tables 3.5 and 3.6 are substituted into the ratios of γ_{tb}/γ_{gb} tabulated in Table 2.1, average values of coherent twin boundary free energy (γ_{tb}) in a number of fcc metals and alloys are measured as given in Table 3.7. Table 3.7 also lists corresponding temperature coefficients for twin boundary free energies in a few cases determined from ratios of γ_{tb}/γ_{gb} measured at various temperatures as illustrated in Fig. 3.27, several of which have been measured from $d\gamma_{gb}/dT$ values (Table 3.6). Measurements of the variation of γ_{tb} with temperature (and consequently $d\gamma_{tb}/dT$) have been made for only a limited number of metals and alloys by the author as illustrated in Fig. 3.28.

It should be apparent from Fig. 3.28 and Table 3.7 that values of $d\gamma_{tb}/dT$ for pure metals are negative as would be expected on thermodynamic grounds, but that

TABLE 3.7
Coherent Twin Boundary Free Energy in FCC Metals and Alloys[a]

Metal or Alloy	Coherent Twin Boundary Free Energy, γ_{tb} (ergs/cm²)	Temperature (°C)	$d\gamma_{tb}/dT$ (ergs/cm² °C)	Ref.
Aluminum	75	450	−0.07	104
Cobalt	13	1354	−	109
Copper	24	800	−0.020	110, 143
Cu–16 a/o Al	10	800	−	144, Murr (unpublished data)
Cu–30 w/o Zn	14	850	−	Murr (unpublished data)
Gold	15	1000	−0.008	110, 143
Nickel	43	1060	−0.020	110, 117
Ni–20 w/o Cr	19	1060	−	102
Platinum	161	25	−0.040	110
Silver	8	950	−0.003	110, 143
Stainless steel (type 304)	19	1060	+0.007	116

[a]Values are measured values or are based directly on measured values of γ_{gb} and γ_{tb}/γ_{gb}.

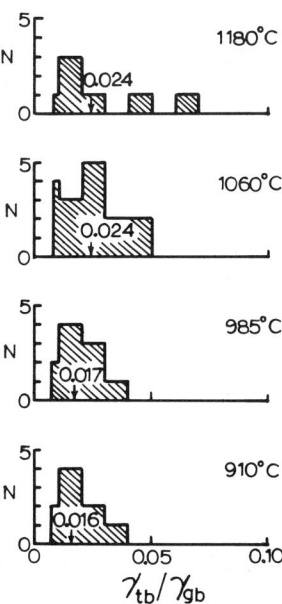

FIGURE 3.27 Histograms of relative interfacial free energy ratios, γ_{tb}/γ_{gb}, in 304 stainless steel at various temperatures. Arrows denote mean values at temperature. (From Murr et al.[116])

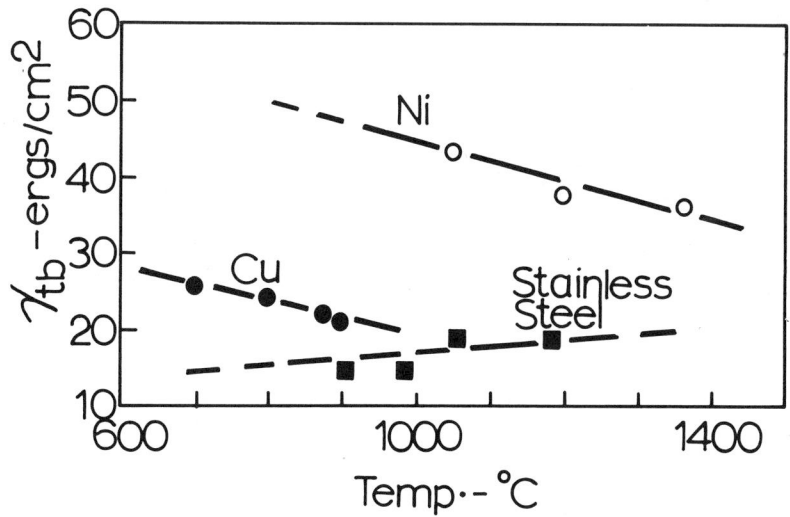

FIGURE 3.28 Temperature dependence of twin boundary free energy in some fcc metals and alloys. (Data from Murr[110,116] and unpublished data.)

values of $d\gamma_{tb}/dT$ for alloys will probably vary from negative to positive or zero, depending on the type and extent of twin boundary segregation, because of the small entropy associated with the coherent twin boundary (coincident with $\{111\}$ in fcc materials). Thus, even though solute segregation to or from twin boundaries may occur in the same proportion as grain boundaries in the same material, the associated temperature coefficient may have an opposite sign depending on the magnitude of the second term on the right-hand side of Eq. (3.47), or the associated entropies of the segregating components which would change the sign of the first term on the right in Eq. (3.47). Generally one would expect that, on considering the implications of Fig. 3.24, the energetics of twin boundary segregation would be decidedly different from those of most grain boundaries because of the expected coincidence structure of the boundary in the $\{111\}$ planes of fcc materials compared with the complex nature of general high-angle grain boundaries. Since this feature has not been experimentally tested for either twin or grain boundaries, it is at this time merely a matter of logical speculation. There have, in addition, been no direct, systematic investigations of the variation of twin boundary free energy with composition, although this variation is expected to follow that expected on the basis of the thermodynamic principles outlined in Chapter 1 (Section 1.3.2). This feature is implicit in the variation of values of γ_{tb}/γ_{gb} with composition in copper alloys as shown in Fig. 3.29 if γ_{gb} varies in the same manner.

Measurement of Noncoherent Twin Boundary Free Energy. A noncoherent twin boundary, as we shall discuss in more detail in Chapter 4, is by its nature similar to a grain boundary. However, because it has a definite, crystallographic coincidence

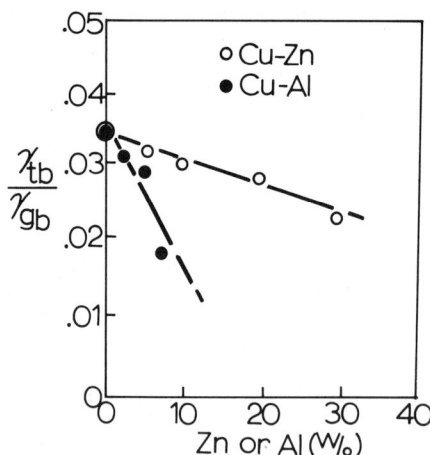

FIGURE 3.29 Variations of mean relative interfacial free energy ratios with composition in copper alloys. (Values for Cu-Zn from ref. 145; values for Cu-Al from ref. 144.)

INTERFACIAL FREE ENERGY

which defines the interface [that is, the noncoherent interface can be associated with a lattice plane, (hkl)], its average energy in fcc metals and alloys would be expected to be less than that for a general high-angle grain boundary as given for example in Table 3.6.

Mean values of noncoherent twin boundary free energy can be measured from intersection equilibria analogous to that described for a single twin boundary intersecting a grain boundary, as demonstrated in the work of Murr et al.[146] Figure 3.30 illustrates some typical examples of noncoherent twin boundaries intersecting a grain boundary, with which a coherent twin boundary also intersects

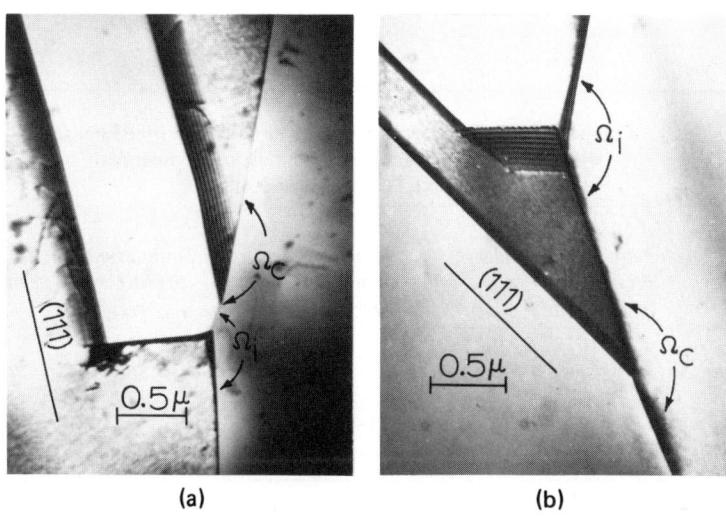

FIGURE 3.30   Coherent-noncoherent twin boundary-grain boundary intersections. (a) Type 304 stainless steel. (b) Cu-5 a/o Al.

in the immediate vicinity. For approximately symmetrical dihedral intersection geometries, the equilibrium conditions implicit in Eq. (2.13) can be rewritten

$$\gamma_{tb}/\gamma_{gb} = 2 \cos (\Omega_c/2) \tag{3.56}$$

and

$$\gamma_{TB}/\gamma_{gb} = 2 \cos (\Omega_i/2) \tag{3.57}$$

where γ_{tb} and γ_{TB} are the coherent and noncoherent twin boundary free energies, respectively, and Ω_c and Ω_i are the corresponding dihedral angles characterizing the intersection as illustrated in Fig. 3.30, determined by treating the intersections as triple junctions and solving Eq. (2.8). The accuracy of Eq. (3.56) as compared with more rigorous calculations utilizing Eq. (2.39) has been checked for coherent twin boundary-grain intersections in nickel and found to agree to within a few percent,[146] lending support to the analysis implicit in Fig. 3.30 and Eqs. (3.56) and (3.57).

Values of γ_{TB}/γ_{gb} obtained as outlined above have been tabulated previously in Table 2.1 for several metals and alloys. Table 3.8 lists the values of γ_{TB} along with corresponding values of γ_{tb} and γ_{gb} at temperature for comparison as determined from Tables 2.1, 3.6, and 3.7. The differences in γ_{TB} and γ_{tb} indicated in Table 3.8 are obvious from the differences in the dihedral angles Ω_i and Ω_c shown in Fig. 3.30.

TABLE 3.8
A Comparison of Mean Noncoherent Twin Boundary Free Energies
with Coherent Twin and Grain Boundary Free Energies
for Some FCC Metals and Alloys

Metal or Alloy	Coherent Twin Boundary Free Energy, γ_{tb} (ergs/cm²)	Noncoherent Twin Boundary Free Energy, γ_{TB} (ergs/cm²)	Grain Boundary Free Energy, γ_{gb} (ergs/cm²)	Temperature (°C)
Copper	21	498	623	950
Cu-5 a/o Al	17	180	560^a	750
Silver	8	126	377	900
Stainless steel (type 304)	19	209	835	1060

$^a\gamma_{gb}$ assumed to be approximately the same as for Cu-16 a/o Al.

3.4.4 Measurement of Stacking-Fault Free Energies in Metals and Alloys

The first direct determination of stacking-fault free energy was reported by Whelan[147] from measurements on extended dislocation nodes[138] in stainless steel observed in the transmission electron microscope. Whelan equated the attractive force, γ_{SF}, on a unit length of the bounding partial dislocation due to the stacking fault, to the force necessary to maintain the curvature, R, of the partial dislocation in the form

$$\gamma_{SF} = G \, |\mathbf{b}_p|^2/2R \tag{3.58}$$

where G is the shear modulus, and $\mathbf{b}_p$ is the partial-dislocation Burgers vector. The variation along the bounding partial of dislocation line energy with character angle was neglected in this equation. Brown[148] has shown that this expression is in error by roughly a factor of 2. A more exact form of the stacking-fault free energy expression for fcc metals and alloys based on isotropic elasticity theory can be written[149,150]

$$\gamma_{SF} = \frac{Ga^2}{6R_0} \left\{ 0.27 - 0.08 \left(\frac{\nu}{1-\nu}\right) \cos 2\alpha + [0.104\left(\frac{2-\nu}{1-\nu}\right) + 0.24\left(\frac{\nu}{1-\nu}\right) \cos 2\alpha] \log_{10}\left(\frac{R_0}{\epsilon}\right) \right\} \quad (3.59)$$

where, with reference to the schematic diagrams of Fig. 3.31,

$$R_0 = R \left[\frac{(\sin^2 \phi + \cos^2 \phi \cos^2 \theta)^{\frac{3}{2}}}{\cos^2 \theta} \right]$$

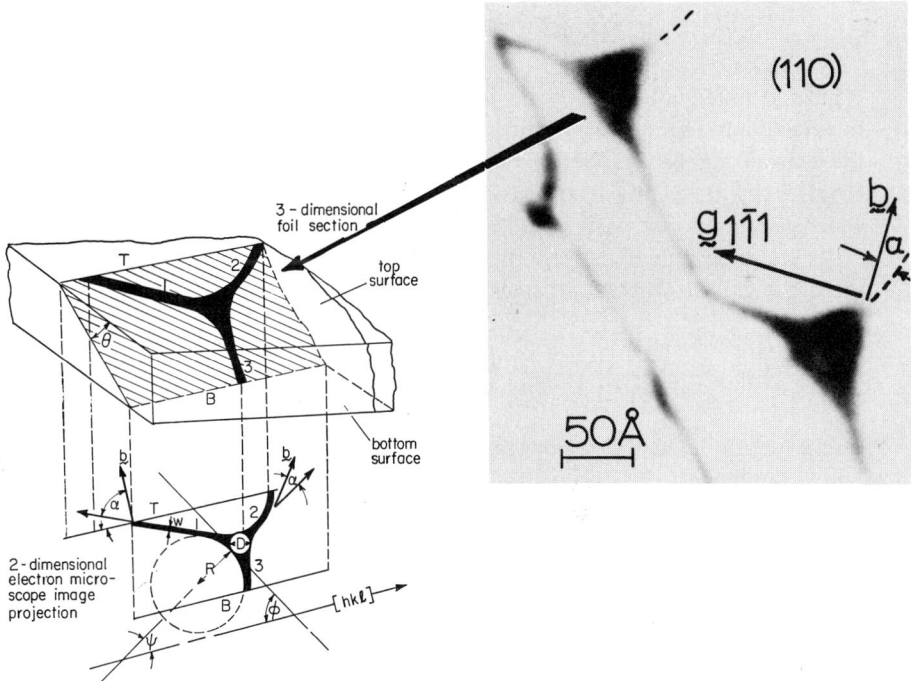

FIGURE 3.31 Schematic section depicting an extended dislocation node on an inclined $\{111\}$ plane in an fcc metal or alloy. R is the radius of curvature as measured directly from a bright-field transmission electron micrograph. The top and bottom foil surfaces are denoted T and B, respectively. The insert shows dislocation nodes in 304 stainless steel on a (111) plane as observed in the transmission electron microscope (TEM).

or

$$R_0 = R \left[\frac{(\cos^2 \psi + \sin^2 \psi \cos^2 \theta)^{\frac{3}{2}}}{\cos^2 \theta} \right]$$

and where a is the lattice parameter, ν is Poisson's ratio, α is the character angle (the angle between the dislocation line and its Burgers vector; $\alpha = 90°$ for a pure edge dislocation and $\alpha = 0°$ for a pure screw dislocation), ϵ is the effective dislocation core radius (or cut-off radius), equal to roughly a Burgers vector ($\epsilon \cong |\mathbf{b}_p|$), and R_0 is the true radius of curvature determined as indicated from the electron transmission (bright-field) image. An alternative expression utilizing the inscribed diameter, D, in Fig. 3.31 (which normally requires no measurable correction) can be written

$$\gamma_{SF} = \frac{Ga^2}{3D} \left\{ 0.055 \left(\frac{2-\nu}{1-\nu} \right) - 0.06 \left(\frac{\nu}{(1-\nu)^2} \right) \cos 2\alpha \right. $$
$$\left. + \left[0.018 \left(\frac{2-\nu}{1-\nu} \right) + 0.036 \left(\frac{\nu}{1-\nu} \right) \cos 2\alpha \right] \log_{10} \left(\frac{R_0}{\epsilon} \right) \right\} \quad (3.60)$$

It must be noted in Eq. (3.60) and with reference to Fig. 3.31 that D is measured to extend only to the inside of the partial dislocation images. Where this is unclear, the diameter can be measured to the outside, where the correct value of D will then be $(D' - 2\xi_g/\pi)$, where ξ_g is the extinction distance for the operating reflection, $\mathbf{g}$, which gives rise to the partial dislocation contrast.

It is also possible to measure the separation, w_0, of partial dislocations as indicated in Fig. 3.31, wherein the stacking-fault free energy can be measured from

$$\gamma_{SF} = \frac{Ga^2}{48\pi w_0} \left[\frac{2-\nu}{1-\nu} \left(1 - \frac{2\nu \cos 2\alpha}{2-\nu} \right) \right] \quad (3.61)$$

for fcc metals and alloys or more generally

$$\gamma_{SF} = \frac{G|\mathbf{b}_p|^2}{8\pi w_0} \left[\frac{2-\nu}{1-\nu} \left(1 - \frac{2\nu \cos 2\alpha}{2-\nu} \right) \right] \quad (3.61a)$$

The measurement of stacking-fault free energy has been reviewed by Saada,[151] Ruff,[152] and Gallagher.[153] In addition to the measurement of extended dislocation nodes or extended dislocations as outlined in Eqs. (3.59) through (3.61) for values of $\gamma_{SF} \lesssim 60$ ergs/cm^2, measurements of the kinetics of loop annealing and stacking-fault tetrahedra have also been utilized where $\gamma_{SF} \gtrsim 60$ ergs/cm^2.[154,155] In addition, indirect methods involving mechanical properties (for example, rolling textures[156]) have been employed. Even field-ion microscopy has been employed in the direct observation of partial dislocations,[157] and Cockayne et al.[158] have devised a technique (weak-beam method) useful in the transmission electron micro-

INTERFACIAL FREE ENERGY

scope which is capable of resolving separations of partials nearly as small as the field-ion microscope.* It must be realized that, although the direct methods of measurement are by far the more reliable, the measurements often include anomalous effects such as the image force in thin films and the undetected segregation of impurities or an unknown component to the stacking fault. The more acceptable values of stacking-fault free energy corresponding to room temperature (25°C) for a number of metals and alloys are listed in Table 3.9. The values listed are averages where more than one accurate measurement has been made. Only measurements from equilibrated dislocation nodes, loops, or stabilized tetrahedra have been included, along with measurements from twin boundary free energy corrected to room temperature (25°C).

TABLE 3.9
Intrinsic[a] Stacking-Fault Free Energies in Metals and Alloys[b]

Metal or Alloy	Stacking-Fault Free Energy, γ_{SF} (ergs/cm²)[c]	Temperature (°C)	$d\gamma_{SF}/dT$ (ergs/cm² °C)	Ref.
Aluminum	166[d]	25	−0.04	104, 159
Al–0.7 w/o Mg	110	170	−	160
Al–1 w/o Si	145	150	−	161
Cadmium	175	25	−	154
Cobalt (fcc)	15	500	+0.03	162
Co–Fe (6 a/o)	12	25	+0.05	163
Co–Fe (8 a/o)	25	25	+0.04	163
Co–Fe (12 a/o)	35	25	+0.008	163
Co–33 w/o Ni	28	500	+0.03	162
Co–32 a/o Ni	11	25	+0.05	163
Copper	78	25	−0.04	110, 155, 164, 165
Cu–5 a/o Al	20	25	−	143
Cu–10 a/o Al	4	25	−	143
Cu–Al (1.11 e/a)	28	25	−	153
Cu–Al (1.17 e/a)	17	25	−	153
Cu–Al (1.20 e/a)	10	25	−	153
Cu–Al (1.32 e/a)	6	25	−	153
Cu–Ga (1.12 e/a)	30	25	−	153
Cu–Ga (1.18 e/a)	10	25	−	153
Cu–Ga (1.24 e/a)	6	25	−	153
Cu–Ga (1.28 e/a)	4	25	−	153

*Both the field-ion and weak-beam techniques are limited in a sense in the same way that the node method is limited, in spite of their characteristically better resolution capabilities. That is, as the separation of partials becomes smaller and smaller, a situation arises in which it is physically impossible to deal with the separation in terms of stacking-fault free energy associated with an interface. In fact, elastic approximations break down as the cores of the partial dislocations overlap, rendering calculations meaningless.

TABLE 3.9 *(continued)*

Metal or Alloy	Stacking-Fault Free Energy, γ_{SF} (ergs/cm²)[c]	Temperature (°C)	$d\gamma_{SF}/dT$ (ergs/cm² °C)	Ref.
Cu–Ge (1.11 e/a)	35	25	–	153
Cu–Ge (1.12 e/a)	23	25	–	153
Cu–Ge (1.20 e/a)	13	25	–	153
Cu–Ge (1.27 e/a)	4	25	–	153
Cu–Si (1.11 e/a)	25	25	–	153
Cu–Si (1.15 e/a)	18	25	–	153
Cu–Si (1.21 e/a)	9	25	–	153
Cu–Si (1.26 e/a)	5	25	–	153
Cu–Sn (1.11 e/a)	33	25	–	153
Cu–Sn (1.19 e/a)	22	25	–	153
Cu–Sn (1.22 e/a)	15	25	–	153
Cu–Sn (1.26 e/a)	11	25	–	153
Cu–Zn (1.10 e/a)	35	25	–	153
Cu–Zn (1.14 e/a)	25	25	–	153
Cu–Zn (1.20 e/a)	18	25	–	153
Cu–Zn (1.30 e/a)	14	25	–	153
Cu–Zn (1.36 e/a)	11	25	–	153
$(Cu_{89.5}Al_{10.5})_{100-x}Mn_x$				
(CuAl)–Mn (0 a/o)	12	25	–	166
(CuAl)–Mn (6 a/o)	15	25	–	166
(CuAl)–Mn (10 a/o)	18	25	–	166
(CuAl)–Mn (15 a/o)	25	25	–	166
Gold	45	25	–0.02	110, 153
Iridium	300	25	–	157, Murr (unpublished data)
Fe–19 w/o Cr–9 w/o Ni (304 stainless steel)	21	25	+0.014	116, 153
Fe–18 w/o Cr–14 w/o Ni	40	25	–	153
Fe–18 w/o Cr–19 w/o Ni	50	25	–	153
Fe–18 w/o Cr–14 w/o Ni				
0.01 w/o Si	50	25	–	153
0.90 w/o Si	32	25	–	153
1.86 w/o Si	26	25	–	153
2.79 w/o Si	22	25	–	153
3.66 w/o Si	21	25	–	153
Magnesium	125	25	–	154
Nickel	128	25	–0.04	110, 153, 167

INTERFACIAL FREE ENERGY

TABLE 3.9 *(continued)*

Metal or Alloy	Stacking-Fault Free Energy, γ_{SF} (ergs/cm²)	Temperature (°C)	$d\gamma_{SF}/dT$ (ergs/cm² °C)	Ref.
Ni–20 w/o Co	120	25	—	153
Ni–40 w/o Co	90	25	—	153
Ni–60 w/o Co	40	25	—	153
Ni–20 w/o Cr	40	25	—	102, Murr (unpublished data)
Ni–Cr–Fe (Inconel 600)	28	25	—	150
Palladium	175	25	—	156
Pd–40 a/o Ag	125	25	—	156
Pd–60 a/o Ag	75	25	—	156
Pd–80 a/o Ag	45	25	—	156
Platinum	322	25	−0.08	110
Silver	22	25	−0.006	110, 153
Ag–Al (1.05 e/a)	22	25	—	153
Ag–Al (1.08 e/a)	16	25	—	153
Ag–Al (1.15 e/a)	8	25	—	153
Ag–Al (1.35 e/a)	2	25	—	153
Ag–In (1.00 e/a)	22	25	—	153
Ag–In (1.02 e/a)	21	25	—	153
Ag–In (1.04 e/a)	22	25	—	153
Ag–In (1.10 e/a)	16	25	—	153
Ag–In (1.15 e/a)	9	25	—	153
Ag–In (1.23 e/a)	6	25	—	153
Ag–1.9 a/o Mn	22	25	—	168
Ag–3.8 a/o Mn	21	25	—	168
Ag–7.5 a/o Mn	21	25	—	168
Ag–13 a/o Mn	22	25	—	168
Ag–18 a/o Mn	22	25	—	168
Ag–Sn (1.00 e/a)	23	25	—	169
Ag–Sn (1.03 e/a)	20	25	—	169
Ag–Sn (1.06 e/a)	19	25	—	169
Ag–Sn (1.12 e/a)	12	25	—	169
Ag–Sn (1.18 e/a)	7	25	—	169
Ag–Sn (1.24 e/a) Ag	5	25	—	169
Ag–9 a/o Sn	2	25	+0.008	170
Ag–12 a/o (hcp)	4	25	−0.006	170
Ag–14 a/o (hcp)	9	25	—	170
Ag–17 a/o (hcp)	18	25	—	170
Ag–Zn (1.00 e/a)	20	25	—	153, 165
Ag–Zn (1.02 e/a)	19	25	—	153, 165
Ag–Zn (1.03 e/a)	19	25	—	153, 165

TABLE 3.9 *(continued)*

Metal or Alloy	Stacking-Fault Free Energy, γ_{SF} (ergs/cm²)[c]	Temperature (°C)	$d\gamma_{SF}/dT$ (ergs/cm² °C)	Ref.
Ag–Zn (1.07 e/a)	18	25	–	153, 165
Ag–Zn (1.09 e/a)	16	25	–	153, 165
Stainless steel (type 304)	21	25	+0.014	116
Tungsten	>50[e]	–	–	171
Zinc	140	25	–	154
Zirconium	240[f]	25	–	172
Zr–0.7 w/o Sn	116[f]	25	–	172
Zr–3 w/o Sn	85[f]	25	–	172
Zr–5 w/o Sn	62[f]	25	–	172

[a] It is generally observed that γ_{SF} (intrinsic) $\cong \gamma_{SF}$ (extrinsic). This has been confirmed in Cu-, Ag-, and Au-base alloys.[153]

[b] In alloys for which the valences of the solute and solvent atoms (V_s and $V_{\bar{s}}$, respectively) are known, it is common practice to denote an alloy with atomic fraction solute content, X, by means of its electron/atom ratio, denoted e/a, where $e/a = (1 - X)V_s + V_{\bar{s}} = 1 + X \Delta V$. Many values given in Table 3.9 are therefore indicated by this normalization.

[c] Note that 1 erg/cm² $\equiv$ 1 mJ/m². Values given in this table are mean values where sufficiently accurate data exist and are based primarily on dislocation node, stacking-fault tetrahedra, or shrinking loop methods.

[d] Values of stacking-fault free energy above about 50 ergs/cm² are usually the mean of extrapolated node data or loop annealing measurements, or are average values obtained by applying Eq. (3.62) at room temperature (25°C), which normally requires a knowledge of $d\gamma_{tb}/dT$ in order to obtain the room temperature value of the coherent twin boundary free energy.

[e] From x-ray stacking-fault probability data.

[f] The value of γ_{SF} was determined by direct observation of partial dislocations under non-ideal conditions which could easily render the value dubious. See footnote p. 145.

The Nature of Stacking Faults. Figure 3.32 illustrates schematically the distinction between intrinsic (i) and extrinsic (e) stacking faults and shows for comparison the geometrical relationship to annealing twins in fcc metals and alloys. The extrinsic fault arises by the superposition of two intrinsic faults, and annealing twins result from the formation of intrinsic stacking faults on every (111) plate within the twinned region. On the basis of this scheme, an extrinsic stacking fault is simply a one-layer twin. Since, as we shall see in more detail in Chapter 4, the interfacial energy is governed to a large extent by short- and longer-range interactions with lattice atoms, the energetics of the intrinsic and extrinsic stacking faults might be expected to differ, and additionally, the annealing twin boundary interfacial free energy would be expected to be related to both intrinsic and extrinsic stacking faults when compared on the same thermodynamic basis—that is, in the same materials at the same temperature, pressure, and equilibrium conditions. Geometrically, we can see from Fig. 3.32 that

INTERFACIAL FREE ENERGY   *149*

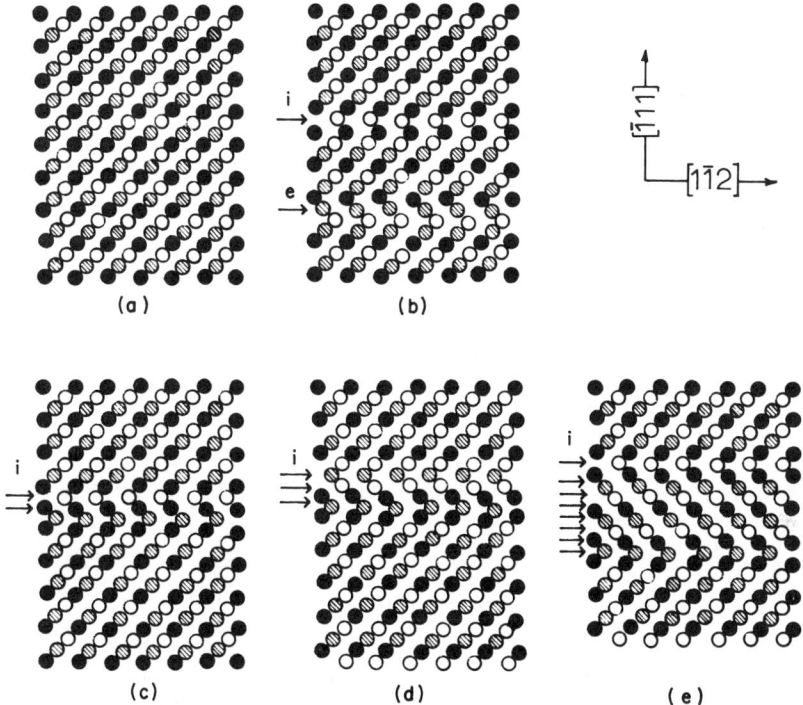

FIGURE 3.32   Idealized view of stacking faults and their development as *n*-layer twins in the close-packed fcc lattice. The line of sight (normal to the page is coincident with [110]. (From Murr.[150])

$$[\gamma_{SF} \cong 2\gamma_{tb}]_T \tag{3.62}$$

Equation (3.62) has important applications in the determination of γ_{SF} in metals or alloys where the value is too high for any measurable node extension to occur. However, it is strictly applicable only at a constant temperature, which normally requires a knowledge of $d\gamma_{tb}/dT$.

Gleiter and Klein[173] have in fact recently shown that the stacking-fault energy in the slip planes next to a coherent twin boundary in copper and a Cu–0.2 a/o Co alloy is only about 2% of the stacking-fault energy of the bulk material—that is, an isolated stacking fault—and that the stacking-fault energy increases with increasing distance from the annealing twin boundary. This feature is illustrated in Fig. 3.33. Although the data of Fig. 3.33 illustrate that the energetics of stacking faults are influenced by the perturbation of an adjacent interface, the results are not exact, since the thermodynamic basis (temperature of formation) differs. Nonetheless,

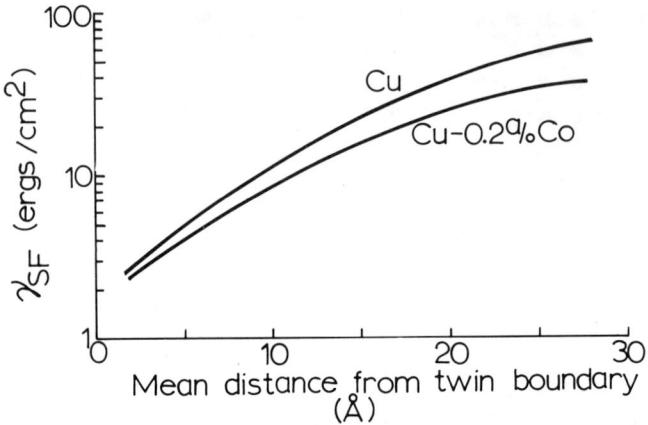

FIGURE 3.33 Measured stacking-fault free energy of copper and Cu–0.2 a/o Co alloy as a function of distance from a coherent annealing twin boundary. (Based on experimental data of Gleiter and Klein.[173])

this interaction allows us to understand the fact that the energy of single stacking faults is higher than that of a coherent twin boundary. The significance of this argument is shown in Fig. 3.34, which illustrates very dramatically the necessity of evaluating interfaces—for example, stacking faults and twins—on a common thermodynamic basis (at the same temperature).

3.4.5 Effects of Temperature and Composition on Stacking-Fault Free Energy

The temperature coefficients for metals and alloys have been measured for only a few cases, and these are indicated in Table 3.9. When the effects of vacancy segregation to a stacking fault are negligible (in the case of a pure metal), the entropy is the controlling factor, and $d\gamma_{SF}/dT$ is negative in accordance with Eq. (3.47). Like a coherent twin boundary, the entropy of a stacking fault would be expected to be low, and only a fraction of that for a high-angle grain boundary. As a consequence of this feature, small additions of an impurity or the occurrence of a high concentration of vacancies (or interstitials) which could systematically adsorb to the stacking-fault interface could become the controlling factor, and the sign of $d\gamma_{SF}/dT$ could become positive.

Suzuki[174] first suggested that a decrease in γ_{SF} with an increase in solute concentration in an alloy system would lead to the presence of a driving force for solute segregation to the stacking-fault interface. However, direct evidence of such an effect is generally lacking, and indeed such a process would also be complicated by temperature. As a result, the value of $d\gamma_{SF}/dT$ for an alloy could conceivably

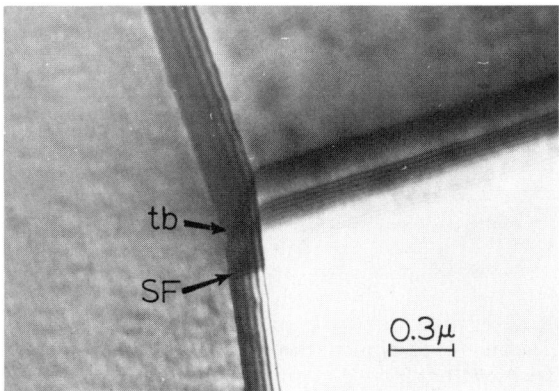

FIGURE 3.34 Stacking fault and coherent annealing twin intersecting a common grain boundary in Cu-5 a/o Al alloy. While the effective stacking-fault free energy would normally be given by Eq. (3.62), the dihedral angles at the intersections (arrows) are not compatible, since the temperatures of formation are different by nearly two orders of magnitude. As a consequence, the stacking fault has little influence on the junction with the grain boundary, since it has not been thermodynamically equilibrated with it. It should be noted in the image that the stacking-fault contrast overlays that of the twin; however, the distance between these interfaces is sufficient to maintain the autonomy of the stacking fault (see Fig. 3.33).

be negative, positive, or in certain instances zero, although it might be expected generally that $d\gamma_{SF}/dT$ would, in the case of solute segregation in alloys, be positive because of the dominance of these terms over the entropy [cf. Eq. (3.47)].

Because of the nature of stacking faults (their instability at high temperature), one of the most effective methods for determining the effect of temperature on γ_{SF} is by direct observations of node response in the transmission electron microscope. As shown generally in Eq. (3.58), an increase or decrease in node size (radius) with temperature can be directly related to a corresponding decrease or increase, respectively, in γ_{SF}. Although this seems straightforward enough, it must be realized that, because of surface effects (image force effects)[150] in thin foils, the observed node response may not accurately reflect either a thermal response (change of entropy) or a segregation effect in the case of an alloy. This response is, nonetheless, demonstrated experimentally in Fig. 3.35; Fig. 3.36 summarizes the node measurements for several fcc alloys.

As we discussed in Chapter 1 (Section 1.3.2), variations of interfacial free energy with composition can be treated by the Gibbs adsorption isotherm. As was pointed out with reference to Eq. (1.37a), there is a dependence of segregation or adsorption on temperature and composition or at fixed compositions, and the variation of adsorption with temperature could account for an increase, for example example, in the interfacial free energy with increasing temperature.

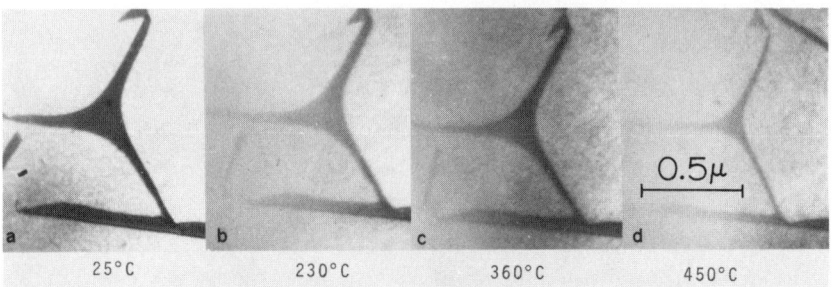

FIGURE 3.35 Variation of dislocation node size (R) in Ag-9 a/o Sn. (After Ruff and Ives,[170] courtesy Dr. A. W. Ruff.)

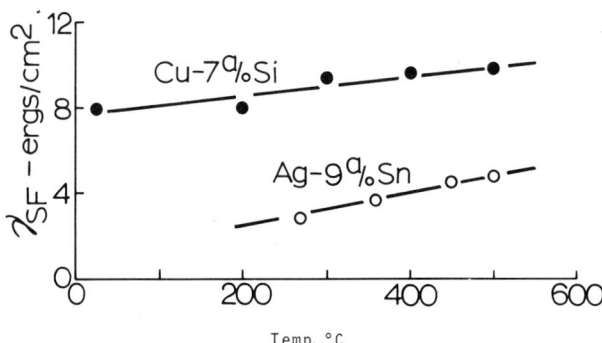

FIGURE 3.36 Temperature dependence of stacking-fault free energy in some fcc alloys. (Data from Ruff and Ives[170] and Nordstrom and Barrett.[175])

It can be seen from Table 3.9 that experimentally γ_{SF} is generally lowered by alloying. A simple expression has been derived relating γ_{SF} and alloy concentration in the form[153]

$$\ln\left(\frac{\gamma_{SF(\text{alloy})}}{\gamma_{SF(\text{solvent})}}\right) = \delta\left[c/(1+c)\right]^2 \tag{3.63}$$

where $\gamma_{SF(\text{solvent})}$ represents the stacking-fault energy of the pure base metal (solvent), δ is a constant, and $c = x/x'$, with x defined as the solute (alloying) concentration, and x' the solubility limit at high temperature. Figure 3.37 illustrates the form of this equation in plotting the data of Nordstrom and Barrett[175] and Gallagher[153] for Cu-Si alloys.

INTERFACIAL FREE ENERGY   *153*

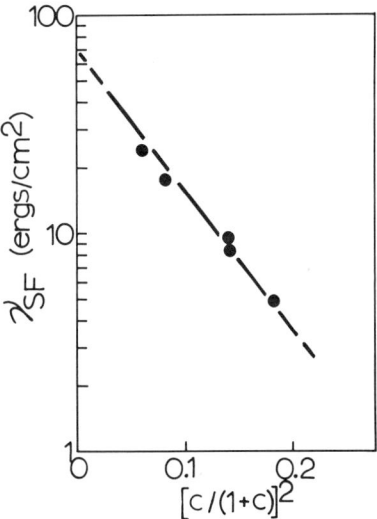

FIGURE 3.37   Semilog plot of γ_{SF} versus $[c/(1 + c)]^2$ for Cu-Si alloys. (Data from Gallagher[153] and Nordstrom and Barrett.[175])

Ideally, on considering the Gibbs adsorption isotherm in a given system, and knowing the requisite thermodynamic data for the matrix phase, μ_2 and μ_v, we can determine the equilibrium Suzuki adsorption to a stacking fault by measuring γ_{SF} as a function of x_2 (solute concentration) at constant temperature [cf. Eq. (1.39)] Although, as was indicated previously, this occurrence of Suzuki segregation seems questionable in many cases, there is some direct evidence of these effects. Perhaps the most convincing proof for segregation or precipitation effects can be demonstrated in the distortion of equilibrium stacking-fault node shapes. As shown in Fig. 3.31, the inward curvature of partial dislocations is implicit in the derivation of Eqs. (3.58) and (3.59). As Venables[176] has pointed out, a basic assumption that is made, either implicitly or explicitly, when this technique is employed for stacking-fault energy measurements, is that the degree of nodal extension is not influenced by precipitation or segregation effects.

Figure 3.38 illustrates dramatically that this feature is not necessarily true, and that segregation and precipitation at stacking faults are indeed real effects. Figure 3.38, although illustrating unambiguous proof for stacking-fault segregation, should also serve as a caution when one is considering node measurements of γ_{SF}.

3.4.6 Measurement of Interphase Interfacial Free Energies in Metals and Alloys

There have been few if any direct measurements of interphase interfacial free energy except in cases where "interphase" is taken to describe the boundary

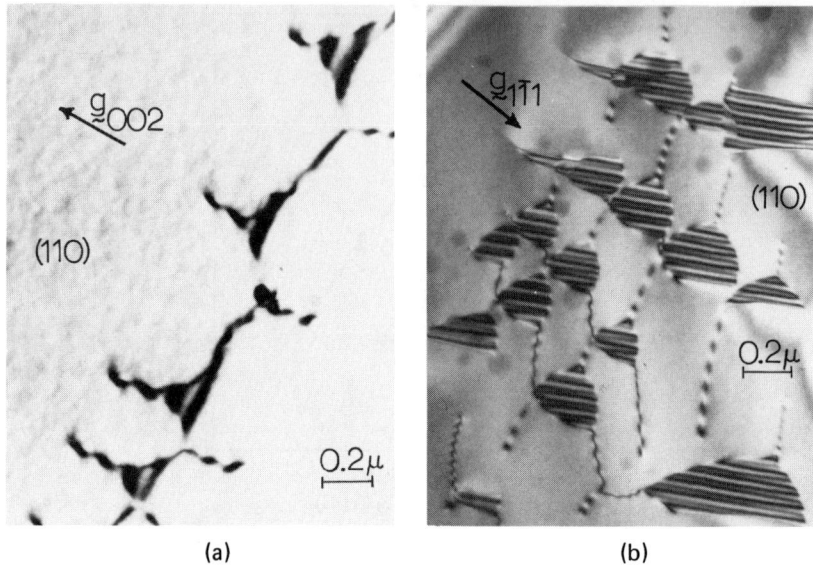

FIGURE 3.38 Comparison of dislocation node shapes in an alloy considered to exhibit small or negligible stacking-fault segregation (stainless steel) (a), and an alloy containing an impurity which segregates to the stacking faults (TiC + B) (b). [(b) is courtesy of J. D. Venables.]

separating a solid from its saturated liquid or vapor. We shall limit the connotation of "interphase interface" to be descriptive of the boundary separating two phases having different chemistry or crystallography, but not in the same state (solid or liquid)—for example, the interface separating a precipitate or dispersoid, etc., from the matrix. The latter examples of interphase interfaces are descriptive of a host of important solid-state interfacial phenomena, which includes eutectic boundaries and related interfaces.

Many of the techniques outlined previously in this chapter for the measurement of interfacial free energies are certainly applicable in the case of interphase interfaces, but experimental attempts are lacking in the literature. Nearly all the reported estimates of interphase interfacial energies have been based on calorimetric determinations of interfacial enthalpy utilizing basically the technique of Åstrom,[103] assuming some value for the associated entropy. Eutectoid phase boundary energies have been measured in this way by Kramer et al.,[177] for example, where the interfacial entropy in all cases was assumed to be $-2/3$ erg/cm^2 °K. Similarly, Hopkins and Kossowsky[178] measured the eutectic interphase boundary energy ($\gamma_{\gamma_1 \alpha_J}$) in Ni-51 w/o Cr, while Boyd and Nicholson[179] have measured the θ'/matrix interfacial free energy in Al-4 w/o Cu and the θ''/matrix interfacial free energy in the same alloy. These and similar measurements are tabulated in Table 3.10.

TABLE 3.10
Interphase-Interfacial Free Energy

Interphase System	Interphase-Boundary Free Energy (ergs/cm²)	Temperature (°C)	Ref.
γ_{SL} (Al)	93[a]	600	180
γ_{SL} (Bi)	61[b]	271	181
γ_{SL} (Cu)	177[a]	1083	180
γ_{SL} (Fe)	204[a]	1535	180
γ_{SL} (Au)	132[a]	1063	180
γ_{SL} (Ni)	255[a]	1455	180
γ_{SL} (Pt)	240[a]	1773	180
γ_{SL} (Sn)	55[a]	232	180
γ_{LS} [Cu-Al]$_L$/[UO$_2$]$_S$[c]	333[b]	1100	182
γ_{LS} [Al-Mg]$_L$/[BeO]$_S$	467[b]	650	182
γ_L (Cu)$_S$(Fe)[c]	430[b]	1100	183
γ_L (Hg)$_S$(Pd)	1717[b]	25	184
γ_L (Hg)$_S$(Al)	1500[b]	25	184
γ_{SS} (Au/Al$_2$O$_3$)	1725[b]	1000	185
γ_{SS} (Ag/Al$_2$O$_3$)	1630[b]	700	185
γ_{SS} (Cu/Al$_2$O$_3$)	1925[b]	850	185
γ_{SS} (Ni/ThO$_2$)	2000[b]	1200	184, 186
γ_{SS} (Ni-20 w/o Cr/ThO$_2$)	2300[b]	1200	186
γ_{SS} (Ni/Al$_2$O$_3$)	2140[b]	1000	185
γ_{SS} (austenitic stainless steel/Al$_2$O$_3$)	1380[b]	1200	184
γ_{SS} (γ-Fe/Al$_2$O$_3$)	2065[b]	1000	185
γ_{SS} (Pt/Al$_2$O$_3$)	1050[b]	1400	120
$\gamma_{\gamma_1\alpha_1}$ (Ni-51 w/o Cr eutectic)	300[d]	700	178
Pearlite/austenite (Fe-C)	700[d]	730	177
Ferrite/cementite (Fe/C)	710[d]	1100	177
$\gamma_{\alpha\beta}$(Pb-An eutectic)	42[b]	–	187
Carbide/austenite (precipitate)	668[b]	1060	116
θ' (Al-4 w/o Cu) precipitate	1530[d]	250	179
θ' (Al-4 w/o Cu) precipitate	530[d]	250	179
γ' (Ni-6.5 w/o Al) precipitate	18.5	–	188

[a] Homogeneous nucleation experiments.

[b] Interfacial equilibrium measurements.

[c] The order of the subscript denotes the phase of the component (S–solid; L–liquid). Undesignated phases are solid state.

[d] Calorimetric measurements.

Table 3.10 shows that in some cases interphase boundary energies appear to be similar in magnitude to those of ordinary grain boundaries (Table 3.6), and it might be expected that the interfacial structures would be similar. As we shall see in Chapter 4, this is a reasonably accurate description in several cases investigated.

Perhaps the most direct measurements of precipitate or similar included phase boundary energy involve the use of equilibrium systems as described previously in Chapter 2 (Section 2.4). Several of these types of measurements have been recorded in Table 3.10, but such measurements are in general sadly lacking. Figure 3.39 illustrates several interfacial equilibrium configurations which can be experimentally utilized to determine interphase interfacial free energies, γ_i, when other related interfacial energies (surface and grain boundary free energies) are known. Such arrangements can be attained by the high-temperature annealing of mixed phases, etc., for sufficient periods to attain equilibrium. This technique has in fact been used by McLean and Hondros[120] and Ahmad and Murr,[186] as shown for their results recorded in Table 3.10. Recently, in addition, Rhee[182] has modified the earlier work of Bashforth and Adams[9] in the determination of liquid-solid interfacial free energies by direct measurements of sessile drops, and at the same time determined the surface free energy of the solid substrate (Table 3.10).

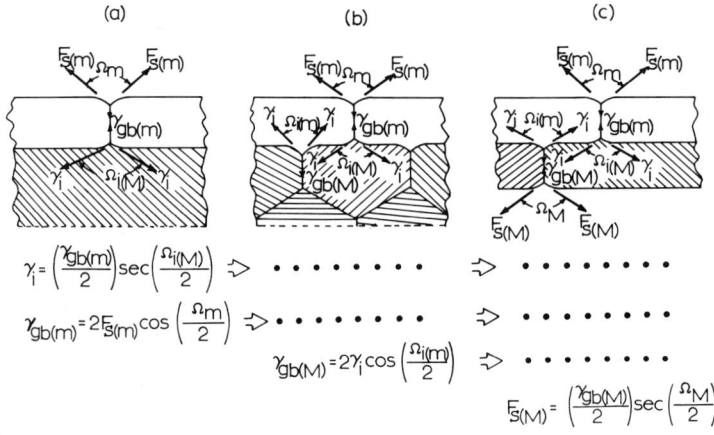

FIGURE 3.39 Equilibrium systems for evaluation of interfacial free energies. The equilibrium solutions are noted for each case, and where they are the same for different cases this is denoted by dotted lines. The interphase interfacial equilibrium is denoted by subscript (*i*). $F_{S(m)}$ denotes either the solid-vapor or solid-liquid surface (interfacial free energy).

It is anticipated that in light of the interest in and development of multiphase systems (composites, precipitation, and dispersion-hardened systems, etc.) a concomitant interest will develop in the measurement of the pertinent interphase boundary free energies. Several aspects of these phenomena will be discussed in Chapters 4 and 5.

INTERFACIAL FREE ENERGY

In dealing with equilibrium at the solid-liquid (or liquid-solid) interface, the condition of melting (or nucleation) forms a useful reference point. In equating the change in energy associated with the solid-liquid transition at the melting point for a small crystal we can write

$$\mu_S - \mu_L = 2\gamma_{SL}(v)/r \tag{3.64}$$

which after appropriate substitutions is observed to be equivalent to Eq. (1.22). In Eq. (3.64) μ represents the corresponding chemical potential, γ_{SL} is the solid-liquid interfacial free energy, v is the molar volume of the particle, and r is the radius of a spherical particle when melting occurs. For equilibrium of the solid particle in the the liquid phase, the temperature must be below the normal melting point, T_m, by some amount, ΔT, such that the increased Gibbs free energy which results is balanced by the decreased free energy of the solid relative to the liquid phase due to this apparent supercooling. A reversal of this process, to be discussed in more detail in Chapter 5, characterizes homogeneous nucleation of a crystal from its melt. The free energy associated with this supercooling effect is $\Delta H_f V \Delta T/T_m$, where ΔH_f is the heat of fusion, V is the particle volume [e.g., Eq. (1.22)], ΔT is the supercooling, and T_m is the melting point. For equilibrium

$$\Delta T = \frac{2\gamma_{SL} T_m}{\Delta H_f r} = \frac{\gamma_{SL} T_m}{\Delta H_f} \left(\frac{1}{r_1} + \frac{1}{r_2} \right) \tag{3.65}$$

where r_1 and r_2 denote general (liquid) surface radii as in Eq. (3.1). In this form, Eq. (3.65) expresses the effective change in melting point at a curved interface.

It can be observed from Eq. (3.65) that the measurement of supercooling for small particles of a liquid of radius r which nucleate as solid crystals can be employed to experimentally measure γ_{SL}. Some of the first experiments of this type were performed by subdividing the particles and keeping them separated by thin oxide films, or by suspending the particles in a suitable fluid and measuring the nucleation rates and the associated temperatures of the aggregate nucleated with a dilatometer.[189,190] Turnbull[180] has accumulated a number of values of pure metal solid-liquid interfacial free energies determined from supercooling data [Eq. (3.65)], and several of these are included in Table 3.10. More recent experiments by Stowell[191] have shown that much lower undercoolings than those observed by Turnbull[180] are possible, and as a consequence many of the experimentally determined values may be in error by as much as a factor of 2.

3.5 SUMMARY

An attempt has been made in this chapter to outline the more reliable techniques for experimentally measuring interfacial energies. In addition, a considerable effort has been made to tabulate the existing experimental data.

It has been pointed out that it is imperative when comparing interfacial energies to do so on a common thermodynamic basis (for example, constant temperature). In connection with this feature, the temperature coefficients have been tabulated where available for various specific interfacial free energies.

In view of the clear distinction recorded for the interfacial free energies of surfaces, grain boundaries, twin boundaries, and stacking faults, a similar distinction would be expected for the corresponding structures and properties of these interfaces. Indeed this feature will be demonstrated in the succeeding chapters. Perhaps the most lacking in attention has been the interphase interface, which constitutes one of the more important interfacial phenomena in contemporary metallurgy and materials science. Although the energetics of these interfaces is not well-known, considerable effort has been expended in the examination of the structure and properties of interphase interfaces. These features will be treated in succeeding chapters.

REFERENCES

1. T. Young, *Proc. Roy. Soc.* (December, 1804); *Phil. Trans.*, p. 65 (1805).
2. P. S. Laplace, "Mechanique Celeste," Suppl. au X Livre, Impr. Imperiale, Paris (1805).
3. K. F. Gauss, "Principae Generalia Theoreticae Figurae Fluidorum in Statu Aequilibrii," Gottingen (1830).
4. S. D. Poisson, "Nouvelle Theorie de l'Action Capillaire" (1831).
5. Leonardo da Vinci, in footnote reference in *Pogg. Ann.*, **101**, 551 (1857).
6. F. Hawksbee, "Physico-mechanical Experiments," London, p. 139 (1709).
7. Lord Rayleigh, *Proc. Roy. Soc., Ser. A*, **92**, 184 (1915).
8. N. K. Adam, "The Physics and Chemistry of Surfaces," 3rd ed., Oxford University Press, London (1941).
9. F. Bashforth and J. C. Adams, "An Attempt To Test the Theory of Capillary Action," Cambridge University Press (1892).
10. J. F. Padday, in "Surface and Colloid Science," E. Matijevic (ed.), Wiley-Interscience, New York, p. 151 (1969).
11. A. Yu. Koshevnik, M. M. Kusakov, and N. M. Lubman, *Zh. Fiz. Khim.*, **27**, 1887 (1953).
12. J. J. Bikerman, "Physical Surfaces," Academic Press, New York (1970).
13. H. Gibson, *Proc. Roy. Soc. S. Australia*, **56**, 51 (1932).
14. A. M. Worthington, *Phil. Mag.*, **20**, 51 (1885).
15. C. Kemball, *Trans. Faraday Soc.*, **42**, 526 (1946).
16. M. E. Nicholas, P. A. Joyner, B. M. Tessem, and M. D. Olson, *J. Phys. Chem.*, **65**, 1373 (1961).
17. M. Humenik and W. D. Kingery, *J. Amer. Ceramic Soc.*, **37**, 18 (1954).
18. W. Bonfield, *J. Mater. Sci.*, **7**, 148 (1972).
19. J. M. Andreas, E. A. Hauser, and W. B. Tucker, *J. Phys. Chem.*, **42**, 1001 (1938).
20. S. Fordham, *Proc. Roy. Soc., Ser. A*, **194**, 1 (1948).
21. C. E. Stauffer, *J. Phys. Chem.*, **69**, 1933 (1965).
22. B. C. Allen, *Trans. AIME*, **227**, 1175 (1963).
23. W. D. Harkins and F. E. Brown, *J. Amer. Chem. Soc.*, **4**, 499 (1919).
24. J. L. Lando and H. T. Oakley, *J. Colloid Interfac. Sci.*, **25**, 526 (1967).
25. M. Cantor, *Ann. Physik*, **47**, 399 (1892).
26. I. E. Verschaffelt, *Verhandel. Koninkl. Ned. Akad. Wetenschap.*, **27**, 208 (1908).

27. R. Schrodinger, *Ann. Physik,* **46,** 413 (1915).
28. V. K. Semenchenko, "Surface Phenomena in Metals and Alloys," Addison-Wesley Publishing Co., Reading, Massachusetts (1962).
29. B. P. Bering and K. A. Ioileva, *Dokl. Akad. Nauk SSSR,* **93,** 85 (1953).
30. J. W. Taylor, *J. Inst. Metals,* **83,** 143 (1954).
31. J. W. Taylor, *Phil. Mag.,* **46,** 867 (1955).
32. N. L. Pokrovskii and M. Saidov, *Zh. Fiz. Khim.,* **29,** 1601 (1955).
33. G. Metzger, *Z. Phys. Chem.,* **211,** 1 (1959).
34. G. Drath and F. Sauerwald, *Z. Anorg. Allg. Chem.,* **162,** 301 (1927).
35. M. Korolkow, *Izv. Akad. Nauk SSSR Otd. Khim. Nauk,* **2,** 35 (1956).
36. V. G. Zhivov, *Trudy Vses. Assoc. Metall. Inzhenerov,* **14,** 99 (1937).
37. Y. Kubitschek, *Izv. Akad. Nauk SSSR Otd. Khim. Nauk,* **2,** 96 (1959).
38. V. de L. Davies and J. M. West, *J. Inst. Metals,* **92,** 208 (1964).
39. Y. V. Naiditsch and V. N. Eremenko, *Fiz. Metal. Metalloved.,* **11,** 883 (1961).
40. V. N. Eremenko, N. Iwaschenko, B. I. Niehenko, and V. Fresenko, *Izv. Akad. Nauk SSSR Otd. Khim. Nauk,* **7,** 144 (1958).
41. E. Pelzel, *Berg. Huettenmaenn. Monatsh.,* **93,** 247 (1949).
42. L. L. Bircumshaw, *Phil. Mag.,* **2,** 341 (1926).
43. L. L. Bircumshaw, *Phil. Mag.,* **12,** 596 (1931).
44. C. C. Addison, J. M. Coldrey, and L. Pulham, *J. Chem. Soc.,* 1227 (1963).
45. V. N. Eremenko, B. I. Nihenko, and L. Tay-Schon-Wej, *Izv. Akad. Nauk SSSR Otd. Khim Khim. Nauk,* **3,** 150 (1960).
46. "Handbook of Chemistry and Physics," Vol. 50, Chemical Rubber Co., Cleveland, Ohio, p. F-19 (1969).
47. D. W. G. White, *Met. Trans.,* **3,** 1933 (1972).
48. H. A. Davies, *Met. Trans.,* **3,** 2917 (1972).
49. V. N. Eremenko and Y. V. Naiditsch, *Izv. Akad. Nauk SSSR, Met. Toplivo,* **2,** 111 (1959).
50. B. C. Allen, *Trans. AIME,* **227,** 1175 (1963).
51. B. C. Allen and W. D. Kingery, *Trans. AIME,* **215,** 30 (1959).
52. S. M. Kaufman and T. J. Whalen, *Trans. AIME,* **230,** 79 (1964).
53. D. A. Belforte and M. P. Lepie, *Trans. AIME,* **227,** 80 (1963).
54. Y. V. Naidich and V. N. Eremenko, *Fis. Metal. Metalloved.,* **11,** 62 (1961).
55. W. Krause, F. Sauerwald, and M. Michalke, *Z. Anorg. Allg. Chem.,* **181,** 353 (1959).
56. B. F. Dyson, *Trans. AIME,* **227,** 1098 (1963).
57. T. J. Whalen and M. Humenik, *Trans. AIME,* **218,** 952 (1960).
58. G. L. Mack, J. K. Davies, and F. E. Bartell, *J. Phys. Chem.,* **45,** 846 (1941).
59. G. Bernard and C. H. P. Lupis, *Met. Trans.,* **2,** 555 (1971).
60. W. D. Kingery and M. Humenik, Jr., *J. Phys. Chem.,* **57,** 359 (1953).
61. W. D. Kingery, *J. Amer. Ceram. Soc.,* **37,** 42 (1954).
62. W. D. Harkins and W. W. Ewing, *J. Amer. Chem. Soc.,* **42,** 2539 (1920).
63. V. N. Eremenko and M. I. Vasiliu, *Russ. Met.,* **2,** 113 (1965).
64. G. Raue, G. Metzger, and H. C. F. Sauerwald, *Metall,* **55,** 605 (1965).
65. R. J. Klein Wassink, *J. Inst. Metals,* **95,** 38 (1967).
66. W. R. Clough, U.S. AEC Report NYO 2159-1 (November, 1965).
67. A. Calverley, *Proc. Phys. Soc.,* **70,** 1040 (1957).
68. J. A. Cahill and A. D. Kirshenbaum, *J. Inorg. Nucl. Chem.,* **27,** 73 (1965).
69. D. W. G. White, *Trans. AIME,* **236,** 776 (1966).
70. A. K. Rayabar and N. N. Gratsiansky, *Ukr. Khim. Zh.,* **28,** 121 (1962).
71. J. D. van der Waals, *Z. Phys. Chem.,* **13,** 657 (1894).
72. R. Von Eötvös, *Ann. Physik,* **27,** 448 (1886).
73. A. V. Grosse, *J. Inorg. Nucl. Chem.,* **22,** 23 (1961)

74. M. Katayama, *Sci. Rep. Tohoku Univ.*, **4**, 373 (1915).
75. E. A. Guggenheim, "Thermodynamics," 4th ed., North-Holland Publishing Co., Amsterdam, p. 195 (1959).
76. D. B. Macleod, *Trans. Faraday Soc.*, **19**, 38 (1923).
77. S. Sugden, *J. Chem. Soc.*, **125**, 32 (1924).
78. A. S. Skapski, *J. Chem. Phys.*, **16**, 389 (1948).
79. R. A. Oriani, *J. Chem. Phys.*, **18**, 575 (1950).
80. D. R. Stull and G. C. Sinke, "Thermodynamic Properties of the Elements," *Advan. Chem. Ser.*, No. 19, American Chemical Society (1956).
81. E. A. Guggenheim, *Trans. Faraday Soc.*, **41**, 50 (1945).
82. T. P. Hoar and D. A. Melford, *Trans. Faraday Soc.*, **53**, 315 (1957).
83. V. V. Fresenko, V. N. Eremenko, and M. I. Vasiliu, "The Role of Surface Phenomena in Metallurgy," Consultants Bureau, New York, p. 31 (1963).
84. V. B. Lazarev, V. S. Arakaleyan, and A. V. Pershikov, *J. Phys. Chem. (USSR)*, **42**, 1 (1968).
85. T. J. Whalen, S. M. Kaufman, and M. Humenik, Jr., *Trans. ASM*, **55**, 778 (1962).
86. M. C. Inman and H. R. Tipler, *Met. Rev.*, **8**, 105 (1963).
87. A. P. Greenough, *Appl. Mater. Res.*, **4**, 25 (1965).
88. J. J. Bikerman, *Phys. Status Solidi*, **10**, 3 (1965).
89. E. D. Hondros, in "Techniques of Metals Research," R. Bunshah (ed.), Vol. IV(A), Chapter 8A, John Wiley & Sons, New York (1970).
90. I. Sawai and M. Nishida, *Z. Anorg. Allg. Chem.*, **190**, 375 (1930).
91. G. Tamman and W. Boehme, *Ann. Phys.*, **12**, 820 (1932).
92. H. Udin, A. J. Shaler, and J. Wulff, *Trans. AIME*, **185**, 186 (1949).
93. H. Udin, *Trans. AIME*, **191**, 63 (1951).
94. A. L. Pranatis and G. M. Pound, *Trans. AIME*, **203**, 664 (1955).
95. E. D. Hondros, *Proc. Roy. Soc., Ser. A*, **286**, 479 (1965).
96. A. P. Greenough, *Phil. Mag.*, **3**, 1032 (1958).
97. A. P. Greenough, *Phil. Mag.*, **43**, 1075 (1952).
98. C. Herring, *J. Appl. Phys.*, **21**, 459 (1950).
99. F. R. N. Nabarro, "Report on Conference on Strength of Solids," Physical Society, London, p. 75 (1948).
100. J. C. Fisher and C. G. Dunn, in "Imperfections in Nearly Perfect Crystals," W. Shockley (ed.), John Wiley & Sons, New York, p. 317 (1950).
101. H. Jones, *Scripta Met.*, **6**, 423 (1972).
102. L. E. Murr, R. J. Horylev, and G. I. Wong, *Surface Sci.*, **26**, 184 (1971).
103. H. V. Åstrom, *Acta Met.*, **4**, 562 (1956).
104. L. E. Murr, *Acta Met.*, **21**, 791 (1973).
105. R. S. Barnes and G. B. Redding, *J. Nucl. Energy*, **10**, 32 (1959).
106. W. W. Mullins, *Acta Met.*, **4**, 421 (1956).
107. B. C. Allen, *Trans. AIME*, **236**, 903 (1966).
108. B. C. Allen, *J. Less-Common Metals*, **29**, 263 (1972).
109. L. F. Bryant, R. Speiser, and J. P. Hirth, *Trans. AIME*, **242**, 1145 (1968).
110. L. E. Murr, *Scripta Met.*, **6**, 203 (1972).
111. M. C. Inman, D. McLean, and H. R. Tipler, *Proc. Roy. Soc., Ser. A*, **273**, 538 (1963).
112. G. Ehrlich, "ASM/AIME Seminar Metal Surfaces: Structure, Energetics, and Kinetics," Chapter 7, ASM, Metals Park, Ohio (1963).
113. F. H. Buttner, H. Udin, and J. Wulff, *Trans. AIME*, **191**, 1209 (1964).
114. A. T. Price, A. P. Greenough, and H. A. Holl, *Trans. AIME*, **230**, 265 (1964).
115. E. D. Hondros and L. E. H. Stuart, *Phil. Mag.*, **17**, 711 (1968).
116. L. E. Murr, G. I. Wong, and R. J. Horylev, *Acta Met.*, **21**, 595 (1973).
117. L. E. Murr, O. T. Inal, and G. I. Wong, in "Electron Microscopy and Structure of

Materials," G. Thomas (ed.), University of California Press, Berkeley, California, p. 417 (1972).
118. E. R. Hayward and A. P. Greenough, *J. Inst. Metals,* **88,** 217 (1959–60).
119. S. V. Radcliffe, *J. Less-Common Metals,* **3,** 360 (1961).
120. M. McLean and E. D. Hondros, *J. Mater. Sci.,* **6,** 19 (1971).
121. E. R. Funk, H. Udin, and J. Wulff, *Trans. AIME,* **191,** 1206 (1951).
122. E. B. Greenhill and S. R. McDonald, *Nature,* **171,** 37 (1953).
123. J. P. Barbour, F. M. Charbonnier, W. W. Dolan, W. P. Dyke, E. E. Martin, and J. K. Trolan, *Phys. Rev.,* **117,** 1452 (1960).
124. A. Bondi, *Chem. Rev.,* **52,** 417 (1953).
125. "Metals Handbook," 8th ed., Vol. 1, ASM, Metals Park, Ohio (1961).
126. A. S. Skapski, *Acta Met.,* **4,** 576 (1956).
127. H. Jones and G. M. Leak, *Acta Met.,* **14,** 21 (1966).
128. P. J. Estrup, in "Modern Diffraction and Imaging Techniques in Materials Science," S. Amelinckx, R. Gevers, G. Remaut, and J. Van Landuyt (eds.), North-Holland Publishing Co., Amsterdam, p. 377 (1970).
129. J. J. Lander, in "Fundamentals of Gas-Surface Interactions," Academic Press, New York, p. 25 (1967).
130. U.S. Government Report ARC-69-0003 (prepared by A. G. Jackson, M. P. Hooker, T. W. Haas, G. J. Dooley, and J. T. Grant) (January, 1969).
131. P. J. Estrup and J. Anderson, *J. Chem. Phys.,* **45,** 2254 (1966).
132. L. A. Harris, *J. Appl. Phys.,* **39,** 1419 (1968).
133. H. P. Bonzel and B. Aaron, *Scripta Met.,* **5,** 1057 (1971).
134. D. McLean, "Grain Boundaries in Metals," Oxford University Press, Oxford (1957).
135. W. R. Bitler, *Scripta Met.,* **5,** 1045 (1971).
136. E. W. Hart, *Scripta Met.,* **2,** 179 (1968).
137. R. Thomson, *Acta Met.,* **6,** 23 (1968).
138. J. P. Hirth and J. Lothe, "Theory of Dislocations," McGraw-Hill Book Company, New York, p. 464 (1968).
139. D. F. Stein, A. Joshi, and R. P. Laforce, *Trans. ASM,* **62,** 776 (1969).
140. M. J. Fleetwood, *J. Inst. Metals,* **90,** 429 (1962).
141. G. Carter, *J. Vac. Sci. Technol.,* **10,** 95 (1973).
142. E. W. Müller, J. A. Panitz, and S. B. McLane, *Rev. Sci. Instrum.,* **39,** 83 (1968).
143. L. E. Murr, "Electron Optical Applications in Materials Science," McGraw-Hill Book Company, New York (1970).
144. L. E. Murr, *Phys. Status Solidi,* **3,** 477 (1970).
145. R. A. Queeney, *Scripta Met.,* **5,** 1031 (1971).
146. L. E. Murr, R. J. Horylev, and W. N. Lin, *Phil. Mag.,* **22,** 515 (1970).
147. M. J. Whelan, *Proc. Roy. Soc.,* **249,** 114 (1959).
148. L. M. Brown, *Phil. Mag.,* **10,** 441 (1964).
149. L. M. Brown and A. R. Tholen, *Discussions Faraday Soc.,* **38,** 35 (1964).
150. L. E. Murr, *Thin Solid Films,* **4,** 389 (1969).
151. G. Saada, in "Theory of Crystal Defects," B. Gruber (ed.), Academic Press, New York, p. 167 (1966).
152. A. W. Ruff, *Met. Trans.,* **1,** 239 (1970).
153. P. C. J. Gallagher, *Met. Trans.,* **1,** 2429 (1970).
154. R. F. Smallman and P. S. Dobson, *Met. Trans.,* **1,** 2383 (1970).
155. T. Jossang and J. P. Hirth, *Phil. Mag.,* **13,** 657 (1969).
156. I. L. Dillamore, *Met. Trans.,* **1,** 2463 (1970).
157. D. A. Smith, T. F. Page, and B. Ralph, *Phil. Mag.,* **19,** 231 (1969).
158. D. J. H. Cockayne, I. L. F. Ray, and M. J. Whelan, *Phil. Mag.,* **20,** 1265 (1969).
159. P. S. Dobson, P. J. Goodhew, and R. E. Smallman, *Phil. Mag.,* **16,** 9 (1967).

160. S. Kritzinger, P. S. Dobson, and R. E. Smallman, *Phil. Mag.*, **16**, 217 (1967).
161. K. H. Westmacott and R. L. Peck, *Phil. Mag.*, **23**, 611 (1971).
162. T. Ericsson, *Acta Met.*, **14**, 853 (1966).
163. T. C. Tisone, *Acta Met.*, **21**, 229 (1973).
164. P. R. Thornton, T. E. Mitchell, and P. B. Hirsch, *Phil. Mag.*, **17**, 1349 (1962).
165. P. C. J. Gallagher and Y. C. Liu, *Acta Met.*, **17**, 127 (1969).
166. B. Pettersson, *Phil. Mag.*, **20**, 831 (1969).
167. A. Howie and P. R. Swann, *Phil. Mag.*, **6**, 1215 (1961).
168. M. A. Quader and R. A. Dodd, *J. Appl. Phys.*, **39**, 4726 (1968).
169. A. W. Ruff and L. K. Ives, *Acta Met.*, **15**, 189 (1967).
170. A. W. Ruff and L. K. Ives, *Phys. Status Solidi (a)*, **16**, 133 (1973).
171. D. A. Smith and K. M. Bowkett, *Phil. Mag.*, **18**, 1219 (1968).
172. D. H. Sastry, M. J. Luton, and J. J. Jonas, *Phil. Mag.*, **30**, 115 (1974).
173. H. Gleiter and H. P. Klein, *Phil. Mag.*, **27**, 1009 (1973).
174. H. Suzuki, *Sci. Rep. Res. Inst. Tohoku Univ.*, *Ser. A*, **4**, 452 (1952).
175. T. V. Nordstrom and C. R. Barrett, *Acta Met.*, **17**, 139 (1969).
176. J. D. Venables, *Met. Trans.*, **1**, 2471 (1970).
177. J. J. Kramer, G. M. Pound, and R. F. Mehl, *Acta Met.*, **6**, 763 (1958).
178. R. H. Hopkins and R. Kossowsky, *Acta Met.*, **19**, 203 (1971).
179. J. D. Boyd and R. B. Nicholson, *Acta Met.*, **19**, 1101 (1971).
180. D. Turnbull, *J. Appl. Phys.*, **21**, 1022 (1950).
181. M. E. Glicksman and C. L. Vold, *Acta Met.*, **17**, 1 (1969).
182. S. K. Rhee, *Mater. Sci. Eng.*, **11**, 311 (1973).
183. L. H. Van Vlack, *Trans. AIME*, **191**, 251 (1951).
184. L. E. Murr, *Mater. Sci. Eng.*, **12**, 277 (1973).
185. R. M. Pilliar and J. Nutting, *Phil. Mag.*, **16**, 181 (1967).
186. U. M. Ahmad and L. E. Murr, *Metallography*, in the press.
187. A. Moore and R. Elliott, *J. Inst. Metals*, **95**, 369 (1969).
188. R. G. Faulkner and B. Ralph, *Acta Met.*, **20**, 703 (1972).
189. B. Vonnegut, *J. Colloid Sci.*, **3**, 563 (1948).
190. D. Turnbull, *J. Appl. Phys.*, **20**, 817 (1949).
191. M. J. Stowell, *Phil. Mag.*, **22**, 1 (1970).

PROBLEMS

(3.1) With a small quantity of mercury in a small glass beaker, use a small (<3 mm inside diameter) glass tube to measure capillary "rise" of the mercury. Compare the measured value with that you calculate using the surface free energy (γ_{LV}) for mercury in Table 3.4. Finally, calculate the surface free energy for mercury by measuring the sessile drop parameters from Fig. 2.27(a). Show your calculations, and compare your result with that in Table 3.4. Discuss any discrepancies you observe.

(3.2) Referring to the principles of interfacial thermodynamics outlined in Chapter 1, discuss on a quantitative basis the variations of liquid surface free energy in the Sn–Cu system, and evaluate $d\gamma_{LV}/dT$ and $d\gamma_{LV}/dx$. Compare the liquid Sn–Cu system with the liquid Sn–Pb system. Evaluate $d\gamma_{LV}/dx$ for the Sn–Pb system at 300°C after constructing a suitable plot of the data.

INTERFACIAL FREE ENERGY

(3.3) Compare the value of F_S (the solid surface free energy) for molybdenum in Table 3.5 with the value calculated at the same temperature from the data in Table 3.4, and discuss any differences. If you observe a difference, would you expect it? Is the difference in the right direction? Why?*

(3.4) Referring to Fig. 3.18, calculate the heats of fusion for silver and copper, and compare with experimental values.

(3.5) The surface free energy of pure nickel at 1300°C is estimated to be 2150 ergs/cm^2. Calculate the weight required for zero creep of a nickel wire at this temperature if the wire diameter is 0.006 cm and $l = 2r$. What will the mean dihedral angle be at the intersection of the grain boundaries with the surface if the ratio of grain boundary free energy to surface free energy is the same at 1300°C as that measured at 1060°C?

(3.6) Calculate the stacking-fault free energy of pure gold at room temperature (25°C) from measured values of twin boundary free energy, and compare it with the value of γ_{SF} measured directly at room temperature. Discuss any differences. Make a simple sketch of the variation of γ_{SF} for gold when alloyed with copper, and discuss any assumptions you make.

(3.7) Referring to Fig. 3.30(b), calculate the value of the noncoherent twin boundary free energy in Cu-5 a/o Al. The sample imaged in Fig. 3.30(b) was characterized by an annealing temperature of 800°C. Assume that, because of geometrical asymmetry and the nature of geometrical corrections, the projected intersection angles are the true dihedral angles. How does this calculated value compare with the experimental value recorded in Table 3.8? What would you expect to measure as the radius of curvature for dislocation nodes observed in Cu-5 a/o Al samples (at 25°C) in the transmission electron microscope if the shear modulus does not differ measurably from that of pure Cu? (Note that Cu-5 a/o Al is an fcc alloy.)

(3.8) A well-annealed thin foil of silver is observed in an electron microscope. During observation of a relatively straight portion of a grain boundary having an inclination angle, θ, of nearly 90°, a stacking fault is seen to emanate from the boundary, nearly normal to the boundary trace (projection). If the temperature of the foil during observation is measured to be 150°C, and the angular resolution is accurate to approximately 1°, will the boundary be observably distorted by the stacking fault intersecting it? Defend your answer using necessary calculations.

(3.9) Describe in detail an experiment for measuring the mean interfacial free energy at a metal-ceramic interface as a function of temperature. How would you determine whether alloying was occurring at the interface with temperature?

(3.10) Make sketches and mathematically document your arguments for the measurement of the interphase interfacial free energy in a eutectic com-

*See Note, p. 164.

posite. Consider the case of eutectic lamellae intersecting the free (solid-vapor) surface. Suppose, in a eutectic composite such as Ni-NiCr an annealing twin extends across the Ni matrix and intersects the NiCr-Ni interface. Show how and under what conditions the interphase interfacial free energy for the NiCr-Ni interface could be determined.

*Note that even though $d\gamma_{LV}/dT$ as well as dF_S/dT might be expected to be constant, there is no reason for these temperature coefficients to be equal. On the contrary, considering the variations in surface structure in the solid state as compared with the liquid state, and the differences in surface atom binding, one might expect them to be different. The value of $d\gamma_{LV}/dT$ might also become nonlinear near T_c.

4

STRUCTURE OF INTERFACES

4.1 SURFACE STRUCTURE IN LIQUID METALS AND ALLOYS

In outlining the concept of the surface tension of liquid metals in Section 3.1, it was pointed out that, unlike ordinary (organic) liquids, metals are not influenced by intermolecular forces and secondary bonds, since only metallic binding applies. A liquid metal is therefore characterized by a state of atomic disorder with reference to a solid crystal, and as a consequence neither interatomic dimensions nor binding energies are explicitly known. Liquid metal surfaces, like those of other liquids, are naturally curved (Fig. 3.1) and characteristically spherical when external forces are zero (a free droplet). The atoms in a liquid surface at the liquid–vapor interface can be thought of structurally perhaps as lacking the degree of bond coordination conceivably present in the bulk where each atom is surrounded by an unsymmetrical array of other atoms. Treated in the context of quantum mechanical wave function overlap, the structure of a liquid metal surface might be envisioned as depicted in Fig. 4.1.*

From a strictly structural point of view, liquid surfaces have no well-defined structure, and as a result of atomic vibration or agitation the surface is continuously changing. However, as noted above, liquid metal surfaces—that is, liquid-vapor interfaces—are characteristically curved as illustrated typically in Fig. 4.2. It can be noted in Fig. 4.2(b) that there is no evidence of faceting or planarity, and this is usually true even for liquid surface areas having dimensions in angstrom units. Somewhat essential to this argument is the presumption that the liquid-vapor interface is characterized by an abrupt change from one phase to the other. However, near the critical point and at very high temperatures for a liquid in equilibrium with its vapor, the interfacial region will become diffuse.

An attempt is made in Fig. 4.1 to depict the electronic structure at the surface as a simple wave-function overlap. It should be recalled from Chapter 3 (Fig. 3.9) that closed valence shells appear to exhibit a level of higher stability as a result of a stronger interatomic interaction. Obviously, this electronic structure is rather

---

*Recent work by F. Abraham and G. M. Pound (Stanford) has shown surface ordering in liquid argon.

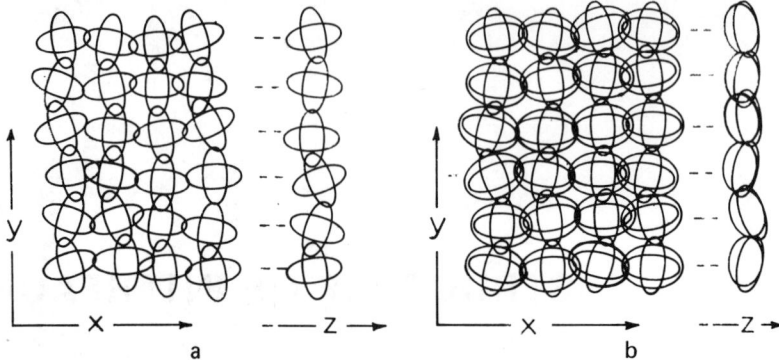

FIGURE 4.1   Simple schematic arrangement for atomic binding in the interior (a) and on the surface (b) of a liquid metal or alloy.

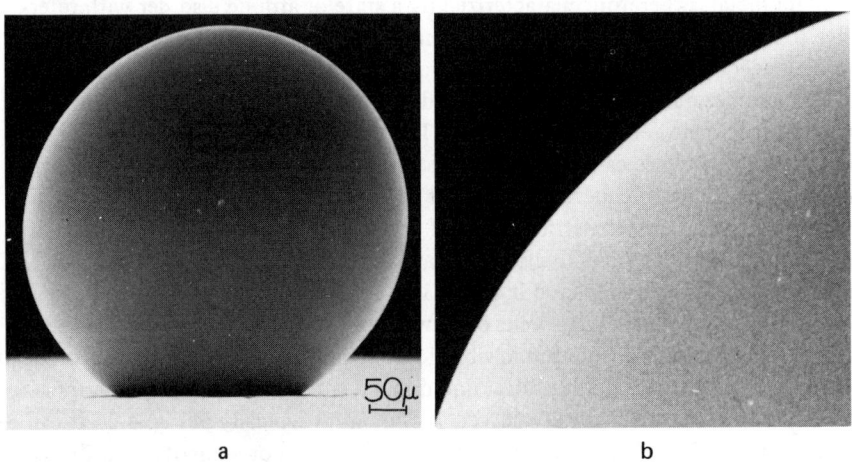

FIGURE 4.2   Mercury droplet on a glass surface as observed in the scanning electron microscope. (a) Low magnification view. (b) Enlarged view of droplet area showing regular curvature.

drastically altered for surface adsorption or reaction, and this is experimentally documented.

In applying the thermodynamic treatment of Chapter 1 to liquid surfaces, it is assumed that the principal radii of curvature (Fig. 3.1) are very large. This is ideally true only for liquids spread over a larger area of a substrate as a result of

STRUCTURE OF INTERFACES

interfacial wetting, or for very small areas of a liquid surface. As a consequence, we might differentiate a liquid metal surface from a solid metal surface in the range of atomic dimensions as one lacking a regular symmetry as a result of constant atomic agitation. Topologically, a liquid surface also generally possesses some curvature. This argument loses significance, as does the interfacial concept, when the radius of a liquid droplet becomes so small as to be comparable with the interfacial thickness. In this regard, on the basis of the changing density of matter in the interfacial state, it has been suggested that the surface tension of a droplet should decrease with increasing curvature.[1-3] Generally, however, an adequate treatment of the effect of curvature on liquid as well as solid interfaces has yet to be achieved.

4.2 LIQUID METAL INTERFACIAL STRUCTURE

When considering the structure of liquid interfaces (liquid–liquid and liquid–solid), it must be pointed out that, just as in the case of the liquid–vapor (surface) interface structure, essentially nothing is definitely known, and there are no truly effective means of illustrating such an interface. It is assumed that there is little if any miscibility and little or no chemical reactivity. This is indeed true for a number of metal (binary) systems,[4] liquid–solid phase systems, or more complex two-phase, multicomponent systems. The concept of an interface as formulated in Chapter 1 is in fact lost when mixing occurs, and the interface thickness, Δt, becomes orders of magnitude larger than the individual atom diameters, forming in reality another phase (Fig. 4.3).

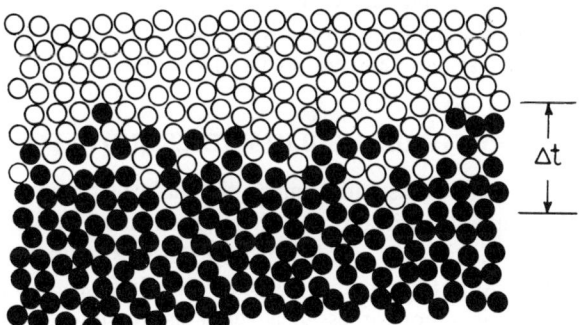

FIGURE 4.3 Liquid–liquid interface in two pure metals exhibiting only slight miscibility. The "interface phase" is denoted Δt.

In dealing with the structure of liquid interfaces, the concept of boundary planarity returns, and the thermodynamic considerations of Chapter 1 are applicable. Such an interface, as a result of the structural uniqueness depicted ideally in Fig. 4.3, would be expected to exhibit a considerable energy difference when compared with surface free energies (surface tensions), because as one can conceptually observe on comparing Fig. 4.1(*b*) and 4.3, the binding considerations are different. Although liquid-liquid interfacial free energies were not discussed specifically in Chapter 3, this point can be demonstrated with reference to the liquid Pb/Zn interface at 420°C measured to be 128 ergs/cm^2,[5] and the Hg/Ga interface at 25°C measured to be approximately 37 ergs/cm^2.[6] These values are considerably less than the values of the corresponding component metal surface (liquid-vapor) free energies at the same or similar temperatures (Table 3.4). In the Hg/Ga system, Hg dissolves only 1.3% of Ga, while the Ga dissolves only 5.8% of the Hg at 25°C, forming an interfacial phase as depicted in Fig. 4.3 having a thickness, Δt, of 2 to 3 atom diameters. It is apparent that whether or not an interface or a new phase exists depends on the limits of Δt, which in turn depends on the miscibility of the contacting phases and/or diffusion phenomena. As a consequence, liquid-liquid interfacial energy will effectively change not only with temperature but also with time at temperature. To a certain extent this may also be true for liquid-solid interfaces. When this occurs, the interfacial structure is being altered. In fact, as indicated above, the interface may become so thick that a new phase exists, at which time the associated interfacial energy must be considered at two new phase boundaries rather than at the original interface.

It must be realized that liquid metal surfaces and interfaces are always in a state of agitation. As a consequence, it is not possible to observe them directly as in the case of solids, because to do so would require an image recording time of less than about 10^{-7} second. Thus any experiments lasting more than a few microseconds would record a smooth surface structure (as in Fig. 4.2) regardless of the image resolution. This is tantamount to attempting to observe a metal end form in a field-ion microscope[7] at high temperatures near the melting point.

4.3 SURFACE STRUCTURE OF SOLID METALS AND ALLOYS

While liquid surfaces are consistently similar, the surface structure of solids varies with service environment, fabrication (including mechanical working or deformation), and heat treatment as well as other features such as composition and bulk morphology (size). As a starting point, we shall initially treat the surface structure of crystalline solids* in a more or less idealized or as-grown form.

4.3.1 Structure of Single Crystals

It may be instructive at the outset to attempt a direct comparison between the structure of a liquid drop and a solid drop, with particular regard to the solid-vapor interface. The fact that single crystals can exist in a form nearly identical in appearance to the liquid metal drop in Fig. 4.2(*a*) is illustrated for vapor-nucleated

---

*See also J. Blakely, "Introduction to the Properties of Crystal Surfaces," Pergamon Press, England (1973).

STRUCTURE OF INTERFACES   169

tin clusters in the electron micrographs of Fig. 4.4.[8] While an apparently identical surface morphology appears on comparison of Figs. 4.2(a) and 4.4(a), Fig. 4.4(b) indicates that, unlike the liquid metal drop which possesses no well-defined bulk or surface structure, the solid tin particles are indeed single crystals.

Figure 4.5 shows schematically the atomic (bulk and surface structure) arrangements which differentiate the liquid [Fig. 4.2(a)] from the solid [Fig. 4.4(a)]. The accommodation to pseudospherical symmetry of the solid particle

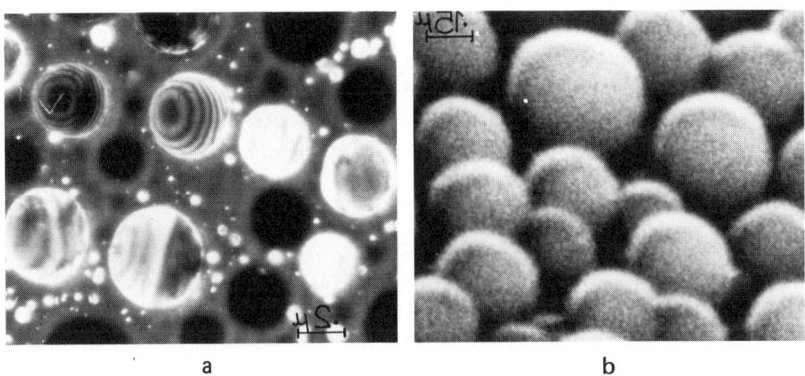

FIGURE 4.4   Solid tin particles nucleated from the vapor onto an NaCl substrate in vacuum and at an elevated substrate temperature.[8] (a) Scanning electron micrograph of tin particles on NaCl substrate. (b) Dark-field transmission electron micrograph of tin particles as in (a) extracted from the NaCl following carbon shadowing to form a support film. Extinction contours and image reversal of particles attest to single-crystal structure.

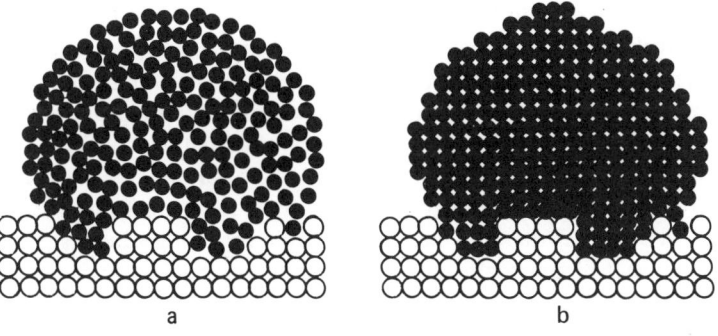

FIGURE 4.5   Schematic representation of liquid ⇌ solid particle phase transition and characteristic surface structure. (a) Liquid particle on solid (crystalline) substrate. (b) Solid (crystalline) particle on solid (crystalline) substrate.

occurs by the formation of steps and systematic facets which vary slightly in surface energy and therefore give rise to a surface-energy anisotropy as discussed in Chapter 1. On the average, however, the transition from the disordered surface structure of Fig. 4.5(*a*) to Fig. 4.5(*b*) (the solid-liquid : liquid-solid phase transition) or vice versa requires a change in energy, ΔE (essentially the heat of fusion), as shown experimentally in Fig. 3.18. Note also the schematic representation of the solid-liquid and solid-solid interfaces in Fig. 4.5. The solid-liquid interface in Fig. 4.5(*a*) is represented as a diffuse interfacial region as a result of the stepped representation of the solid substrate, synonymous with the equilibrium particle structure depicted in Fig. 4.5(*b*). The growth thermodynamics of the phase transition depicted structurally in Fig. 4.5 can be treated in a more or less contemporary manner as nucleation and interface kinetics,[9] while the faceting which finally characterizes the solid equilibrium particle in Fig. 4.5(*b*) has been discussed in Section 1.2. The residual surface structure of single crystals is therefore ideally the result of growth by surface mechanisms and, in the case of small particles (Section 1.2), is dictated by equilibrium shape.

Direct Observations of Surface Structure. One of the most effective ways to observe the atomic features of surface structure is by field-ion microscopy.[7] When fine metal wires are chemically etched and further "smoothed" to an equilibrium shape (pseudohemispherical) by systematic removal of surface atoms in the field-ion microscope by field evaporation,[7] the resulting surface structure resembles that shown schematically in Fig. 4.5(*b*). These features are emphatically illustrated in Fig. 4.6. It should be noted that the atomic surface structure is essentially identical to that expected on considering the crystallography of bulk crystal structure. That is, there is little measurable distortion of the atom positions in the exposed planes of the solid-vapor interface (surface) as compared with the bulk. This particular feature is illustrated in Fig. 4.7. Even where differences in interatomic distances do occur, the expected symmetry in atom positions is preserved.

The description of solid surface structure has to this point either explicitly or implicitly involved pure metals. However, in the case of alloys, particularly solid solutions resulting in an ordered phase, the surface structure will be similarly characterized by the expected atomic symmetry deduced from inspection of a unit cell of the alloy as in Fig. 4.7. This has been demonstrated for the FePt, PtCo, and Ni_4Mo systems, to name but a very few.[7,10-13] The connection between the surface structure and surface energy changes for pure metals and alloys as demonstrated in Section 3.3.3 therefore manifests itself in the changes in local binding energies as a result of systematic substitutions of alloying elements on symmetry (lattice) sites, and/or a systematic change in crystal (lattice) structure as shown in Fig. 4.8. This phenomenon as well as the distinction of the surface interfacial binding from that in the bulk is illustrated schematically in Fig. 4.9. A comparison of Fig. 4.9 with Fig. 4.1 will also illustrate the secular distinction between liquid and solid surface structure in terms of wave-function overlap.

STRUCTURE OF INTERFACES   *171*

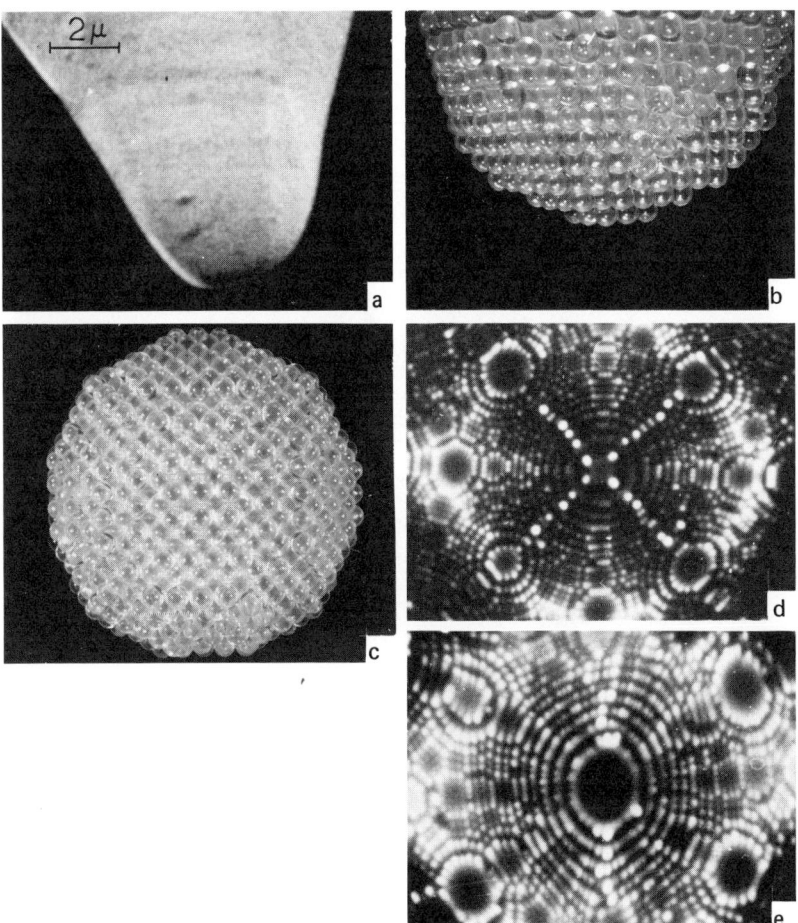

FIGURE 4.6 Observation of surface structure of metal end forms in the field-ion microscope. (a) Scanning electron micrograph of iridium end form following field evaporation. Note crystallographic features (faceting). (b) Crystallographic marble model corresponding to (a). (c) End view of (a) looking down [00$\bar{1}$] axis. (d) Field-ion image of iridium end form (78°K). Atomic (image) prominence occurs preferentially at plane edges. [001] orientation. (e) Field-ion image of tungsten and form having [011] axial symmetry. Note difference in image symmetry between (d) and (e) as a result of the viewing along different zone axes ([001] and [011]).

In retrospect, then, in perfect metal and alloy crystals, the surface structure, as illustrated for example in Figs. 4.5(*b*) and 4.6(*d*) and (*e*), consists of either steps or smooth plane areas (facets) in any combination. In real crystals and polycrystalline

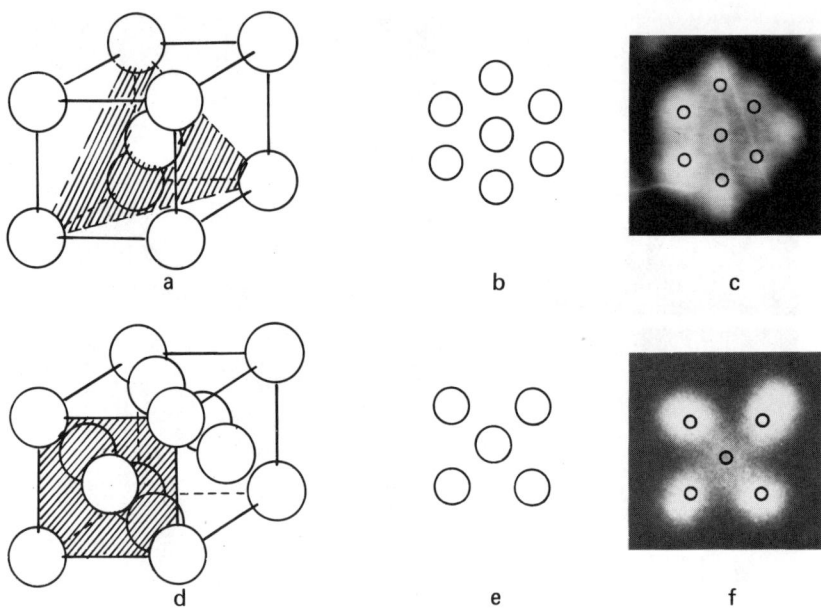

FIGURE 4.7   Correspondence of observed atomic surface structure with idealized (expected) lattice symmetries. (a) BCC unit cell with the (111) plane indicated (shading). (b) Ideal atomic arrangement on the (111) plane. (c) Field-ion microscope image of tungsten showing (111) arrangement of surface atoms. (d) FCC unit cell with the (001) plane indicated (shading). (e) Ideal atomic arrangement on the (001) plane. (f) Field-ion microscope image of iridium showing (001) arrangement of surface atoms.

metals and alloys (real materials), the surface structure in the absence of physical, chemical, or mechanical treatment usually contains defects. Except for grain boundaries or related interfaces which intersect a solid–vapor (surface) interface, these defects, in low concentrations, cause only a slight disturbance in the surface regularity. Such defects are in most cases necessary in nucleation, growth, and solid–vapor interface kinetics. Therefore, they form an integral part of the surface structure.

Steps, Kinks, Adsorbed Atoms, and Vacancies in Solid Surfaces. There are essentially two types of surface structure characterizing the solid–vapor interface which result primarily from more or less two-dimensional growth processes;[9] this structure can also apply to the solid–liquid interface in a metal or alloy melt from which solidification occurs, or in related growth processes such as electrodeposition and physical and chemical vapor deposition. The first is a simple step-plane structure described previously and illustrated, for example, in Figs. 4.5(b) and 4.6, and the second is a modification involving irregular steps and surface vacancies or vacancy aggregates (voids). These features are shown schematically in Fig. 4.10. It can be

STRUCTURE OF INTERFACES

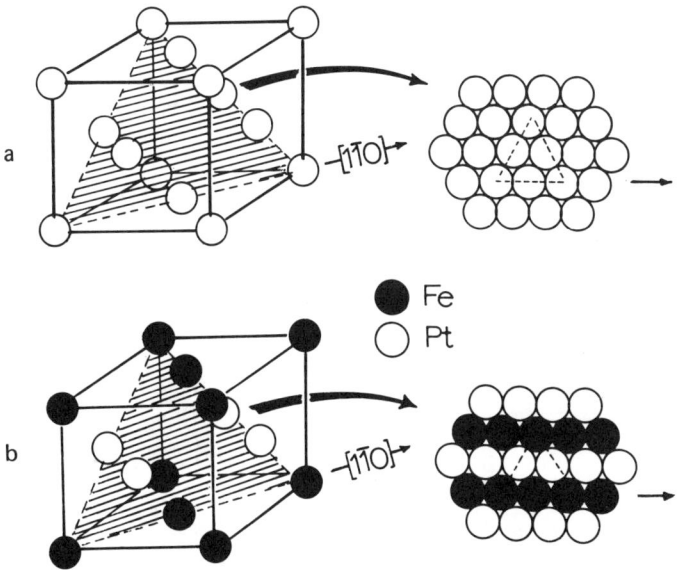

FIGURE 4.8 Comparison of (111) atomic surface structure in fcc Pt (a) and fcc FePt alloy (b).

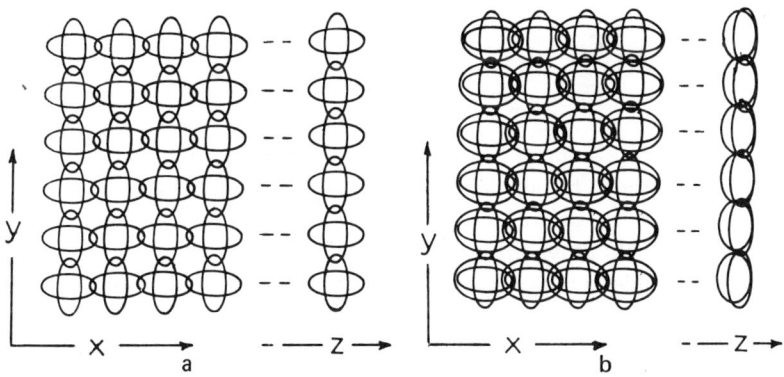

FIGURE 4.9 Simple schematic arrangement for atomic binding in the interior (a) and surface (b) of a solid (crystalline) metal or alloy.

observed in Fig. 4.10(a), and with reference to Fig. 4.6, that atomically flat crystal facets such as those apparent in Fig. 4.6(d) and (e) will be retained as long as the angle σ [Fig. 4.10(a)] is near unity:

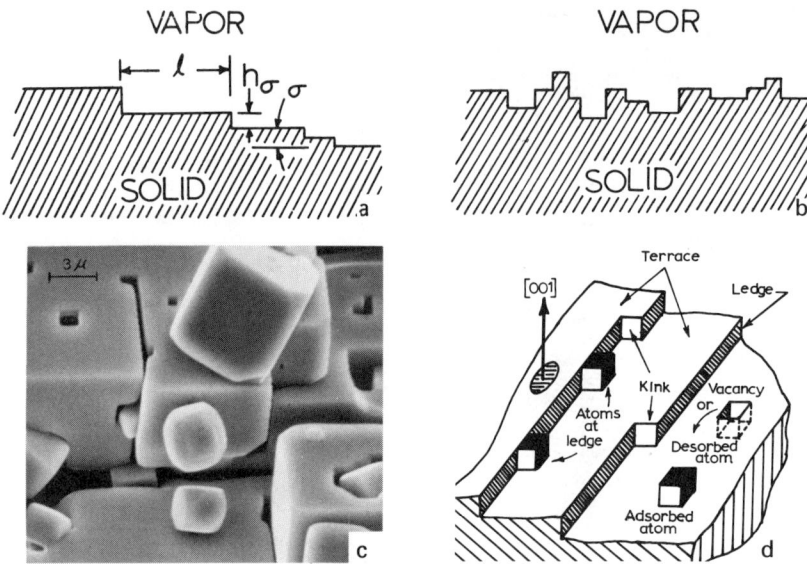

FIGURE 4.10 Step and modified step structures of solid surfaces. (a) Step-plane (flat) surface. (b) Diffuse interface or modified step structure. (c) Exaggerated surface structure of NiO crystals observed in the scanning electron microscope. (d) Schematic representation of (c) showing combined surface structure of (a) and (b). [(c) is courtesy of C. Miglionico.]

$$\sigma = h_\sigma/l \tag{4.1}$$

where h_σ is the step height, and l is the step spacing. Macroscopically flat facets, on the other hand [Fig. 4.10(c)], are retained as long as σ in Eq. (4.1) is small. These features are illustrated by comparing Fig. 4.6(d) and (e) with Fig. 4.11. In particular the enormous sizes of the $\{001\}$ and $\{011\}$ facets observed in Fig. 4.11(c) are to be compared with those of Fig. 4.6(d) and (e), respectively (the central planes of the respective field-ion images, perpendicular to the [001] and [011] zone axes).

The fact that atoms nucleate on a surface as suggested in Fig. 4.10(d) has been demonstrated experimentally by Murr et al.[14] in the field-ion microscope, and these features are seen to occur characteristically as surface structure. Surface atoms cluster either at ledges (C1), on the terraces between ledges (C2), or on the facets characterizing crystal planes (C3). These features are illustrated in the experimental observations shown in Figs. 4.12 and 4.13 for the vapor deposition of platinum atoms onto perfect tungsten and iridium surfaces, respectively. It can be noted in Fig. 4.13 that large facets (plane areas) are formed in the (001), $\{103\}$, $\{011\}$, and $\{111\}$ planes as a result of high-temperature anneal in vacuum. From a crystallo-

STRUCTURE OF INTERFACES   175

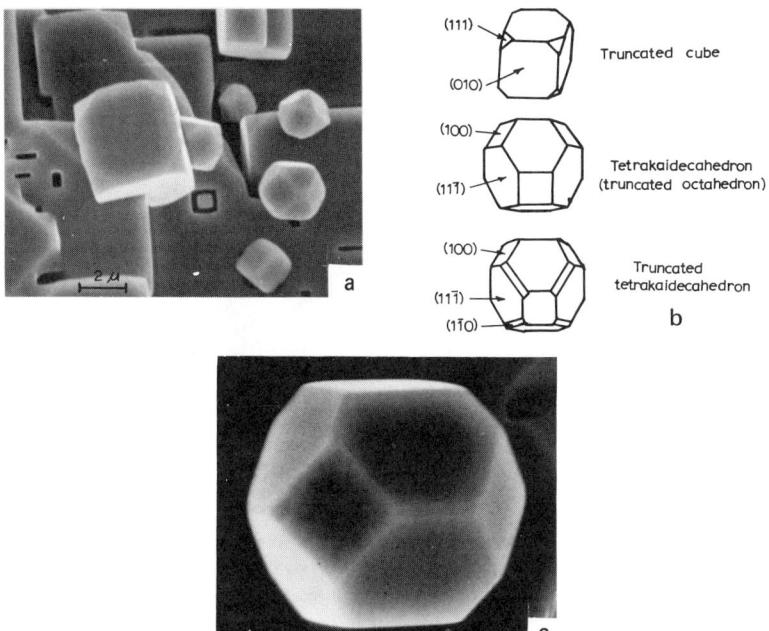

FIGURE 4.11 Faceting and surface structure of vapor-grown NiO single crystals. (a) Surface structure containing single-crystal polyhedra (scanning electron micrograph). (b) Schematic representation of polyhedral facet crystallography. (c) Scanning electron micrograph of individual polyhedra. Facet crystallography in order of decreasing area is $\{111\}$, $\{001\}$, $\{011\}$. (Scanning micrographs courtesy of C. Miglionico.)

graphic standpoint, the iridium end form shown in Fig. 4.13(b) would resemble the polyhedral structure shown in Fig. 4.11(c), neglecting the prominence of the $\{103\}$ planes in Fig. 4.13(b).

Dislocations and Surface Structure: Growth Spirals and Steps. Well-annealed polycrystalline metals and alloys as well as single crystals in the undeformed state are known to contain dislocation densities which can range from approximately 10^2 to 10^7 cm^{-2}. When a screw dislocation emerges at a surface, the Burgers vector is essentially normal to the interface plane, and a spiral ramp results on the surface whose maximum height is $|\mathbf{b}|$. When an edge dislocation line intersects a surface, there is no disturbance normal to the interface plane; however, if an edge dislocation moving parallel to the surface emerges on it, a step is formed. Again **b** is perpendicular to the surface, and the step height is $|\mathbf{b}|$. Consequently, for any

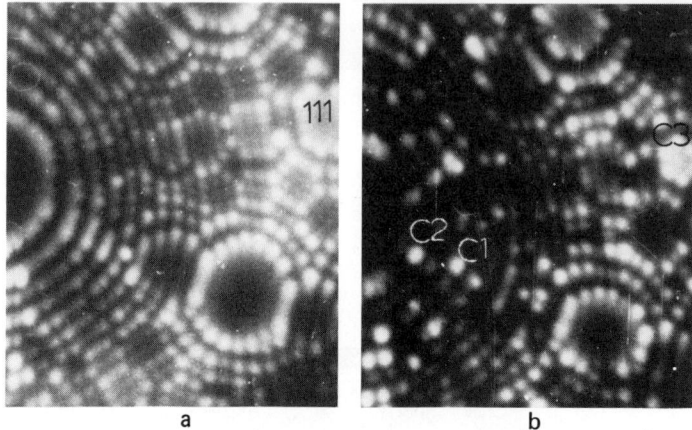

FIGURE 4.12 Change in atomic surface structure by vapor deposition of approximately a monolayer of platinum onto a tungsten end form in the field-ion microscope. (a) Perfect tungsten end form before platinum vapor deposition. The nucleation and clustering of platinum surface atoms at ledges, terraces, and on crystal planes are denoted by C1, C2, C3, respectively.

dislocation at a surface for which **b** is perpendicular to the surface, a step is formed as illustrated in Fig. 4.14.*

Frank[15] first pointed out that a surface step formed by the emergence of a screw dislocation in a surface constitutes an alternative to two-dimensional nucleation and growth at a solid–liquid interface. Such a step is additionally self-perpetuating, and the surface effectively grows up a spiral staircase, causing the step in Fig. 4.14(a) to rotate continuously about the point of emergence of the screw dislocation line as shown, for example, in Fig. 4.15(a). Figure 4.15(b) shows the formation of macrosteps associated with a growth spiral on the surface of electro-deposited silver formed by a step-bunching mechanism originally described by Frank[16] and Pick et al.,[17] and Burton et al.[18] have shown a detailed treatment of a crystal surface by the advancement of growth spirals, which followed the classical theory for two-dimensional nucleation.[9,19-21]

Surface roughness tends to increase generally with temperature, since at $T = 0°K$ steps tend to be as straight and smooth as possible. Frenkel[23] has pointed out that thermal fluctuations will produce an increasing number of kinks which may be positive, forward jumps of an edge, or negative, backward drops which may

---

*Many workers have called emergent dislocations spiral dislocations, which may be preferable because a mixed dislocation or an edge dislocation with its Burgers vector normal to the surface can in fact act as a source of spiral ledge.

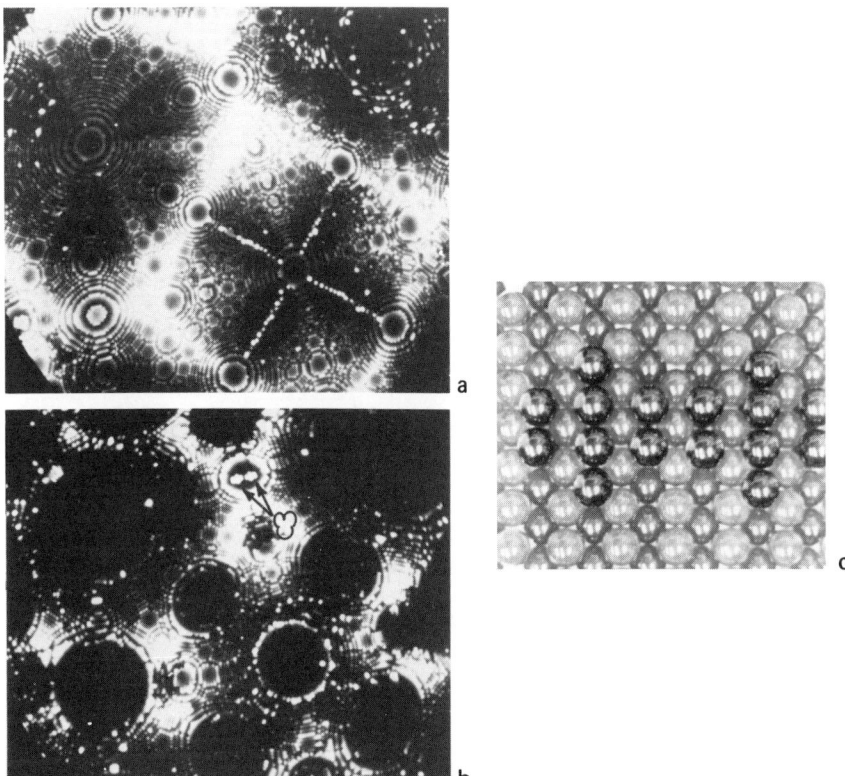

FIGURE 4.13 Structure of platinum clusters on the (011) plane of iridium as observed in the field-ion microscope. (a) Initial, perfect iridium end form. (b) Surface structure of iridium end form in (a) following a 0.25-hour anneal at 1300°K and vapor deposition of approximately a monolayer of platinum onto the end form outside the field-ion microscope. Note thermally induced faceting of prominent crystal planes. (c) Marble model of platinum clusters on the (011) plane indicated by C3 on (b). Slightly darker marbles denote platinum atoms. Clusters accommodate epitaxially on the iridium crystal surface. (After Murr et al.[14])

be single or multiple lattice distances (atomic diameters). Dunning[24] has discussed the roughness of steps and surfaces and has indicated that progressive roughening may be viewed as a cooperative disordering of the surface as illustrated in Fig. 4.16.

4.3.2 Structure of the Solid–Liquid Interface

While the previous discussions of surface-step and spiral-step structure were concerned primarily with the ideal features of the solid–vapor interface, they originate in most instances as a result of crystal growth, and therefore characterize the solid–liquid interface in solution or the growth surface in the melt. Although a

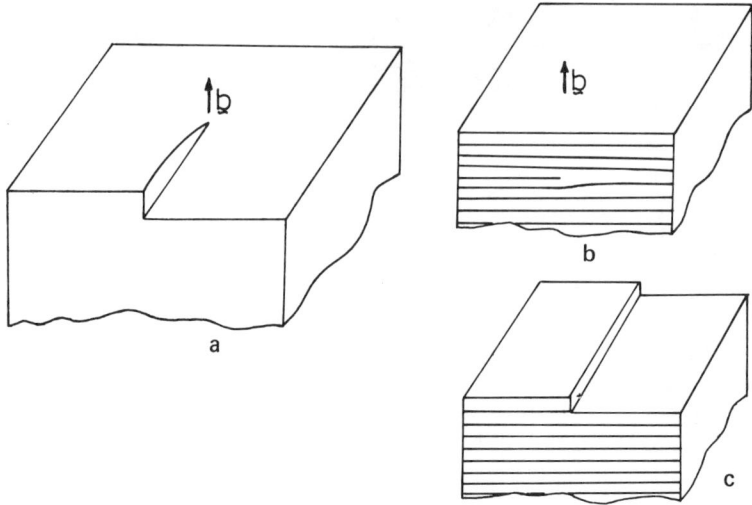

FIGURE 4.14   Dislocations emerging at a solid surface. (a) Screw dislocation forming a surface ramp. (b) Crystal containing an edge dislocation moving toward the surface. (c) Edge dislocation emerging on the surface forming a continuous step (slip step).

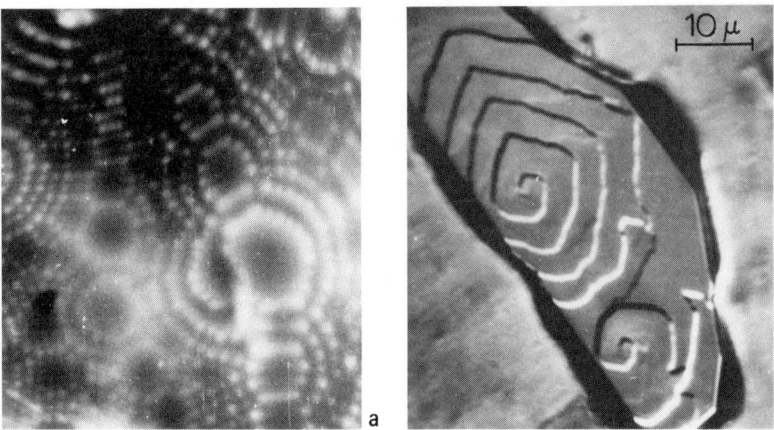

FIGURE 4.15   Spiral-step structure on metal surfaces. (a) Atomic structure of spiral on the surface of an iridium crystal observed in the field-ion microscope. (b) Multilayer spiral steps on electrodeposited silver surface observed by replication-transmission electron microscopy. [(a) is courtesy of Dr. O. T. Inal; (b) is courtesy of Dr. Toshio Suzuki.[22]]

STRUCTURE OF INTERFACES

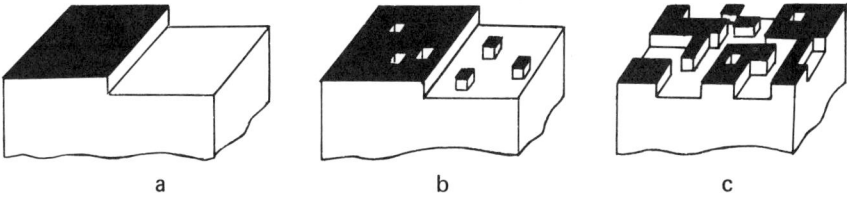

FIGURE 4.16 Progressive roughening of a simple (001) surface. (a) Flat surface containing a monatomic half-layer (step); $T = 0$. (b) $T > 0$; atoms desorb from upper half-layer and are adsorbed in lower half-layer. (c) $T \gg 0$; progressively rougher surface. (After Dunning.[24])

solid–liquid surface can exhibit flat features, it is characteristically diffuse as illustrated ideally in Fig. 4.5(a).

The solid–melt interface is indeed a special and most apparent solid–liquid interface. The interface is continually changing (dynamic) and possesses special features such as spirals and steps or other nucleation mechanisms which facilitate continual change in structure. There are, however, numerous cases of liquid metal–solid metal interfacial systems which as a result of immiscibility are essentially steady-state structurally. The example shown in Fig. 4.2(a) illustrates this situation conceptually, while Fig. 4.5(a) shows the interface schematically (ideally). This feature is particularly true of solid–liquid alloy and multiphase systems. A number of successful models for dealing with the structure of the solid–liquid interface are treated somewhat systematically in a recent book by Woodruff.[25]

4.3.3 Surface Structures Associated with Adsorption, Oxidation, and Related Surface Reactions

Our discussions of surface structure, particularly the solid–vapor interface, have explicitly or implicitly involved clean surfaces. It has been indicated that metals and alloys as a rule show the expected lateral periodicity for surface structure as a parallel plane in the bulk. Although there may be at least two exceptions—namely, the (001) surfaces of platinum[26] and gold[27] as examined by low-energy electron diffraction (LEED)—direct observations of the atomic surface structure in the field-ion microscope (FIM) do not support this contention unambiguously.[7,28]

In adsorption (physisorption and chemisorption) as well as catalysis and oxidation, the adlayers are also observed to form regular periodic or superlattice structures which may show sequences of structural transitions as adsorption and/or reaction proceeds.[29-34] Additionally, adsorption is influenced by the concentration of steps and kinks associated with crystallographic planes. Figure 4.17 illustrates schematically the surface structure typical of hydrogen adsorption on (001) tungsten and many similar crystallographic situations. Adsorption and other

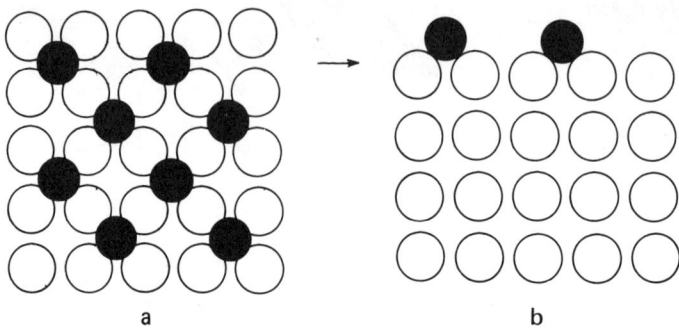

FIGURE 4.17 Model of (001) metal or alloy surface with an adsorbate superstructure. (a) Lateral periodic array. (b) Surface profile, 90° to (a).

reactions are sometimes favored on specific crystallographic surfaces which, because of their atomic structure, are more conducive to accommodation. It seems apparent that surface structure—that is, the atomic arrangement involving channels (systematic or apparent vacant zones in a surface orientation), steps, and kinks—is an important consideration in adsorption or reaction at the surface.

Oxidation and related reactions typify surface interactions which lead to the formation of new compounds and new surface structures. The interaction of oxygen with nickel is a specific example in which oxygen enters the nickel substrate to form crystalline NiO as a new surface. The NiO surface assumes a lateral periodicity identical to the bulk structure, and as a result the surface energy is no longer governed by the structural integrity of pure nickel but rather is governed by that for NiO. That NiO possesses the characteristics of surface structure described generally by steps, kinks, etc., is evident from Figs. 4.10 and 4.11. In addition to changes in surface stoichiometry, the surface structure and orientation are likely to be altered by oxidation or surface reaction. In the specific case of the oxidation of nickel, the NiO (001) plane forms parallel to the (110) nickel surface.[35-37] Numerous examples of metal surface structure research can be found in any current issue of the journal *Surface Science*.

Liquid Surface Structure Associated with Adsorption and Reaction. Liquid surface structure, like solid surface structure, will be altered with adsorption of an atomic species and subsequent reactions which include oxidation. Since the liquid surface is not periodic, the distribution of an adsorbate will also not be periodic. If a reacted surface remains in the liquid state, then the liquid–vapor interface will remain nonperiodic, and it can be visualized generally as depicted previously in Fig. 4.3 where mixing occurs at the liquid–liquid interface. When a liquid-surface transition occurs in such a way that a solid-state reaction follows, the surface

STRUCTURE OF INTERFACES

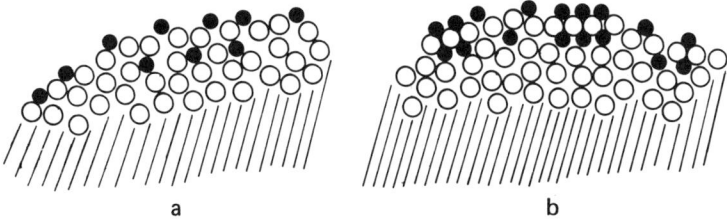

FIGURE 4.18 Residual surface structure following (a) adsorption and surface reaction (for example, oxidation) of a liquid metal or alloy; (b) solid-state reaction at a liquid surface.

structure will be composed of small periodic (crystalline) regions. These features are illustrated schematically in Fig. 4.18.

4.4 CALCULATION OF SURFACE TENSIONS AND SURFACE ENERGIES

Mullins[38] has argued that the work, $E(\infty)$, required to separate two adjacent atomic planes of area A can be estimated as the surface-free energy or the elastic work required to move them through a distance x, with the result

$$F_S \cong \gamma = \frac{E(\infty)}{2A} = \frac{Y}{2} \int_0^{10^{-8}} \left(\frac{x}{a_0}\right) dx = \frac{Y \cdot (10^{-8})^2}{4a_0} \qquad (4.2)$$

where Y is the elastic modulus (Young's modulus) and a_0 represents the atomic radius; the limit of integration has been set at 1 Å (10^{-8} cm). Substituting typical values of $Y \cong 10^{12}$ dynes/cm^2 and $a_0 = 2 \times 10^{-8}$ cm, representative of many solid metals and alloys, into Eq. (4.2) yields values of $\gamma \cong 10^3$ ergs/cm^2. These values agree in order of magnitude with the experimental results discussed previously in Chapter 3.

There have been numerous attempts to calculate the surface energy of metals on both an empirical[39-48] and a theoretical[49-55] basis, but the discrepancies between calculated and experimentally determined surface energies have been so great in many instances as to make the calculational schemes nearly impractical as a general formulation for all metals. Theories of metal surfaces have been lacking primarily because of the difficulties in dealing with the decrease of electron density near the surface and the loss of translational symmetry at the surface. It is apparent that, on considering the structure (or lack of it) for liquid metals, the surfaces of different metals are fundamentally distinguishable only by differences in electron density. As a consequence, the more successful theories of surface energy have followed a treatment of the electron theory of metals, specifically a free-electron or jellium model[53] with supplemental adjustments to the model by first-order pseudopotential calculations.[54,56]

4.4.1 Empirical Calculations of Metal Surface Energies

The more successful empirical and semiempirical theories developed to calculate the surface energy or surface tension of metals rely on the relationship of physical properties of the metal such as the atomic arrangement, the electron work function, and the heat of sublimation or heat of fusion.[40,43,48] In applying the original theoretical formalisms of Skapski,[40] it was shown in Chapter 3 that the solid surface free energies at a temperature, T, below the melting point, T_m, could be calculated from

$$F_S \cong 1.2(\gamma_{LV})_m + 0.45(T_m - T) \tag{3.51}$$

where $(\gamma_{LV})_m$ is the liquid-metal surface energy at the melting point. It is obvious, however, that either $(\gamma_{LV})_m$ must be known experimentally, or it must be amenable to calculation with reasonable certainty in order for Eq. (3.51) to be valid.

Avraamov and Gvozdev[43] have calculated values of the surface energy and the temperature coefficient of surface energy for bcc metals from the general formulas

$$F_S = \frac{3}{2}RT\left\{\frac{k_s\Delta[\Delta H_f(T) - \ln 2]}{f_{(hkl)}N^{1/3}(Ak_v/\rho)^{2/3}k}\right\}$$

and

$$(dF_S/dT) = -3RT \ln 2(n\Delta/k) \tag{4.3}$$

where R is the gas constant; T is the temperature; ρ is the density; A is the atomic weight; $\Delta H_f(T)$ is the heat of fusion at temperature; k_s is the number of atoms for a two-dimensional unit cell of area S ($n = k_s/S$); k is the coordination number of atoms in the metal; k_v is the number of atoms belonging to a three-dimensional unit cell; N is Avogadro's number (for a cubic metal); $a^3 = k_v A/\rho N$; $S = a^2$ (where a is the lattice parameter), and $f_{(hkl)}$ is the packing factor which describes the density of atoms on a crystal plane (hkl). Equation (4.3) is derived on the basis that the surface energy of a metal is that necessary for an atom to withdraw from the interior to the surface, and is predicated on the assumption that the heat of fusion at the absolute zero of temperature, $\Delta H_f(0)$, characterizes the strength of the bond between metal atoms in the form

$$F_S(0) = \frac{1}{2}n\Delta[\Delta H_f(0)/k] \tag{4.4}$$

where Δ is the number of broken bonds at an atom on the surface.

Table 4.1 shows for comparison the calculated and experimental values of surface energy and temperature coefficient of surface energy for a number of bcc metals derived from Eqs. (4.3). Calculated and experimental values of the surface energy have been adjusted to the temperature shown by using the corresponding calculated or experimental temperature coefficients. The experimental values have been taken from Table 3.5 for solid metals. The calculated values are from reference 43. Calculated values of dF_S/dT given in Table 4.1 are the averages for values determined for (100) and (110) surface orientations. It is noted from Table 4.1 that it is difficult to draw any meaningful conclusions regarding the applicability of Eqs. (4.3) to solid metals. However, the temperature coefficients appear to be too small by at least a factor 2, even on considering experimentally determined liquid metal values as previously presented in Table 3.4.

TABLE 4.1
Calculated and Experimental Surface Energies
and Temperature Coefficients for Solid BCC Metals

Metal	Temperature (°C)	F_S (calculated) (ergs/cm²)	F_S (experimental) (ergs/cm²)	dF_S/dT (calculated) (ergs/cm² °C)	dF_S/dT (experimental) (ergs/cm² °C)
Chromium	1400	1440	2300	−0.14	−
Iron (α, δ)	1450	1715	1950	−0.14 (α)	−0.9 (δ)
Molybdenum	2350	2120	2400	−0.12	−0.2
Niobium	2250	1230	2100	−0.10	−
Tungsten	2000	2640	2800	−0.12	−

The relationship between the heat of fusion and the electron work function has been discussed by Frenkel[49] and modified by Belogurov[57] in the form

$$E_w = \frac{2h}{\pi a} \sqrt{\frac{\Delta H_f}{Nm}} \tag{4.5}$$

where E_w is the work function, h is Planck's constant, a is the lattice parameter (for cubic unit cells), N is Avogadro's number, and m is the electron mass. Conversely, the connection between the surface energy and the work function has been described in the form

$$F_{S(hkl)} = CE_w f_{(hkl)} \tag{4.6}$$

where C is a proportionality constant, and $f_{(hkl)}$ is the atom density of a particular surface orientation (hkl).

Considering the free electron theory applied to a metal surface (or the electron in a box)[45] and the effective number of free electrons per atom of metal,[44] Missol[48] has recently determined a formula for the liquid metal surface energy in the form

$$\gamma_{LV} = 1530 E_w \left(\frac{\rho_L}{M}\right)^{\frac{2}{3}} Z^{\frac{2}{3}} \qquad (4.7)$$

where ρ_L is the liquid density, M is the atomic mass, and Z is the effective number of free electrons (valence electrons) per atom. The units of Eq. (4.7) are in ergs per square centimeter (millijoules per square meter) when E_w is expressed in electron volts, ρ_L in grams per cubic centimeter, and M in grams. For solid metals, Missol[48] derives an expression of the form

$$F_S = \frac{1530}{q} E_w \left(\frac{\rho_S}{M}\right)^{\frac{2}{3}} Z^{\frac{2}{3}} \qquad (4.8)$$

where ρ_S is the solid metal density, and q is a structural coefficient which takes account of the atom arrangement. Values of q for bcc and fcc metals are calculated from the ratios of specific surface densities of solid and liquid metal atoms to be 0.93 and 0.80, respectively;[48] they vary from 1 to 2 for other crystal structures such as hcp and tetragonal. On comparing Eqs. (4.7) and (4.8), it is seen that

$$F_S = \frac{\gamma_{LV}}{q} \left(\frac{\rho_S}{\rho_L}\right)^{\frac{2}{3}} \qquad (4.9)$$

where for a liquid metal $q = 1$.

Equations (4.7) and (4.8) suffer the disadvantage, when one is calculating the surface energy, of lack of temperature reference, and the values determined will correspond to the temperature at which E_w and ρ apply. Table 4.2 compares calculated[48] and experimentally determined values of the surface energy of solid fcc and bcc metals at the melting point, T_m. The experimentally determined values are taken from Table 3.4 for liquid metals at T_m and should be considered low when compared with the calculated solid surface energies because of the required heat of fusion separating the liquid from the solid state, amounting to roughly 10% of the value of γ_{LV} at the melting point. That is, corresponding experimental values of F_S at T_m should be approximately $1.1\gamma_{LV}$. When this correction is made in Table 4.2, the calculated values are observed to be in good agreement.

4.4.2 Theoretical Calculation of Metal Surface Energies

As we noted in Section 4.4.1, empirical attempts to calculate the surface energy of metals have relied on the relationship of surface energy to physical properties such as surface work functions, density, the heat of fusion, and in the case of solid metals the surface crystallography or a related symmetry parameter.

Generally speaking, the surface of a metal must be treated in terms of electron-electron interactions as well as ion-electron interactions. This means that a successful metal surface theory must involve a calculation of the electron density at the surface as well as appropriate pseudopotential calculations.[56] Bardeen,[50] following

TABLE 4.2
Comparison of Calculated[48] and Experimentally Determined Values
of Surface Energy of FCC and BCC Metals at the Melting Point

Metal	T_m (°C)	F_S (calculated) (ergs/cm²)	γ_{LV} (experimental) (ergs/cm²)
Aluminum	600	1070	866
Chromium	1950	2500	1590
Copper	1083	1540	1300
Gold	1063	1330	1140
Iron	1535	2000	1880
Lead	327	538	450
Molybdenum	2620	2210	2250
Nickel	1455	1930	1780
Niobium	2473	1940	1900
Palladium	1549	1490	1500
Platinum	1773	2010	1800
Silver	1000	1180	895
Tantalum	3020	2510	2150
Tungsten	3410	2670	2400

the original suggestions of Frenkel,[49] was the first to perform an approximately self-consistent calculation for sodium, while Smith[53] utilized more recent general formulations of electron theory applied to systems of inhomogeneous electron density[51,52] to perform calculations of the surface energies of sodium, potassium, and lithium by replacing the metal surface ions by a model which considers the surface to be a uniform semi-infinite positive charge density. Although good qualitative agreement is obtained for the surface work functions with this formulation,[53,58] the calculated surface energies are too small by a factor of 2 or more, when compared with corresponding experimental values.

Lang and Kohn[54] have used the so-called jellium or free-electron model of a metal surface to obtain self-consistent electron density distributions and to calculate metal surface energies by replacing the uniform positive background representing surface ions by a pseudopotential model of the ions (an ion lattice model). The total surface energy in the ion lattice model is given by[54]

$$(F_S, \gamma_{LV}) = \gamma = \gamma_u + \delta\gamma_{cl} + \delta\gamma_{ps} \qquad (4.10)$$

where γ_u is the surface energy in the uniform background model[51,52] and $\delta\gamma_{cl}$ is the classical cleavage energy:

$$\delta\gamma_{cl} = \alpha Z\bar{n} = E(\text{lat, lat}) + \frac{1}{2}E(-, \text{lat}) - \frac{1}{2}E_\infty(\text{lat, lat})$$
$$+ \frac{1}{2}E_\infty(-, \text{lat}) \qquad (4.11)$$

where α is a dimensionless constant, Z is the ionic charge,[59] $\bar{n}$ is the interior (lattice) value of the electron density distribution, and $E(\text{lat}, \text{lat})$ and $E(-, \text{lat})$ refer to the interaction energy of the lattice (lat) and the uniform negative background. The subscript ∞ indicates that the charge distributions of the negative background and the lattice fill all space, while the unsubscripted terms denote half-space occupancy, $x < 0$, where x denotes the direction normal to the unit area of surface. The term $\delta\gamma_{ps}$ in Eq. (4.10) represents the change in the electron density from its step-function form in the uniform background model to its actual form $n(x)$:

$$\delta\gamma_{ps} = \int_{-\infty}^{\infty} \delta v(x)[n(x) - n_+(x)] \, dx \qquad (4.12)$$

where $\delta v(x)$ is the average, over the y–z plane, of the sum of the ionic pseudopotentials of the half-lattice ($x < 0$), minus the potential due to the semi-infinite uniform charge background. Using a simple model (Thomas-Fermi approximation), Rouhani and Schattler[55] have shown that the main contribution to the metal surface energy comes from a positive charge near the surface.

Lang and Kohn[54] calculated γ in Eq. (4.10) for both fcc and bcc lattice structures, assuming the coordination numbers in the liquid metal state to lie between the coordination numbers of 8 and 12 for fcc and bcc lattices, respectively.[40] The calculations were made by selecting crystal faces of maximum packing density of atoms [(111) and (110) for fcc and bcc lattices, respectively] and for mean electron densities, $\bar{n}$, appropriate to the particular solid metal. Table 4.3 shows for comparison the calculated solid metal surface energies and the corresponding values at the melting point taken from Table 3.4.

TABLE 4.3
Comparison of Theoretical and Experimental Values of Metal Surface Energies[a]

Metal	$\gamma_{theo}(111)$ (ergs/cm²)	$\gamma_{theo}(110)$ (ergs/cm²)	$\gamma_L V(T_m)_{exp}$ (ergs/cm²)
Aluminum	730	1030	866
Cesium	100	100	60
Lithium	360	380	398
Magnesium	540	640	540
Potassium	140	140	101
Rubidium	110	120	76
Sodium	210	230	191
Zinc	440	580	770

[a]Experimental surface energy values refer to the melting point, T_m.

Table 4.3 illustrates that reasonably good agreement is found for the metal surface energies calculated from Eq. (4.10). It must be recognized that a comparison of experimental values at T_m may not be applicable, since the theoretical values refer to the solid state, and the exact temperature applicable to the theoretical calculations is not specified.[54,60] However, the experimental values seem to be in excellent agreement at T_m.

We observe retrospectively, on comparing Tables 4.1, 4.2, and 4.3, that good agreement can be attained between calculated and experimentally determined values of the surface energy of metals, particularly in the vicinity of the melting point (at high temperatures in the solid state). More important, however, is the fact that both experimental (Tables 4.1, 4.2, and 4.3) and theoretical values of metal surface energies have magnitudes approximating those values derived from Eq. (4.2), and it can only be concluded that 10^2 to 10^3 ergs/cm^2 is indeed the true range for surface free energies of metals and alloys.

It was noted in Chapter 1 (Section 1.2) and Chapter 2 (Fig. 2.15) that because of equilibrium conditions governing surface formation, and consequently atomic surface structure, surface free energy anisotropy exists. This feature is demonstrated in the theoretical calculations of Table 4.3. From a practical point of view, the experimental data of Chapter 3 (Table 3.5) show only the relative surface free energy, or the average value. It is apparent, in view of the crystallographic differences associated with solid metal surfaces, that differences in the corresponding surface free energies would occur. This has been demonstrated experimentally in Fig. 2.15.

4.4.3 Theoretical Calculations of the Solid–Liquid Interfacial Free Energy

In his theory of the surface tension of metals, Skapski[40] attempted a calculation of γ_{SL} based on the alternative nearest-neighbor approach. Combining Eqs. (2.52) and (3.50), he obtained

$$\gamma_{SL} = \frac{Z_0}{Z_i} \frac{\Delta H_f}{A_S} + \left[\left(\frac{\rho_S}{\rho_L}\right)^{\frac{2}{3}} - 1\right] \gamma_{LV} + \frac{T_m}{A_S}\left(\Delta S_L - \Delta S_S\right) \quad (4.13)$$

where $Z_0 = Z_i - Z_a$, and ΔS_L and ΔS_S are the associated entropy changes. Skapski's calculations utilizing Eq. (4.13) compare very favorably with experimental results from undercooling experiments[61,62] as demonstrated in Table 4.4.

4.5 INTERFACIAL STRUCTURE IN SOLID METALS AND ALLOYS

Solid interfaces—that is, solid–solid interfaces in metals and alloys—are, as outlined in Chapters 1 through 3, characterized by grain boundaries, twin boundaries, phase boundaries, stacking faults, and related transformation boundaries. Because solid–solid metal interfaces are associated with crystal transitions or metal crystal structure transformations, their structure would be expected to be systematically characterized either crystallographically or geometrically or both. Early models of grain boundary structure envisioned the interface as an amorphous region characterized as an undercooled liquid, having a thickness of more than 10 atom diameters.[64-68]

TABLE 4.4
Comparison of Theoretical and Experimental Values
of the Solid–Liquid Interfacial Free Energies for Pure Metals[a]

Metal	γ_{SL} (theoretical)[b] (ergs/cm²)	γ_{SL} (experimental)[c] (ergs/cm²)
Sodium	15	20[d]
Lithium	27	30[d]
Lead	32	33
Silver	101	126
Gold	121	132
Copper	128	177
Platinum	236	240

[a] Values apply generally at T_m.
[b] From Skapski.[40]
[c] From Turnbull.[61,62]
[d] From Taylor.[63]

In the general case of a grain boundary, or crystal boundary, the grain boundary is an interface where two crystals of different orientation contact. Because the crystals (or grains) are periodic arrays of atoms, the atoms at the interface would also be expected to form a periodic array differing from the neighboring crystals over an accommodation distance of not more than a few atom diameters (the interfacial thickness, Δt, in Fig. 1.5). This feature is in fact observed experimentally [Fig. 1.7(a)].

4.5.1 Grain Boundary Structure

Intense efforts have been made by a large number of investigators in a variety of disciplines to elucidate the structure of grain boundaries. The development of models for the periodic boundary structure has included dislocation models, island models, and coincidence lattices, and grain boundaries have been described autonomously by these models or combinations and modifications of them as outlined in numerous reviews over the past several decades.[69-77] An attempt will be made in this section to review some of the more successful (correct) descriptions of grain boundaries and associated models and transition lattice theories.

Dislocation Models. Since, as was stated above, a grain boundary represents a deviation from an atomic periodicity associated with specific contacting crystals, the interfacial phase can be depicted as a transition lattice[78] which in its simplest form can be envisioned as an array of dislocations. Such dislocation models were first proposed for simple low-angle grain boundaries by Bragg[79] and Burgers,[80] who visualized symmetrical arrays of perfect edge dislocations parallel to the tilt axis, which defines the misorientation, Θ, as described in the geometrical considerations outlined previously in Section 2.2.2 for a 1-degree-of-freedom system. The dislocation models, depicted schematically in Fig. 4.19, are based on the original

STRUCTURE OF INTERFACES   189

considerations of Read and Shockley.[81] If the lattices are rotated about an axis normal to the boundary plane a twist boundary results. Generally for a polycrystalline metal or alloy both tilt and twist components are accommodated by more complex dislocation arrays.

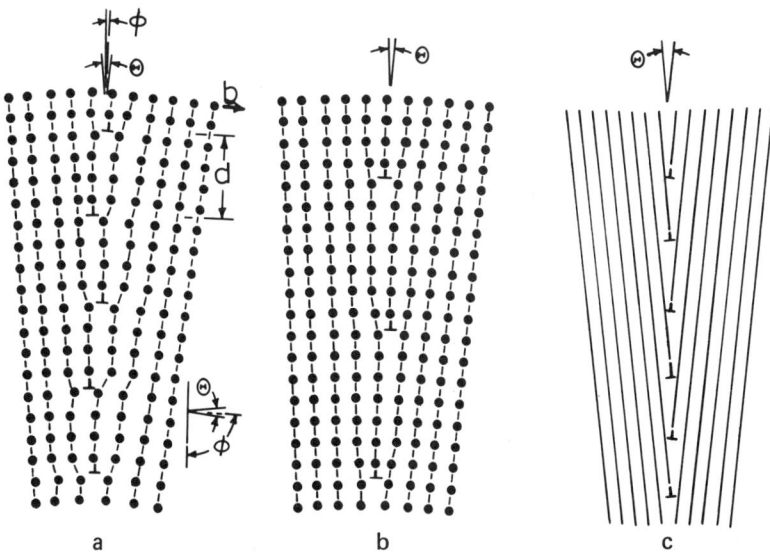

FIGURE 4.19   Dislocation models of grain boundaries. (a) Asymmetrical (two-degree-of-freedom) tile boundary (general case). (b) Symmetrical ($\phi = 0$) tilt boundary. (c) Symmetrical tile boundary having one dislocation per lattice plane ($\Theta = 20°$).

Direct evidence of the dislocation structure of small-angle boundaries ($\Theta \leqslant 2°$) was obtained by etch-pit techniques[82,83] to reveal the positions of dislocations in the boundary plane as depicted in Fig. 4.19, and by transmission electron microscopy which revealed the dislocations in associated planar boundary arrays. These features are illustrated in Fig. 4.20. Figure 4.20(a) and (b) illustrate tilt boundary arrays consisting of rows of parallel edge dislocations at appropriate spacings to accommodate to the required misorientation and asymmetry:

$$d \cong |\mathbf{b}|/\Theta \sin \phi \tag{4.14}$$

or

$$d = |\mathbf{b}|/\Theta \cos \phi \tag{4.14a}$$

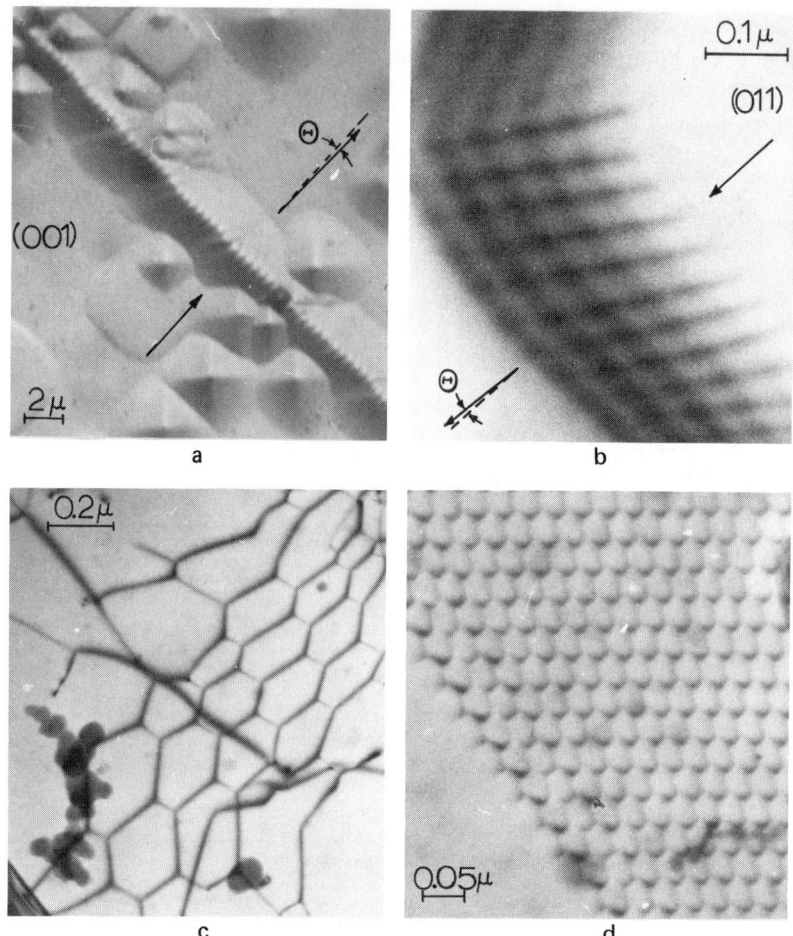

FIGURE 4.20 Dislocation structure of low-angle grain boundaries. (a) Tilt boundary structure revealed by etch pits in LiF crystal. (b) Tilt boundary structure in stainless steel similar to (a) revealed by transmission electron microscopy. (c) Twist boundary structure in beryllium consisting of a crossed-grid of screw dislocations for $\Theta \cong 0.5°$. (d) Dislocation twist boundary array in beryllium for $\Theta < 1°$.

for small Θ values,* where d is the dislocation spacing in the boundary [Figs. 4.19 and 4.20(a) and (b)], **b** is the Burgers vector of the boundary dislocations, Θ is the misorientation angle (Figs. 2.8 and 4.19), and ϕ is the asymmetry angle (Figs. 2.8 and 4.19), or the deviation from the symmetrical case. It can be observed in Fig.

*See also W. T. Read, "Dislocations in Crystals," McGraw-Hill Book Co., New York, p. 174 (1953).

STRUCTURE OF INTERFACES   191

4.20(b) that, as the trace of the plane of the boundary varies (ϕ changes), the dislocation spacing varies. Read and Shockley[81] found that the energy of a low-angle dislocation grain boundary depends on the misorientation angle, Θ, in the form

$$\gamma_{gb} = E_0 \Theta (A_0 - \ln \Theta) \qquad (4.15)$$

where E_0 depends on the elastic constants of the material, and A_0 is related to the core energy of a dislocation. The total energy of a twist boundary [Fig. 4.20(c) and (d)] in an isotropic material is

$$\gamma'_{gb}(\text{twist}) = (G|\mathbf{b}|/2\pi) \Theta (A_0 - \ln \Theta) \qquad (4.16)$$

The energy of a corresponding tilt boundary is

$$\gamma'_{gb}(\text{tilt}) = [G|\mathbf{b}|/4\pi(1-\nu)] \Theta (A_0 - \ln \Theta) \qquad (4.17)$$

where G is the shear modulus, and ν is Poisson's ratio. For a symmetrical low-angle tilt boundary in a 1-degree-of-freedom system, Eq. (4.17) can be expressed approximately by

$$\gamma'_{gb}(\text{tilt}) = \frac{G|\mathbf{b}|\Theta}{4\pi(1-\nu)} \left[\frac{4\pi(1-\nu)\gamma_c}{G|\mathbf{b}|^2} - \ln \frac{\epsilon\Theta}{b} \right] \qquad (4.18)$$

where γ_c is the dislocation core energy, and ϵ is the core cut-off radius for Θ expressed in radians. Murr et al.[84] have measured average misorientations, Θ, of 1.7° and 1.9° in 304 stainless steel and nickel, respectively, by utilizing direct observations in the transmission electron microscope [Fig. 4.20(b)]. By measuring an average value for $\gamma'_{gb}(\text{tilt})$ from intersection equilibrium situations described in Chapter 2 for low-angle boundaries intersecting with high-angle grain boundaries [$\gamma'_{gb}(\text{tilt})/\gamma_{gb} = 0.455$ and 0.432, or $\gamma'_{gb}(\text{tilt}) = 383$ and 342 ergs/cm^2, for nickel and 304 stainless steel at 1060°C, respectively]:

$$\bar{\gamma}_c \cong \frac{|\mathbf{b}|}{\bar{\Theta}} \bar{\gamma}'_{gb}(\text{tilt}) + \frac{G|\mathbf{b}|\bar{\Theta}\ln(\epsilon\bar{\Theta}/|\mathbf{b}|)}{4\pi(1-\nu)}$$

where $\bar{\gamma}_c$, $\bar{\gamma}'_{gb}(\text{tilt})$, and $\bar{\Theta}$ represent average values, and ϵ is taken as $|\mathbf{b}|$. The average dislocation core energy in nickel and 304 stainless steel was found to be 0.55×10^{-4} erg/cm and 0.59×10^{-4} erg/cm, respectively, representing roughly 6% of the average edge dislocation self-energies in agreement with theoretical predictions.[85,86]

Relative grain boundary energies can be determined by the experimental fabrication of equilibrium intersection systems resulting from the intergrowth of oriented tricrystals as outlined in Chapter 2. In addition, the misorientation angle, Θ, for a recognizable low-angle boundary can be determined by direct analysis in the transmission electron microscope [Fig. 4.20(b)]. The associated energy ratios can be calculated for such a boundary intersecting a high-angle grain boundary, or the average values of γ'_{gb} can be obtained (as illustrated in Fig. 4.21) from

$$\gamma'_{gb} \cong 2\gamma_{gb} \cos (\Omega/2) \tag{4.20}$$

where γ_{gb} is the average high-angle grain boundary free energy determined as described in Section 3.4.1, and Ω is the true dihedral angle determined as described in Section 2.2.1. Experimental results of this kind can then be compared with

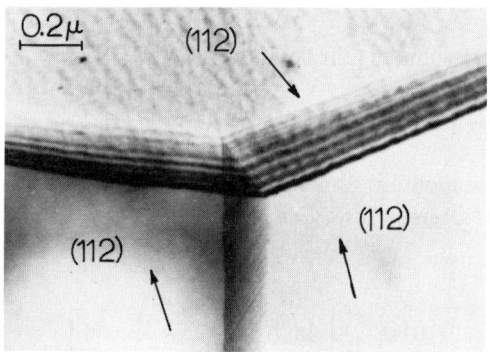

FIGURE 4.21   Intersection of a low-angle grain boundary with a high-angle grain boundary in 304 stainless steel. The [1$\bar{1}$0] directions common to all grains are indicated by the arrows. Note structural differences of two boundaries.

theoretical predictions [Eq. (4.15)]. Figure 4.22 illustrates for comparison several experimental plots of average grain boundary energy or relative energy ratios as a function of misorientation, Θ, which demonstrate the general validity of the dislocation model as implicit in Eq. (4.15).

It is apparent that the dislocation model, while originally developed for low-angle boundaries, may be extended on a purely geometrical basis to high-angle boundaries [Eq. (4.17), Figs. 4.19 and 4.22]; however, a uniform dislocation spacing in a high-angle boundary results only when the dislocation spacing is an integral number of the lattice planes terminating at the boundary as illustrated in Fig. 4.19(c). If the misorientation changes in the case of a symmetrical tilt boundary, the dislocation spacing ideally varies from uniform to nonuniform as shown schematically in Fig. 4.23. Gleiter[76] has shown that geometrically the boundaries depicted in Fig. 4.23 may be described as a uniform dislocation array and a superimposed nonuniform array; that is, Fig. 4.23(b) can be described as a 53° tilt boundary and a superimposed 7° tilt boundary. The variation in the energy

STRUCTURE OF INTERFACES   193

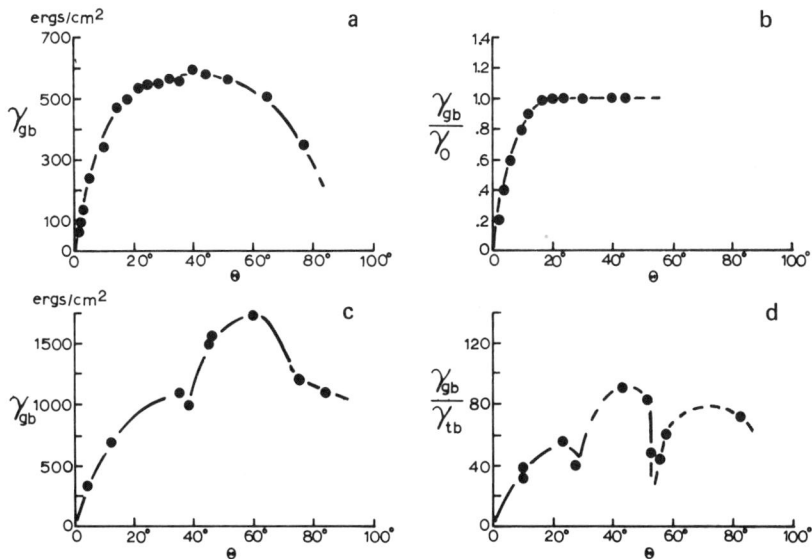

FIGURE 4.22   Grain boundary energy or relative energy ratios as a function of misorientation. (a) Copper <100> tilt boundaries.[87] (b) Germanium <100> tilt boundaries.[88] (c) Nickel misorientations of (110) grains (L. E. Murr, unpublished data). (d) Cu–16% Al, misorientations of (110) grains.[89]

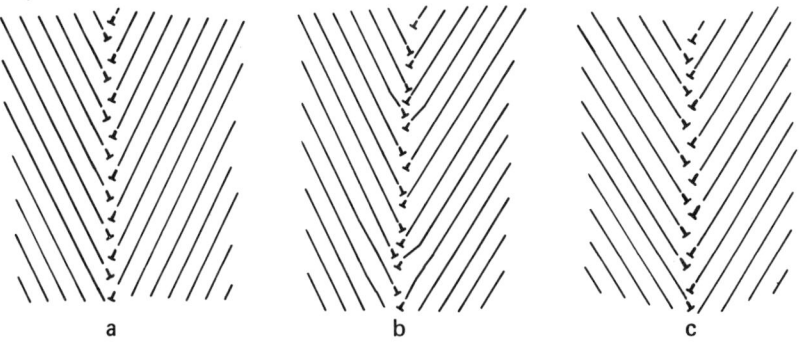

FIGURE 4.23   Dislocation models of symmetrical tilt boundaries showing variation of dislocation spacing uniformity with misorientation angle. (a) 53° misorientation, (b) 60° misorientation, (c) 62° misorientation.

of such symmetrical tilt boundary arrays, assuming the energy of a boundary to consist of two portions—that due to the uniform geometrical array, plus the energy of a small-angle boundary necessary to make up the angular difference between the uniform and the nonuniform dislocation distribution [Fig. 4.23(b) and (c) compared with (a)] —is then calculated to vary with misorientation as illustrated in Fig. 4.24.[90] The energy cusps in Fig. 4.24 correspond to the symmetrical, uniform dislocation boundary arrays—for example, Figs. 4.19(c) and 4.23(a) and (c). The experimental data of Fig. 4.22(c) and (d) in fact exhibit such energy cusps.

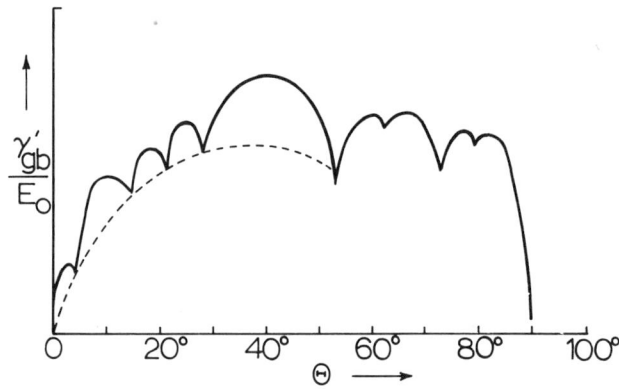

FIGURE 4.24 Relative grain boundary energy versus misorientation angle for symmetrical tilt boundaries which vary in uniformity of dislocation arrays. Cusps correspond to uniform dislocation arrays. (After Gleiter.[76])

Li[91] examined the variation of dislocation core parameters and spacing as an extension of the original Read-Shockley model [Eq. (4.15)] over a range of misorientation angles, Θ, assuming a core radius characterized by a disappearance of stress at the surface of the core, and linear elasticity theory to apply over a large misorientation range. In such a model, a symmetrical tilt boundary consists of a single dislocation array, while for increasing misorientation angles the dislocation cores increase in size and become elliptical, finally touching and combining to form a continuous boundary-phase region as depicted schematically in Fig. 4.25 for $\Theta \geqslant 40°$.

Generally, dislocation structure, even complex dislocation structure, is insufficient to explain the appearance of random, high-angle grain boundaries in a metal or alloy ($\Theta \geqslant 5°$). In some cases, particularly in the range of misorientations between 2° and 8°, grain boundaries possess a complex, transitional structure consisting of dislocations, ledges having coincident boundary portions, sloping double ledges, serrations, protrusions, and microfacets—some of which tend to decrease in frequency with increasing misorientation.[84] Many of these features are illustrated in Fig. 4.26, which shows a grain boundary section in nickel. Note that

STRUCTURE OF INTERFACES   195

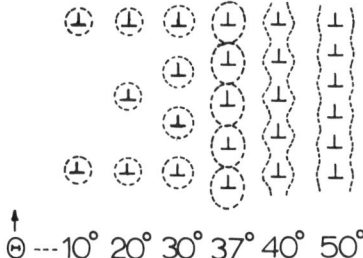

FIGURE 4.25   Calculated shape of dislocation cores in symmetrical tilt boundaries of different misorientation angles indicated. (After Li.[91])

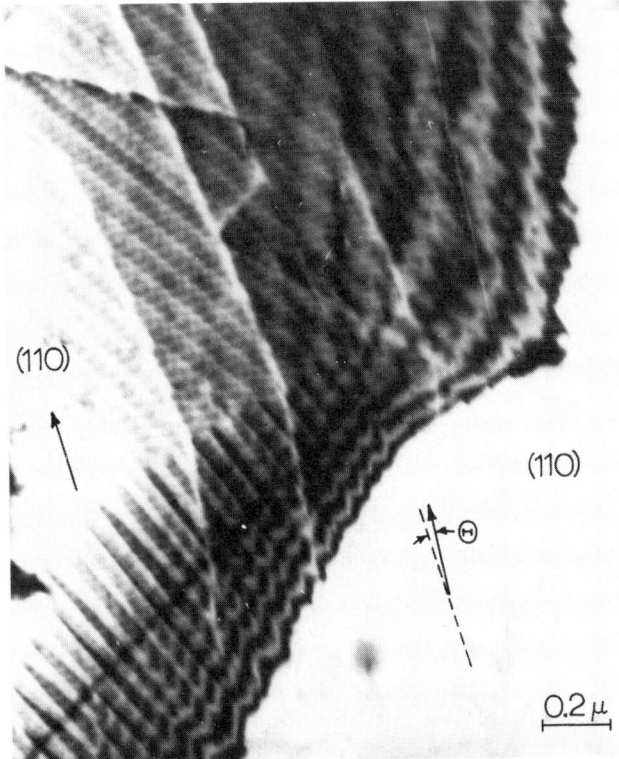

FIGURE 4.26   Complex grain boundary structure in nickel. Arrows denote the corresponding [1$\bar{1}$0] directions in each grain, from which the misorientation, Θ, is observed to be 6°. (After Murr et al.[84])

the boundary possesses only 3 degrees of freedom, but that asymmetry, ϕ, and boundary tilt (or inclination with respect to the section surfaces), θ, vary along the interface. Obviously such a boundary, and other more general boundaries, require modifications to, or additions to, the dislocation model such as the dualistic treatment of Bollmann,[92,93] and the concept of the coincidence site lattices developed originally by Kronberg and Wilson[94] and Frank,[95] and elaborated upon by Brandon et al.,[96] Brandon,[97] and Ranganathan.[98]

Coincidence Site Models. As demonstrated in Fig. 4.22(c) and (d) and Fig. 4.24, certain crystallographic-geometric situations exist that involve lattice "coincidence" which, in the case of symmetrical boundaries, produce uniform structural arrays—specifically dislocations. There are, additionally, some misorientations which permit certain atoms to belong to the lattices of both grains—coincident lattice sites—forming an interface of a kind of superlattice common to the two grains which were originally referred to as Kronberg-Wilson[94] boundaries.

Two crystal lattices related by a rotation about an axis have certain common lattice sites which can be located on a single lattice of larger cell dimensions. The larger lattice is referred to as a coincidence-site lattice (CSL), and its characteristics will generally depend on the axis, $\langle hkl \rangle$, about which the rotation is made, and the angle of rotation (the misorientation angle, Θ). However, the only atoms that are common to the two lattices are those of the boundary itself.

Ranganathan[98] has shown that CSL's may be generated by using the function

$$\Sigma = x^2 + Ny^2 \qquad (4.21)$$

where Σ is the ratio of the area of the coincidence lattice cell to that of the original lattice, commonly referred to as the multiplicity, or the reciprocal of the density of common lattice sites; x and y are nonnegative integers representing the Cartesian coordinates of the lattice point joined to an origin; and $N = h^2 + k^2 + l^2$. A CSL is always generated for a rotation of 180° about a rational direction $\langle hkl \rangle$, and when the value of Σ determined from Eq. (4.21) is even, it must be repeatedly divided by 2 until an odd number results. The angle of misorientation is

$$\Theta = 2 \tan^{-1}(y/x)\sqrt{N} \qquad (4.22)$$

and $\Theta = 180°$ corresponds to $x = 0, y = 1$. Thus, the rotation of 180° around $\langle hkl \rangle$ in the cubic system gives rise to a coincidence site lattice of $\Sigma = h^2 + k^2 + l^2$ if N is odd, or $N/2$ if N is even.

Since the same values of Σ may be generated by Eq. (4.21) from different sets of x, y, and N, different CSL's giving the same value are distinguished by labeling them a, b, etc. The corresponding axis-angle pairs for coincidence boundaries in the cubic lattice system for Σ values from 1 to 19 calculated from Eq. (4.21) are tabulated in Table 4.5.[99] Values of Θ in Table 4.5 correspond to "ideal" coincidence boundaries.

TABLE 4.5
Axis-Angle Pairs for Coincidence Boundaries in the Cubic Lattice System[a]

Θ°	Σ	Θ°	Σ	Θ°	Σ	Θ°	Σ
<100> Axis		<221> Axis		<410> Axis		<522> Axis	
22.62	13a	61.93	17b	107.92	13b	160.25	17b
28.07	17a	90.00	9	152.73	9		
36.87	5	112.62	13b	180.00	17a	<530> Axis	
53.13	5	143.13	5			142.14	19b
61.93	17a	180.00	9	<411> Axis		180.00	17a
67.38	13a			93.37	17a		
112.62	13a	<310> Axis		129.52	11	<531> Axis	
118.07	17a	76.66	13b	153.47	19b	99.59	15
126.87	5	93.02	19a	180.00	9	126.22	11
143.13	5	115.38	7			160.81	9
151.93	17a	144.90	11	<421> Axis			
157.38	13a	180.00	5	113.58	15	<532> Axis	
				155.38	11	180.00	19b
<110> Axis		<311> Axis					
26.53	19a	50.70	15	<430> Axis		<533> Axis	
38.94	9	67.11	9	118.07	17b	105.35	17a
50.48	11	95.74	5	157.38	13b	130.83	13b
70.53	3	117.82	15			162.66	11
86.63	17b	146.44	3	<431> Axis			
93.37	17b	180.00	11	137.17	15	<551> Axis	
109.47	3			180.00	13b	110.01	19b
129.52	11	<320> Axis				134.43	15
141.06	9	71.59	19b	<432> Axis		164.06	13a
153.47	19a	100.48	11	121.76	19a		
		121.97	17b	158.96	15	<553> Axis	
<111> Axis		149.00	7			137.33	17a
27.80	13b	180.00	13a	<433> Axis		165.16	15
38.21	7			142.14	19a		
46.83	19b	<321> Axis		180.00	17b	<610> Axis	
60.00	3	86.18	15			161.33	19a
73.17	19b	123.75	9	<441> Axis		161.33	19a
81.79	7	150.07	15	160.25	17a		
92.20	13b	180.00	7			<611> Axis	
147.80	13b			<510> Axis		180.00	19a
158.21	7	<322> Axis		137.17	15		
166.83	19b	107.92	13a	180.00	13a	<711> Axis	
180.00	3	152.73	9			110.01	19a
		180.00	17b	<511> Axis		134.43	15
<210> Axis				73.17	19a	164.06	13b
48.19	15	<331> Axis		92.20	13a		
73.40	7	63.82	17b	120.00	9	<731> Axis	
96.38	9	82.16	11	158.21	7	137.33	17b
131.81	3	110.92	7			165.16	15
180.00	5	154.16	5	<520> Axis			
		180.00	19a	121.76	19b	<733> Axis	
<211> Axis				158.96	15	139.74	19b
62.96	11	<332> Axis					
78.46	15	99.08	19a	<521> Axis		<751> Axis	
101.54	5	133.81	13a	139.88	17b	166.83	19a
135.58	7	180.00	11	180.00	15		
180.00	3						

[a] From Pumphrey and Bowkett.[99]

198   INTERFACIAL PHENOMENA IN METALS AND ALLOYS

It should be pointed out that, whereas the density of coincidence sites depends only on the orientation relationship, the density of coincidence sites at a boundary depends on the trace direction of the boundary plane intersecting with the coincidence lattice. This feature is apparent in Fig. 4.27, especially Fig. 4.27(b) which represents a bcc lattice with a coincidence site lattice corresponding to a 50.5° rotation about a <110> axis. Figure 4.27(c) represents a typical example of a bubble raft model. Bubble rafts can be employed very effectively in producing almost any of the more common, close-packed coincidence-site boundary structures possible (Table 4.5). An excellent experimental method for producing bubble raft

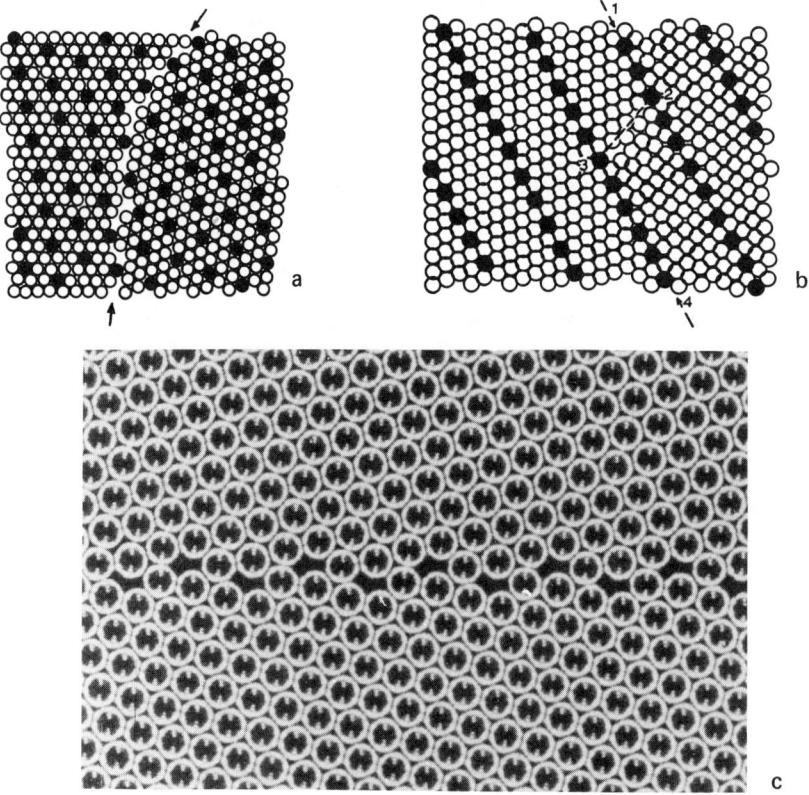

FIGURE 4.27   Two-dimensional coincident-site grain boundary models. (a) Cubic (111) grains separated by a grain boundary having a misorientation of 38° (38° rotation about <111>; $\Sigma = 7$).[94] (b) Cubic (110) grains separated by a grain boundary having a symmetric coincidence portion 1–2, 3–4, $\phi = 0$, and an unsymmetric coincidence portion 2–3, $\phi = 68°$. The misorientation is 50.5° (50.5° rotation about <110>; $\Sigma = 11$).[96] (c) Bubble raft model of coincidence boundary, $\Theta = 27.8°$, $\phi = 0$ (27.8° rotation about <111>; $\Sigma = 13$). (Courtesy of Y. Ishida.[100])

STRUCTURE OF INTERFACES

models has been demonstrated by Ishida.[100] Figure 4.28 illustrates the change in the boundary structure in the same CSL of different atomic dimensions.

Figures 4.27 and 4.28 illustrate that CSL boundary configurations occur as regular repeat structures (long-range order boundary) and that the boundary structure will depend on the coincidence plane (hkl) referred to the crystal structure characterizing the metal or alloy.[84] In addition, it is observed that perfect and imperfect dislocations can occur in long-range order boundaries and manifest themselves as steps in the boundary plane. Obviously, as we indicated in referring to Figs. 4.22 and 4.24, the energy of the more perfect (low Σ) boundaries will be

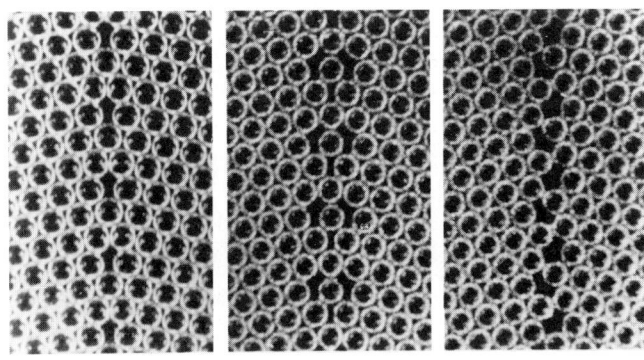

FIGURE 4.28 Variations in the boundary configurations of bubble raft boundaries in $\Sigma 7$ coincidence orientation relationship which result by varying bubble size. (Courtesy of Y. Ishida.[100])

lower than that of the more imperfect (high Σ) boundaries; this is ideally demonstrated with reference to the two boundary regions shown in Fig. 4.27(b). It is therefore more important to consider the complexities of the actual atomic configurations along the boundary rather than the orientation relationships of the ordered boundary (as implicit in Table 4.5), since the boundary energy is determined by the atomic fit which may be only indirectly related to lattice fit. Traditionally, the concept of atomic matching along a boundary has been treated by the so-called island models,[68,74,101] which are based on the assumption that a boundary consists of islands of good atomic matching separated by regions of poor matching. Figure 4.26 lends a certain credibility to such a model. Brandon et al.,[96] on the basis of their work, suggested that when the macroscopic orientation of the coincidence grain boundary deviates from the ideal planes, the boundary consists of a series of steps made up of regions of good and bad fit. Thus, it is assumed that the boundary will have a minimum energy if it follows closely packed planes in the coincidence site lattice because this constitutes the condition of best fit. Consequently, deviations from the close-packed planes in the coincidence site lattice will be taken up by the boundary taking up a stepped structure. In addition, the boundary has superimposed upon it a dislocation network constituting a subboundary in the coincidence lattice which accommodates deviations from perfect coincidence. Many of these features are apparent in the transmission electron micrographs of Fig. 4.29. Figure 4.29(a) illustrates a coherent [(111) coincidence] annealing twin boundary in fcc Inconel 600 (76% Ni, 16% Cr, 8% Fe) containing noncoherent steps [portions not coincident with (111)]. The noncoherent steps are, nonetheless, associated with coincidence planes; that is, they are coincident with (hkl) lattice planes. The noncoherent twin boundary portions are, as noted in Section 3.4.3 (see Fig. 3.30 and Table 3.8), of higher energy than the coherent twin portions as a consequence of the larger Σ value and the poorer atomic matching at the interface. Figure 4.29(b), on the other hand, illustrates a dislocation network accommodating the deviation from a perfect coincidence plane.

The concept of accommodation by dislocation networks is analogous to that presented previously for deviations from symmetric (regular) dislocation boundary arrays, and as a result the coincidence boundary model[97] is observed to be simply an extension and combination of the dislocation model of Read and Shockley[81] and the lattice coincidence model of Kronberg and Wilson.[94]

Grain Boundary Dislocations. Many experimental observations of grain boundary structure in the transmission electron microscope and the field-ion microscope have demonstrated the existence of grain boundary dislocations (GBD)[102-104] when the

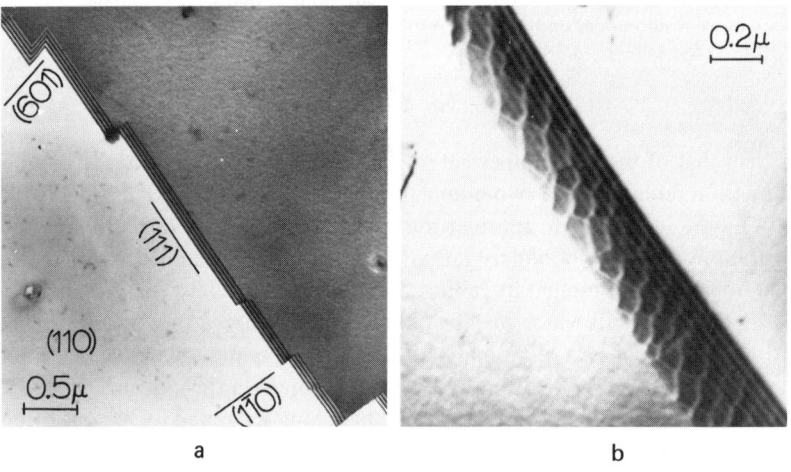

FIGURE 4.29 Coincidence boundary structure observed in the transmission electron microscope. (a) Coherent-noncoherent twin boundary structure in Inconel 600 alloy. Coincidence planes of the boundary are shown. (b) Dislocation network in a boundary plane deviating slightly from a coincidence orientation in stainless steel.

orientation relationships deviate from ideal coincidence orientations.[105-113] The GBD has been observed to be distinct—that is, unique to the grain boundary. It is not a lattice dislocation, and it requires the assumption of a periodic grain boundary structure to define its Burgers vector. The possible Burgers vectors of perfect boundary dislocations are equal to the difference vectors between the two participating crystal lattices:

$$\mathbf{B} = \mathbf{b}_A \pm \mathbf{b}_B \qquad (4.23)$$

where $\mathbf{b}_A$ and $\mathbf{b}_B$ are lattice vectors of the A and B grains composing the boundary. In general a GBD occurs as a step in the boundary whose height varies with the misorientation. That is, the Burgers vector is determined only by the misorientation of grains A and B:

$$|\mathbf{B}| = 2 b \sin(\Theta/2) \qquad (4.24)$$

where b is the magnitude of the lattice vector. GBD's can form by deformation or grain boundary migration, but the Burgers vectors are invariant irrespective of the mode of formation. Recently, Ishida and McLean[114] derived a set of general equations of GBD Burgers vectors for simple cubic, fcc, and bcc crystal structures for coincidence systems with rotation axes [100], [110], and [111]; the general equations of the unit Burgers vectors for boundary dislocations they derived are given in Table 4.6. The unit Burgers vectors of the principal coincidence systems calculated from the equations in Table 4.6 are listed in Table 4.7. With reference to Table 4.7, Eq. (4.23) can be rewritten

$$\mathbf{B} = m\mathbf{b}_1, n\mathbf{b}_2; \quad 1 \equiv A, 2 \equiv B \qquad (4.23a)$$

or

$$\mathbf{B} = m\mathbf{b}_1 + n\mathbf{b}_2 \qquad (4.23b)$$

where m and n are all integers. In addition, when considering the coincidence boundary, the vector component parallel to the rotation axis is given by[114]

$$\mathbf{b}_3 = \mathbf{b}_3^* + \mathbf{d} \qquad (4.25)$$

where $\mathbf{b}_3^*$ is the component normal to the rotation axis, and $\mathbf{d}$ is the spacing of the plane. It is apparent, therefore, that grain boundary dislocations are possible with Burgers vectors that are integral multiples of the unit Burgers vectors listed in Table 4.7. Figure 4.30 illustrates two examples of grain boundary dislocations associated with an ideal coincidence (twin) boundary and an adjoining nonideal boundary section in a thin film of Inconel 600 alloy observed in the transmission electron microscope.

TABLE 4.6
General Equations of Unit Burgers Vectors for GBD's in fcc and bcc Crystal Lattices[a]

Rotation Axis	bcc Structure	fcc Structure
[100]	$\dfrac{a}{k^2 + l^2}[0kl], \dfrac{a}{k^2 + l^2}[0l\bar{k}]$ $\dfrac{a}{2(k^2 + l^2)}[k^2 + l^2, k+l, l-k]$	$\dfrac{a}{2(k^2 + l^2)}[0, k+l, l-k]$ $\dfrac{a}{2(k^2 + l^2)}[0, l-k, k+l]$ $\dfrac{a}{2(k^2 + l^2)}[k^2 + l^2, k, l]$
[110]	h even: $\dfrac{a}{2(2h^2 + l^2)}[h+l, h+l, l-2h]$ $\dfrac{a}{2(2h^2 + l^2)}[l-h, h-l, 2h+l]$ $\dfrac{a}{2(2h^2 + l^2)}[2h^2 + l^2 + h, 2h^2 + l^2 - h, l]$ h odd: $\dfrac{a}{2(2h^2 + l^2)}[hhl], \dfrac{a}{2h^2 + l^2}[ll2h]$ $\dfrac{a}{2(2h^2 + l^2)}[2h^2 + l^2 + l, 2h^2 + l^2 - l, 2h]$	$\dfrac{a}{2(h^2 + l^2)}[ll2h], \dfrac{a}{2h^2 + l^2}[hhl]$ h even: $\dfrac{a}{4(2h^2 + l^2)}[2h^2 + l^2 + 2h - l, 2h^2 + l^2 + l - 2h, 2h - 2l]$ h odd: $\dfrac{a}{4(2h^2 + l^2)}[2h^2 + l^2 + l, 2h^2 + l^2 - l, 2h]$
[111]	$\dfrac{2a}{h^2 + k^2 + l^2}[hkl], \dfrac{2a}{h^2 + k^2 + l^2}[klh]$ $\dfrac{2a}{h^2 + k^2 + l^2}[lhk], \dfrac{a}{6(h^2 + k^2 + l^2)}$ $[h^2 + k^2 + l^2 + 4h - 4k,$ $h^2 + k^2 + l^2 + 4k - 4l,$ $h^2 + k^2 + l^2 + 4l - 4h]$	$\dfrac{a}{h^2 + k^2 + l^2}[hkl], \dfrac{a}{h^2 + k^2 + l^2}[klh],$ $\dfrac{a}{h^2 + k^2 + l^2}[lhk]$ $\dfrac{a}{3(h^2 + k^2 + l^2)}[h^2 + k^2 + l^2 + h - k,$ $h^2 + k^2 + l^2 + k - l, h^2 + k^2 + l^2 + l - h]$

[a]From Ishida and McLean.[114]

TABLE 4.7
Unit Burgers Vectors for Grain Boundary Dislocations
in bcc and fcc Crystal Structures for Principal Coincidence Systems[a]

Σ Values	bcc Crystal Boundary			fcc Crystal Boundary		
	b_1	b_2	b_3	b_1	b_2	b_3
3	$\frac{a}{6}[1\bar{1}1]$	$\frac{a}{3}[1\bar{1}2]$	$\frac{a}{3}[21\bar{1}]$	$\frac{a}{6}[1\bar{1}2]$	$\frac{a}{3}[1\bar{1}1]$	$\frac{a}{6}[21\bar{1}]$
5	$\frac{a}{5}[012]$	$\frac{a}{5}[02\bar{1}]$	$\frac{a}{10}[531]$	$\frac{a}{10}[03\bar{1}]$	$\frac{a}{10}[013]$	$\frac{a}{10}[512]$
7	$\frac{a}{7}[12\bar{3}]$	$\frac{a}{7}[2\bar{3}1]$	$\frac{a}{14}[3\bar{1}5]$	$\frac{a}{14}[12\bar{3}]$	$\frac{a}{14}[2\bar{3}1]$	$\frac{a}{14}[536]$
9	$\frac{a}{18}[1\bar{1}5]$	$\frac{a}{6}[1\bar{1}1]$	$\frac{a}{9}[452]$	$\frac{a}{18}[1\bar{1}4]$	$\frac{a}{9}[2\bar{2}1]$	$\frac{a}{6}[211]$
11	$\frac{a}{22}[1\bar{1}3]$	$\frac{a}{11}[3\bar{3}2]$	$\frac{a}{11}[74\bar{1}]$	$\frac{a}{22}[3\bar{3}2]$	$\frac{a}{11}[1\bar{1}3]$	$\frac{a}{22}[74\bar{1}]$
13a	$\frac{a}{13}[023]$	$\frac{a}{13}[03\bar{2}]$	$\frac{a}{26}[13\ 5\ 1]$	$\frac{a}{26}[051]$	$\frac{a}{26}[0\bar{1}5]$	$\frac{a}{26}[13\ 2\ 3]$
13b	$\frac{a}{13}[13\bar{4}]$	$\frac{a}{13}[3\bar{4}1]$	$\frac{a}{26}[391]$	$\frac{a}{26}[13\bar{4}]$	$\frac{a}{26}[3\bar{4}1]$	$\frac{a}{26}[8\ 11\ 7]$
15	$\frac{a}{5}[102]$	$\frac{a}{10}[3\bar{1}\bar{5}]$	$\frac{a}{15}[052]$	$\frac{a}{5}[10\bar{2}]$	$\frac{a}{10}[2\bar{5}1]$	$\frac{a}{30}[354]$
17a	$\frac{a}{17}[014]$	$\frac{a}{17}[04\bar{1}]$	$\frac{a}{34}[17\ 5\ 3]$	$\frac{a}{34}[053]$	$\frac{a}{34}[0\bar{3}5]$	$\frac{a}{34}[17\ 1\ 4]$
17b	$\frac{a}{34}[5\bar{5}\bar{1}]$	$\frac{a}{34}[\bar{1}17]$	$\frac{a}{17}[10\ 7\ \bar{2}]$	$\frac{a}{34}[3\bar{3}4]$	$\frac{a}{17}[2\bar{2}3]$	$\frac{a}{34}[985]$
19a	$\frac{a}{38}[3\bar{3}1]$	$\frac{a}{19}[1\bar{1}6]$	$\frac{a}{19}[10\ 9\ \bar{3}]$	$\frac{a}{38}[1\bar{1}6]$	$\frac{a}{19}[3\bar{3}1]$	$\frac{a}{38}[10\ 9\ \bar{3}]$
19b	$\frac{a}{19}[23\bar{5}]$	$\frac{a}{19}[3\bar{5}2]$	$\frac{a}{38}[7\ 1\ 11]$	$\frac{a}{38}[23\bar{5}]$	$\frac{a}{38}[3\bar{5}2]$	$\frac{a}{38}[13\ 10\ 15]$

[a] Values computed from equations in Table 4.6; after Ishida and McLean.[114] Burgers vectors for GBD's are discussed also in a series of articles as follows: Y. Ishida and M. McLean, *Phil. Mag.*, **30**, 453 (1974); M. A. Fortes, *ibid.*, 457; D. H. Warrington and H. Grimmer, *ibid.*, 461.

Grain Boundary Ledges. As we noted in the previous discussion of grain boundary dislocations, and as is shown, for example, in Fig. 4.30, GBD's can assume the geometry of a step in the grain boundary. The agglomerations of GBD's can thus lead to the formation of large steps or ledges in the grain boundary. Ledge formation can be envisioned as two distinct agglomeration processes involving GBD's. In the first, GBD's introduced into the grain boundary as a result of thermal migration during recrystallization and grain growth are induced to glide along the boundary

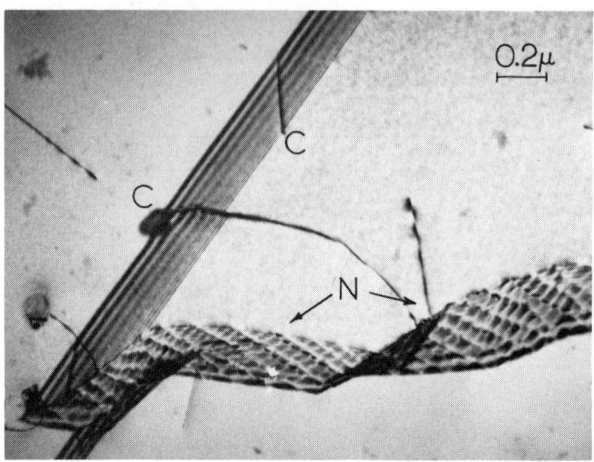

FIGURE 4.30 Grain boundary dislocations in Inconel 600. Dislocations in the coherent twin boundary which produce a small step are observed at *C*. Complex arrangement of GBD's in nonideal boundary portion is observed at *N*.

plane. The gliding GBD's, on encountering a suitably oriented glide plane in either one or the other lattice composing the interface, form a ledge or ledges in the boundary as illustrated in Fig. 4.31(a) and (b). In the second process, glide dislocations in grain A with Burgers vector $\mathbf{b}_A$ pass through the grain boundary into grain B in response to an applied stress. On passing through the grain boundary, the Burgers vector is altered to $\mathbf{b}_B$, and, as a consequence of this transboundary glide, a disturbance is left in the interface which can be visualized as a GBD or an agglomeration of GBD's forming a ledge as shown in Fig. 4.31(c) and (d). Marcinkowski[105] has also discussed the creation of grain boundary ledges by dislocation glide loops. In addition, dislocations gliding through a grain boundary along various glide planes can create individual GBD's which, under certain thermal or mechanical treatment, agglomerate to form grain boundary ledges.

It is observed in Fig. 4.31 that the size of the grain boundary ledges will depend on the number of dislocations causing heterogeneous shear of the boundary plane, or the number of GBD's coalescing or agglomerating within the interface to form a particular ledge, and on the misorientation of the grains composing the interface plane (grains A and B in Fig. 4.31). As a result, the effective Burgers vector of a grain boundary ledge can be represented as

$$\mathbf{B}_L = 2n\mathbf{b} \sin(\Theta/2) \tag{4.24a}$$

where n is an integer representing the number of glide dislocations of Burgers

STRUCTURE OF INTERFACES 205

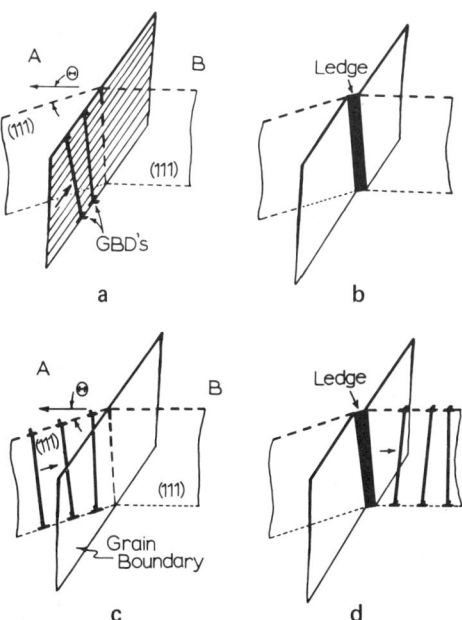

FIGURE 4.31   Creation of grain boundary ledges. (a) Movement of GBD's along boundary plane (arrow). (b) Coalescence and agglomeration of GBD's to form ledge. (c) Movement of lattice dislocations from grain A into the boundary plane. (d) Heterogeneous shear of the grain boundary causing ledge formation.

vector, **b**, or more generally in the context of Eq. (4.24),

$$\mathbf{B}_L = m\mathbf{B} \tag{4.24b}$$

where m is the number of GBD's composing the ledge. Values of $\mathbf{B}_L$ are therefore implicit in Table 4.7 when either n or m is known. The modulus of $\mathbf{B}_L$, $|\mathbf{B}_L|$, therefore represents the ledge height when reference is made to the boundary plane, and this ledge height is generally larger in magnitude than any single lattice dislocation.

Li,[91] in his original treatment of grain boundary ledges, noted that, since the free energy of formation of a ledge is strongly dependent on the grain boundary misorientation (being small for high-angle boundaries), ledges should constitute a prominent structural feature of high-angle boundaries. Indeed, ledge structure is prominent over a range of misorientations in metals and alloys as illustrated typically in Fig. 4.32. Bernstein and Rath[115] have determined that, as the misorientation increases, the ledge density (calculated as the number of ledges per

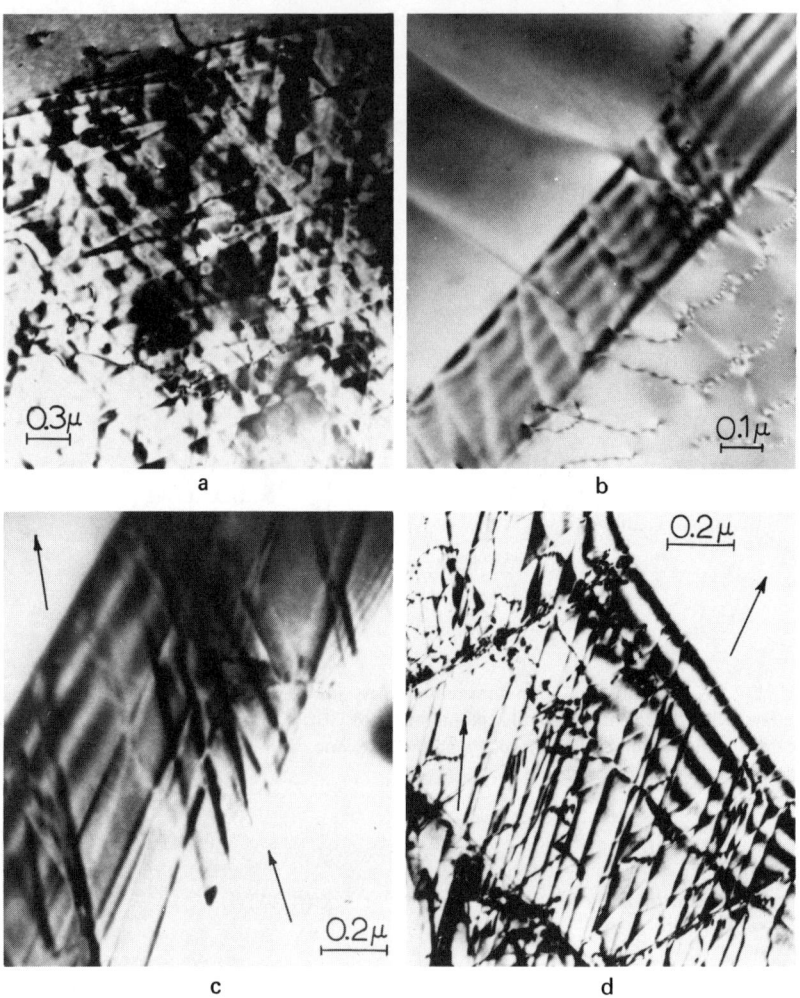

FIGURE 4.32   Grain boundary ledges observed in the transmission electron microscope. (a) Complex ledge structure in aluminum. (b) Ledges in a beryllium grain boundary. (c) Ledge structure in a nickel grain boundary. (d) Ledges in a 304 stainless-steel grain boundary. Arrows denote common directions.

unit length of boundary) increases markedly in a Fe-0.15 Ti steel. More recently, Murr[116] has found that the ledge density in pure nickel and type 304 stainless steel varies in a similar manner as illustrated in Fig. 4.33.

Figure 4.33 shows a consistent increase in ledge density in grain boundaries in the range $0° < \Theta < 50°$. This response was in fact predicted by Li and Chou.[117]

STRUCTURE OF INTERFACES

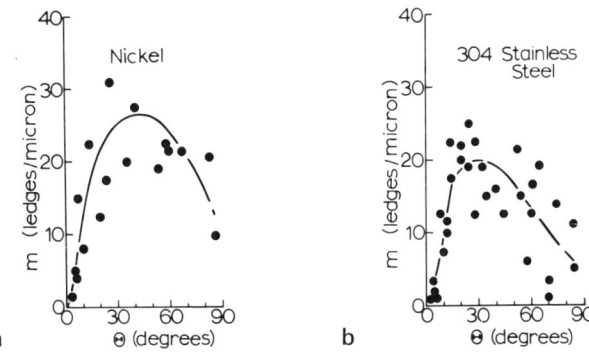

FIGURE 4.33 Variation of grain boundary ledge density with misorientation in (a) nickel and (b) 304 stainless steel as determined for boundaries separating identical (110) grain orientations observed in the transmission electron microscope.[116]

One of the important features of grain boundary ledges is that they are, as originally proposed by Li,[91] most effective sources of dislocations. Numerous examples of dislocation emission from grain boundary ledges have been made;[84,116,118,119] several typical occurrences are reproduced in Fig. 4.34. The implications of grain boundary ledges as sources of dislocations in dealing with mechanical properties will be discussed in detail in Chapter 5.

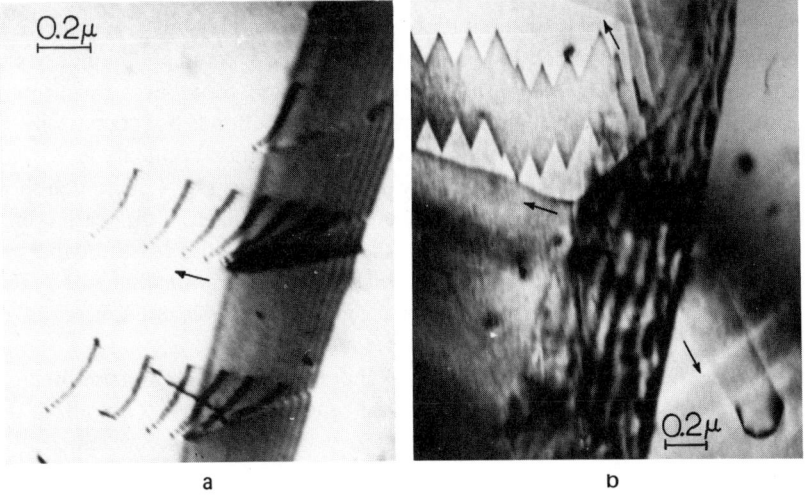

FIGURE 4.34 Emission of dislocations at grain boundary ledges. (a) Ledge emission at grain boundary in 304 stainless steel. (b) Ledge emission at grain boundary in aluminum.

Grain boundary ledges have been observed in a wide variety of metals and alloys in both field-ion microscope and transmission electron microscope studies.[84,106,120-123] Ledge structure is thus a prominent feature of grain boundary structure, as illustrated in Figs. 4.26 and 4.32, and apparent to a lesser degree in other examples of grain boundary structure presented in this section.

4.5.2 Interpretation and Description of Grain Boundary Structures

The models of grain boundary structure discussed in Section 4.5.1 are obviously limited in their application and, like many theories, possess a considerable lack of causality because one single model cannot accurately describe all grain boundaries or even complete interfacial regions in engineering metals and alloys. Many of the model features discussed are based on geometrical considerations to derive the boundary structure and are therefore not necessarily atomic descriptions of boundary structure but of boundary geometry and crystallography. The descriptions are also based, for the most part, on simple symmetry arrays which are 1-degree-of-freedom models (Figs. 4.19, 4.23, 4.27, and 4.28); they are not descriptive of a grain boundary plane, or the structure of an interface, but rather depict the intersection of the boundary plane with a free surface (the solid-vapor interface). This is another of the limitations of grain boundary studies in the field-ion microscope, except where field evaporation is employed to systematically reveal, by atom layer removal, the sequential arrangement of the grain boundary plane. Studies of this type have in fact been reported[124,125,*] in which the boundary plane structure was characterized by facets which, ideally, could be described in terms of ledges and sloping double ledges, which Ryan and Suiter[120] have referred to as protrusions. These features are all apparent in the complete view of the grain boundary plane section shown in Fig. 4.26.

Figure 4.35 illustrates the limitations implicit in the previous discussion of field-ion microscope observations as compared with the more complete boundary plane observations possible in the transmission electron microscope. Even the electron microscope view of a corresponding grain boundary section is not statistically significant, however, since only a small portion of one grain boundary is represented. This feature is emphatically demonstrated in Fig. 4.36, which shows a section of a polycrystalline tin film. There is a distinct lack of dislocation or related boundary structure in Fig. 4.36, and the grain boundaries form triple points where they intersect, characterized by dihedral angles approximating 120°. The boundaries are therefore primarily high-angle grain boundaries but are certainly not uniquely describable by any single structural model.

Figures 4.35 and 4.36 illustrate that a canonical description of grain boundary structure is possible only when the structural concepts of many, if not all, of the proposed models and theories of grain boundary structure are integrated or extended as necessary to obtain a qualitatively useful interpretation of grain

---

*D. A. Smith, Proceedings of the Institute of Physics Anniversary Meeting, private communication (1971).

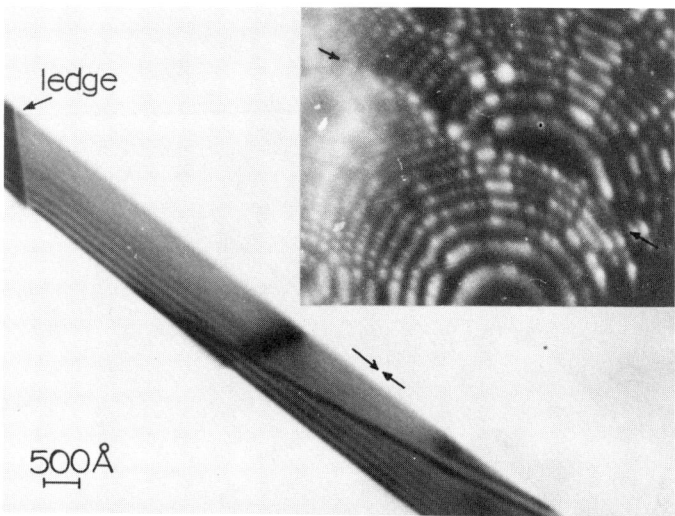

FIGURE 4.35 Comparison of grain boundary region observed in the field-ion microscope (insert) with a corresponding grain boundary section observed in the transmission electron microscope. The examples are from iridium. The arrows along the grain boundary plane indicate the equivalent length. Note that the field-ion image boundary portion is only commensurate in size with a ledge along the boundary projection observed in the electron microscope.

boundary structure—that is, a description of structure which has some applications in understanding residual properties of metals and alloys. Bishop and Chalmers,[126] recognizing this feature, originally proposed that the structure of grain boundaries in metals (with particular reference to high-angle grain boundaries) can be described equally well in terms of a two-dimensional array of coincidence atoms in the boundary plane, atomic ledges in the boundary, and a dense dislocation array. Similarly, Gleiter[76] has argued that the structure of a grain boundary may be described as periodic arrays of (small) structural units (structural units of low energy) for a few specific orientation relationships and boundary misorientations ($\Theta \lesssim 15°$), while for all other orientation relationships ($\Theta > 15°$) the boundary maintains these units of low energy at the expense of lattice strain and faceted interfacial structure. More recently, however, Sadananda and Marcinkowski[127] proposed a canonical theory of grain boundary structure in which the structure of both low- and high-angle grain boundaries is examined from a dislocation approach, and it is shown that coincidence site lattice theory can adequately account for the structure of boundaries of any angle of misorientation. This so-called canonical theory also gives a physical basis for ledges and other structural units which are

FIGURE 4.36 Grain boundary structure in polycrystalline tin film. Arrow denotes boundary exhibiting recognizable structural features (dislocations).

included in the models of Bishop and Chalmers[126] and Gleiter.[76] Like the other models,[76,126] however, the model of Sadananda and Marcinkowski[127] suffers from an oversimplification identical to that described above. As a conceptual basis for the formation of grain boundary dislocations, ledges, and coincidence plane portions, the shuffle model of Murr et al.[84] is perhaps a reasonably effective approach to understanding the origin of grain boundary structural features. It is, nonetheless, the only three-dimensional model of grain boundary structure, as illustrated in Fig. 4.37.

The O-Lattice Interpretation of Grain Boundary Structure. Bollmann[128-130] has devised a geometrical treatment which allows the description of the near-neighbor relations of lattice points in interpenetrating lattices at a grain boundary. The theory is a mathematical method for calculating the detailed structure of the interface between two arbitrary crystals in some relative orientation (or misorientation) and with an arbitrary path of the boundary between the two crystals. Points of best match are defined as a coincidence of points in equivalent positions in both lattices. Such points can include, in addition to atomic positions, the whole continuous space of the crystal. Consequently, they can be the coincidence centers of

STRUCTURE OF INTERFACES   211

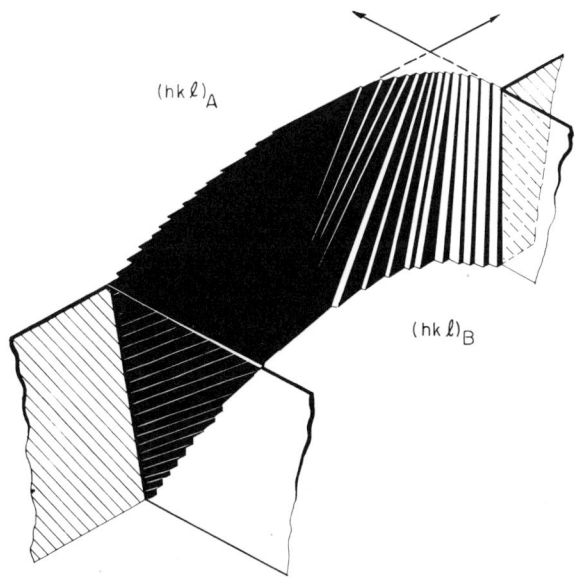

FIGURE 4.37   Phenomenological scheme of grain boundary structure. The two crystallographic planes having trace directions indicated by arrows can be associated with either grain A or grain B, or one in each grain. Grain boundary steps and ledges result from adjustments or shuffling of these planes or dislocation glide on these planes as in Fig. 4.31. Systematic adjustments give rise to coincident boundary portions. Boundary inclination and curvature occur simply by ledge and step adjustments. (From Murr et al.[84])

unit cells or generally points with the same internal coordinates—that is, the same coordinates within the respective unit cells of the two lattices. Points fulfilling this condition are called O-points. Such O-points can form a point lattice within the interpenetrating lattices or a lattice consisting of parallel lines or of parallel planes. The O-lattice within the interpenetrating crystal lattices constitutes the sum of all locations of best match.

The mathematical procedure for calculating the O-lattice first involves a point-to-point correlation between the two lattices composing the grain boundary:

$$\mathbf{x}^{(BL)} = \mathbf{T}\mathbf{x}^{(AL)} \tag{4.26}$$

where BL and AL refer to lattice points in grains A and B, and $\mathbf{T}$ represents a linear homogeneous transformation which may involve a rotation, a translation, or a combination of both. Once $\mathbf{T}$ is given, the whole O-lattice is determined by the solution of

$$(\mathbf{I} - \mathbf{T}^{-1})\mathbf{x}^{(O)} = \mathbf{b}^{(L)} \tag{4.27}$$

where L refers to a lattice point and $\mathbf{I}$ is a unit transformation (= 1). This equation, as Bollmann pointed out,[128] is a generalization of Frank's formula,[95] where

$$\mathbf{b} = \mathbf{x}^{(B)} - \mathbf{x}^{(A)} \tag{4.28}$$

or

$$\mathbf{b} = (\mathbf{I} - \mathbf{T}^{-1})\mathbf{x}^{(B)} \tag{4.29}$$

Under the condition that $\mathbf{b}$ is a translation vector of the A-lattice, $(\mathbf{t}^{(A)})$, $\mathbf{x}^{(B)}$ becomes an O-point, $\mathbf{x}^{(O)}$:

$$\mathbf{t}^{(A)} = (\mathbf{I} - \mathbf{T}^{-1})\mathbf{x}^{(O)} \tag{4.30}$$

Thus, once the relative position and orientation of two crystals joining in a boundary are fixed, the interpenetrating lattices A and B as well as the O-lattice are given.

For transformations, $\mathbf{T}$, deviating slightly from the unit transformation, $\mathbf{I}$, the cells of the O-lattice become large compared with the unit cells of the A- and B-lattices. Consequently, for a fixed boundary within the two lattices, the misfit will be concentrated at the intersection of the boundary with the O-cell walls. Simply considering two equivalent lattice arrangements of point atoms to be superimposed and rotated or translated or both will then result in geometrical moiré bands which can contract to distinct dislocations, and the resulting boundary is therefore depicted by a dislocation network. The formation of a dislocation network instead of a continuous change in the boundary structure, mathematically expressed, is a replacement of a homogeneous transformation by a translational sequence where each dislocation initiates a translation having a magnitude of its Burgers vector.

Thus, by certain translations of lattice B with respect to lattice A, the resulting "pattern" can be shown to be representative of observed grain boundary structure, particularly with regard to dislocation structure. Several examples of this model in interpreting grain boundary structure are shown in Fig. 4.38. Similar examples have been observed by Balluffi et al.[113] and by Levy.[108]

The O-lattice theory is not necessary for understanding the structure of grain boundaries, particularly in interpreting dislocation structure of grain boundaries. However, it is a reasonable description of grain boundary structure involving dislocations, and it affords a means of computing structures or images of grain boundary structure which might occur, particularly in the transmission electron microscope.

Plane Matching Theory and Periodic Grain Boundary Structures. In many observations in the transmission electron microscope, periodic misfit lines have been shown to be associated with high-angle (or high-energy) grain

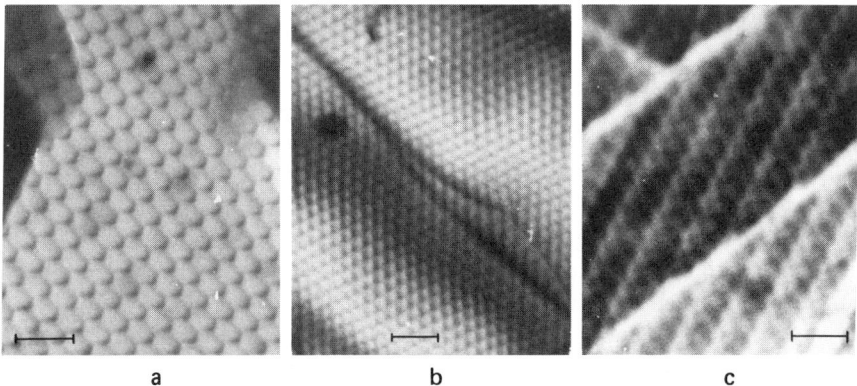

FIGURE 4.38 Dislocation structure of grain boundaries. (a) Simple twist boundary (subboundary) in beryllium. (b) Twist boundary structure in beryllium. (c) Interacting twist and tilt boundary structure and moiré pattern in complex grain boundary in nickel. Markers = 0.1 micron.

boundaries.[107,108,110,111] Such grain boundary structures, as illustrated typically in Fig. 4.39, are not associated with coincidence-related boundaries—that is, boundary planes slightly deviating from a CSL plane—since, as was discussed previously, CSL deviation is accommodated by the superposition of a dislocation network [Fig. 4.29(b)] on the boundary plane. In addition, such periodic linear features are not

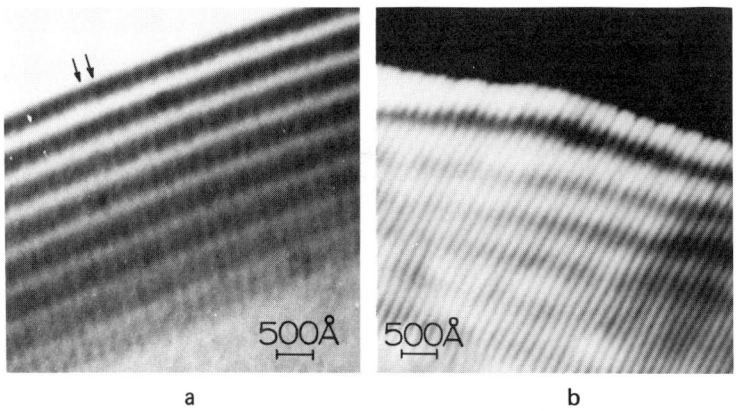

FIGURE 4.39 Periodic misfit defects in high-angle grain boundaries. (a) Parallel periodic defects (arrows) in high-angle grain boundary portion. (b) Dark-field electron transmission image of misfit lines. Samples are 304 stainless steel.

moiré fringes, since their orientation and spacing are independent of the operating reflection in the electron microscope.

Pumphrey[131,132] has demonstrated that, when a set of $\{200\}$, $\{220\}$, or $\{111\}$ planes are mismatched across a high-angle grain boundary in an fcc metal or alloy, the boundary has a periodic structure. If at least one set of crystal planes of relatively high density—for example, (111) in fcc—are slightly mismatched across a high-angle grain boundary, the traces in the boundary plane of the sets of mismatched planes on either side of the boundary plane correspond to rows of atoms on either side of the grain boundary slightly rotated with respect to each other, and the spacings between rows of atoms will be different. These rows of atoms correspond to optical diffraction gratings. If relaxation occurs in the lattice surrounding each dense line of atoms, the strain fields will be observable in the electron microscope, and the resulting lines will be observable using operating reflections in either grain composing the interface [Fig. 4.39(b)]. While such lines are intrinsic structure components in the interface, their spacing has the same dependence as moire fringes, but they are independent of the structure of the interface.*

The implications of plane matching are primarily that periodicity and near-periodicity are, under certain conditions, important in determining grain boundary structure. Recent evidence in support of this contention has been presented in the field-ion microscope observations of ordered grain boundary structures in tungsten.[134] Marcinkowski et al.[135] and Tseng et al.[136] have also demonstrated that a unique set of grain boundary dislocations are associated with a unique set of coincidence site lattices which depend on the rotation axis characterizing the boundaries. Plane-matching concepts are therefore similar to those contained in the generalized theory of grain boundaries proposed earlier by Marcinkowski and Das[137] whereby the grain boundary structure is described as comprising various combinations of crystal lattice dislocations, denoted CLD, associated with the grains composing the interface.[138-140]

Disclination Model of Grain Boundary Structure. It should be apparent from the foregoing descriptions of grain boundary structure that, although the dislocation model is strictly applicable to low-angle grain boundaries, it does not adequately describe the structure of high-angle grain boundaries. Since many grain boundaries in metals and alloys are characteristically rotational defects, they may be describable by disclinations rather than dislocations because disclinations are in fact rotational defects whereas dislocations are translational defects.[141,142] There have

---

*Thölen[133] has shown that the contrast obtained from closely spaced dislocations is indistinguishable from moire fringes when the dislocation spacing is much smaller than one extinction distance. In such cases parallel lines (misfit lines) are observed regardless of the complexity of the dislocation network composing the interface. When dislocations in networks become spaced closer than $\sim 0.3\xi_g$ (ξ_g = extinction distance) the individual dislocations are not resolvable in the electron microscope. This is not due to a resolution limitation but rather is a result of the image-forming (contrast) process.

STRUCTURE OF INTERFACES   215

been several attempts to model disclination structures of grain boundaries, as illustrated in Fig. 4.40.[105,143,144] Each dislocation in the simple symmetric tilt boundary in Fig. 4.40(c) is transformed into a wedge disclination dipole in Fig. 4.40(d). The rotation axis of the disclination is the common direction of the two grains composing the interface. If the low-energy plane of the disclination dipole does not coincide with the boundary plane, the boundary may show a ledge structure as discussed previously. Consequently, grain boundary ledges may be an indication of disclination structure in the grain boundary. It can also be observed that a wedge disclination array is equivalent to a tilt dislocation island; that is, a disclination dipole is a tilt island. Consequently, if the separation of disclination dipoles is wide, they should be observable in the electron microscope. For small separations, however, the contrast would appear identical to that for a dislocation.

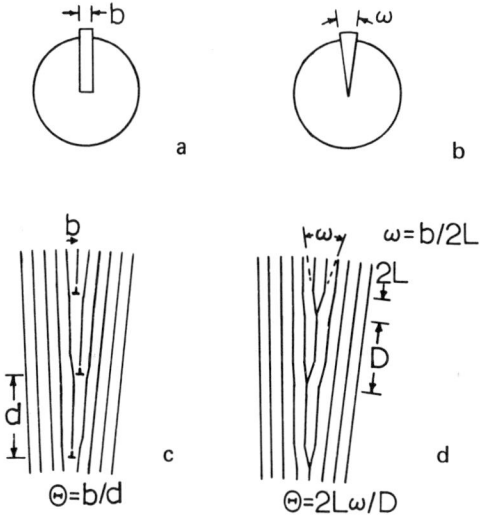

FIGURE 4.40   Disclination model of symmetric tilt boundary. (a) Edge dislocation. (b) Wedge disclination. (c) Simple dislocation (tilt) boundary. (d) Simple wedge disclination boundary. (After Li.[143])

That is, a disclination dipole is equivalent to a dislocation for small separation, and to a dislocation wall when the separation is large. Figure 4.41 illustrates some of these features for a low angle grain boundary in 304 stainless steel.*

*It should be emphasized that disclinations and dislocations are alternative linear elastic approximations of the same entity. As such they predict the same elastic fields, rotations, etc., and differ only in "core" structure (Fig. 4.40) where linear elasticity breaks down anyway.

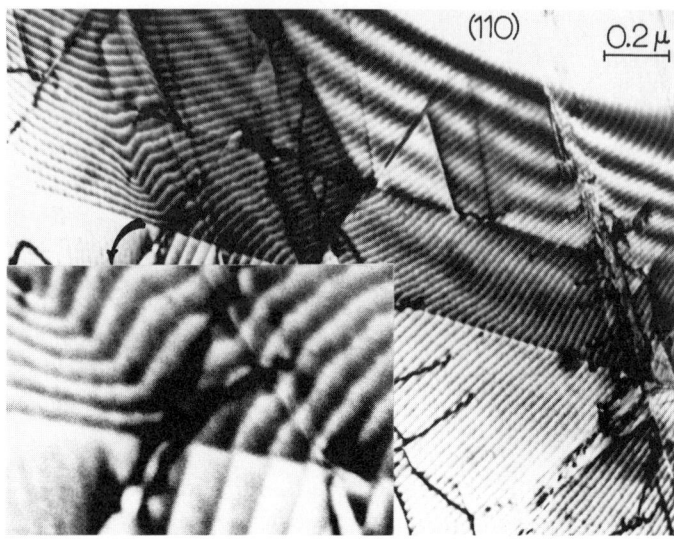

FIGURE 4.41 Structure of low-angle grain boundary in 304 stainless steel resembling disclination structure. Insert shows disclination-type arrangement in boundary portion. Small ledges are observable in the boundary composition.

Li[143] has shown that a general grain boundary consists of regions of low energy separated by disclinations. In addition, Li has calculated the elastic energy caused by the pressure of these disclinations and proposed that this energy represents the excess energy between cusps in the energy misorientation relations (Figs. 4.22 and 4.24).

Practical Description of Grain Boundary Structure. It should be apparent, on considering the many models and structural features of grain boundaries, that interfacial structure can be visualized in terms of discrete structural elements such as dislocations, coincidence units, ledges, steps, protrusions, and more-complex interacting sets of these elements. While these features are observable in thin metal and alloy sections in the transmission electron microscope, many grain boundaries possess no discernible structure (Fig. 4.36). Most low-angle grain boundaries favor a recognizable dislocation structure, while the majority of high-angle grain boundaries in well-annealed materials possess little structure. However, it is difficult to determine simply from this criterion the structural distinction between low-angle (or low-energy) and high-angle (or high-energy) grain boundaries as demonstrated in Fig. 4.36. A distinction is apparent on the basis of structure and misorientation as illustrated in Fig. 4.42.

STRUCTURE OF INTERFACES

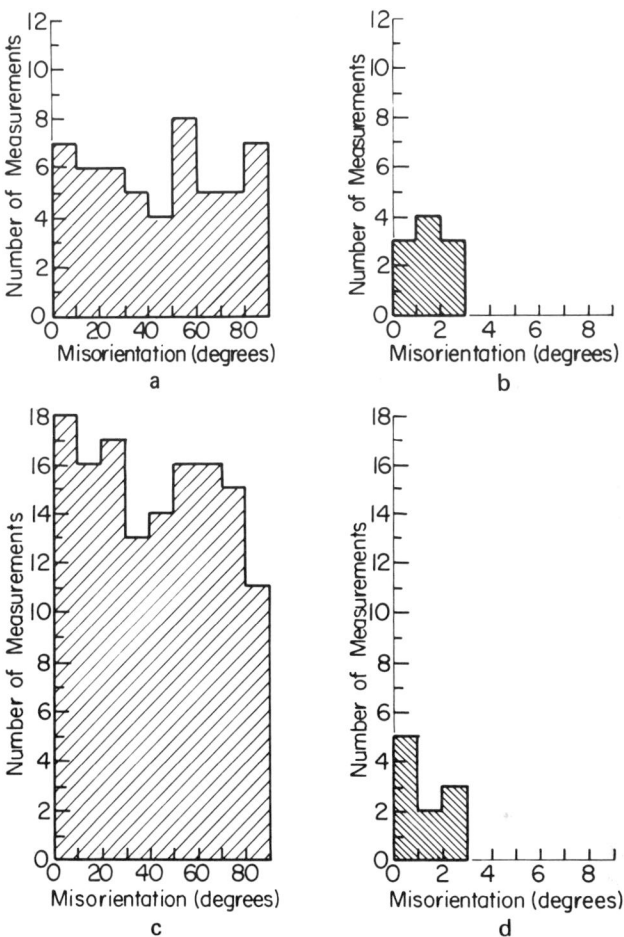

FIGURE 4.42   Histograms of grain boundary misorientation distributions. (a) Distribution of regular grain boundary misorientations in nickel. (b) Distribution of low-angle (dislocation) grain boundary misorientations in nickel. (c) Regular grain boundary misorientations in 304 stainless steel. (d) Low-angle (dislocation) grain boundary misorientations in 304 stainless steel.

Because of the randomness of grain boundary geometry and crystallography in polycrystalline metals and alloys—that is, any boundary portion may be defined by variations in the five degrees of freedom describing the interface—the structural features of a grain boundary can be expected to vary regularly or irregularly along an interface, and to be altered abruptly as the crystallography or geometry varies abruptly. These features are illustrated typically in Fig. 4.43 and in many of the

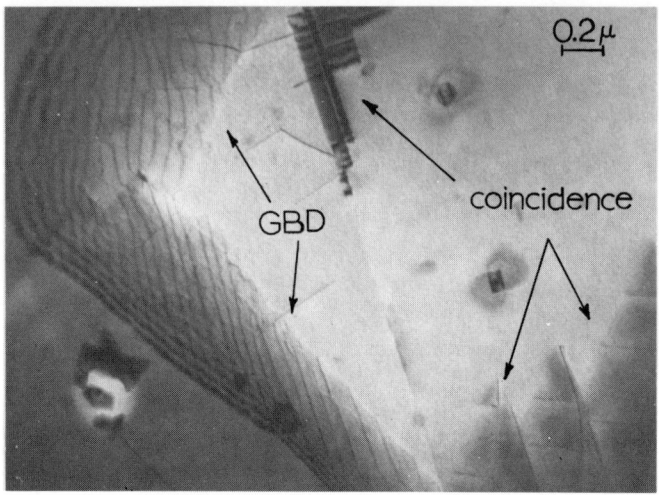

FIGURE 4.43 Variations in grain boundary structure with changes in interfacial crystallography and geometry. Electron transmission microscope image in Inconel 600.

other structural examples of grain boundaries presented previously. In the practical sense, grain boundaries are generally not simple symmetric arrays depicted in the ideal model systems, and therefore the models, while helpful in understanding structural units, are not strictly applicable to practical descriptions of grain boundary structure. For example, Loberg et al.[112] have shown that the coincidence site model is not applicable in the majority of grain boundary structural observations, and therefore the lack of structure cannot be taken to indicate coincidence. This conclusion must apply generally for other structural models or structural units where lack of structure is apparent, as indicated in Fig. 4.41. Aaron and Bolling[145] have discussed the criterion for the applicability of grain boundary models on the basis of grain boundary free volume, V_F [defined as the parameter expressing the change in volume of a polycrystalline material with grain boundary area (A) at constant temperature, pressure, number of atoms (n_i), and composition x] as

$$V_F = (\partial V/\partial A)_{T,P,n_i,x} \cdots \text{cm}^3/\text{cm}^2 \tag{4.31}$$

which for the simple tilt boundary becomes

$$V_F = V_0 \Theta (B - \ln \Theta) \tag{4.32}$$

where V_0 is derived from the calculations of Seeger and Haasen,[146] and B is an unknown constant. Their conclusion is that free volume is the best criterion for calculation of grain boundary structures. However, even this condition is inadequate in detailed descriptions of all possible complex grain boundary structures.

It is clear that grain boundary structure accommodates the change from one orientation to the other. This can involve translation, rotation, or translation-rotation of one lattice with respect to the other at the interface. The accommodation can obviously be performed by simple or complex dislocation arrays (which include ledges and GBD's), steps, and coincidence portions. These structural features are consistent with those characterizing a solid-vapor or solid-liquid interface as discussed previously in Sections 4.3.1 and 4.3.2. That is, crystal lattice structures associated with solid-vapor and solid-liquid interfaces were characterized by dislocations emerging on the surface, steps, or terrace (planar) portions which can be compared with coincidence planes or regions of good fit. Since these structural units were observed to form more or less integrated and complex units on the free surface or at the solid-liquid interface (Figs. 4.6, 4.10, and 4.14), it is logical that solid-state interfaces would exhibit similar features as demonstrated in this section. Spiral structures in the grain boundary similar to those illustrated in Fig. 4.14 have also been observed in the solid-solid interface.[147]

Finally, it should be pointed out that the accommodation from one lattice orientation to another must be accomplished according to the same binding characteristics which bind the perfect lattice. This will necessitate the grain boundary having a width of the order of an atom diameter in regions of good fit such as a coincidence region (for example, a twin boundary) or of not more than roughly three atom diameters in poorly accommodated (high-energy) boundaries. This feature is apparent in the field-ion micrograph of Fig. 4.44.

The models alluded to above not only are idealized, but also they refer in the main to a regular lattice—that is, that of a pure metal. It should be apparent that the accommodation at an interface will be greatly complicated when the lattice atoms vary in size as in an alloy. In addition, even an idealized boundary which satisfies some coincidence relationship will be expected to differ in energy when one is comparing the same geometries in a pure metal lattice and a complex alloy lattice such as nickel and stainless steel, for example. We shall elaborate upon this phenomenon when discussing the calculation of grain boundary free energies in Section 4.8.

We shall observe in Chapter 5 that the structural models, and in fact the observed structural features of grain boundaries, will form the basis for describing a whole range of properties and applications of solid-solid interfaces including boundary migration and creep, melting, segregation and embrittlement behavior, precipitation, and a number of additional features; in a quantitative manner.

4.5.3 Structure of Stacking-Fault and Twin Interfaces

A stacking fault, as illustrated in Fig. 3.32, can be considered a singular interface plane in a regular crystal lattice. Disregistry therefore occurs only at the interface, and the crystallography and composition on either side of the interface are identical (Fig. 4.45). Obviously, the stacking fault as such is a special coincidence

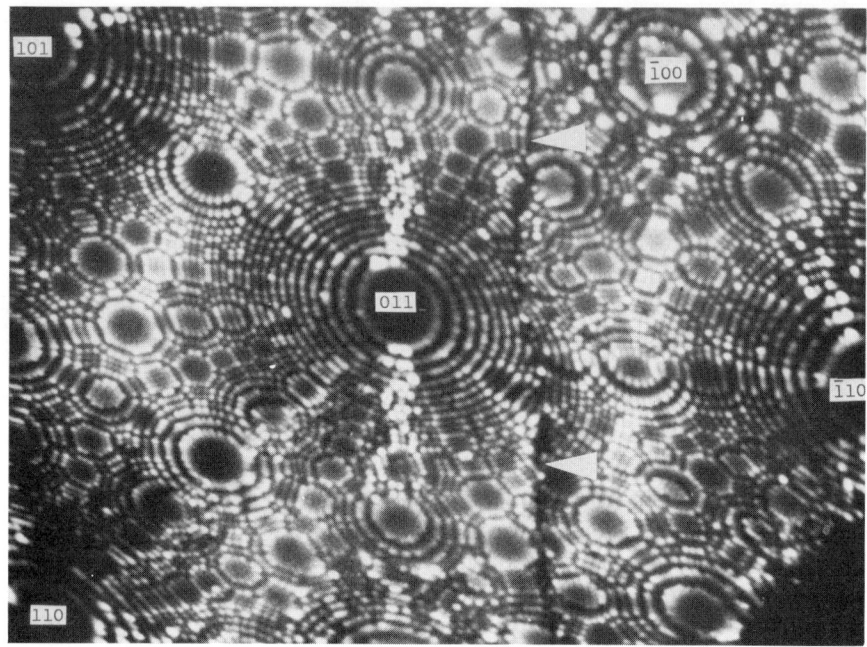

FIGURE 4.44   Grain boundary in tungsten. (Courtesy of Dr. E. W. Müller.)

"boundary" whose energy, in the context of Fig. 4.24, is characterized by a deep cusp. The stacking-fault plane will normally be extremely short in length, determined by the equilibrium separation of partial dislocations.[148] However, in many fcc metals and particularly in alloys of low stacking-fault free energy, stacking faults can extend across grains and thereby attain dimensions commensurate with those of other solid–solid interfaces—for example, a grain boundary or twin boundary.

The fcc system is a particularly simple one in which to examine the structure of stacking faults and coherent twin boundaries because of their intimate relationship. As was indicated above, the intrinsic stacking fault in fcc metals and alloys is an interfacial singularity in an otherwise perfect lattice. On the other hand, a coherent twin boundary, while crystallographically coincident with the same singularity, is a dividing surface separating a lattice from its twin. The twin boundary is therefore structurally similar to an intrinsic stacking fault, but energetically different [Eq. (3.62)]. The structural similarities are illustrated in Fig. 4.46. It can be noted in Fig. 4.46(a) and (b) that the intrinsic stacking fault can be visualized as two coherent twin boundaries separated by a single atom layer (a zero-layer twin).

STRUCTURE OF INTERFACES   221

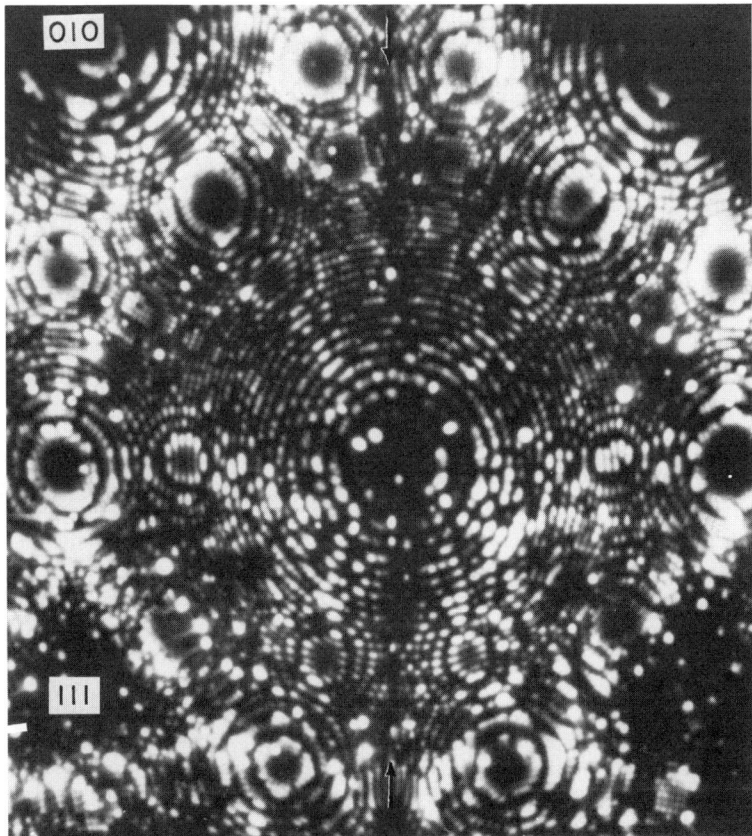

FIGURE 4.45   Coherent twin boundary in a carbon steel observed in the field-ion microscope. (Courtesy of Dr. O. Nishikawa and Dr. E. W. Müller.)

Conversely, an extrinsic stacking fault in fcc (Fig. 3.32) formed by two overlapping intrinsic stacking faults is equivalent to two twin planes separated by two atomic layers (a one-layer twin).

Twin-Fault and Deformation Twin Interfaces. We shall observe in Chapter 5 that, as is implicit in Fig. 4.46(a) and (b), stacking faults and consequently twins in fcc can be produced by shearing operations on the $\{111\}$ planes. If these operations are irregular or erratic with reference to a continuous interface plane (coincident with a specific $\{111\}$ plane), the resulting fault may not be a perfect twin, and consequently the interface may be invisioned as multistepped, resembling a

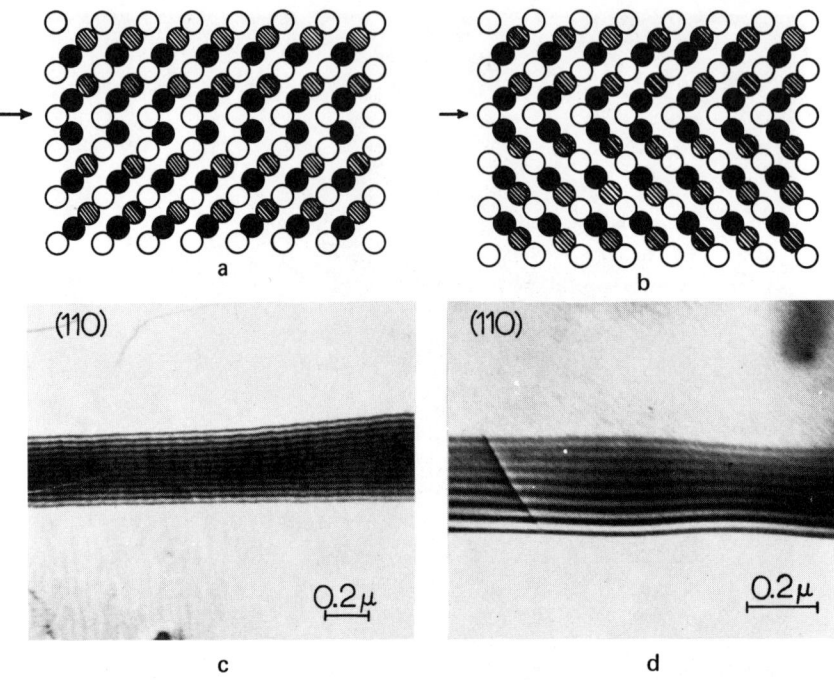

FIGURE 4.46 Stacking-fault and coherent twin interfaces in fcc metals and alloys. (a) Schematic view of intrinsic stacking-fault plane. (b) Schematic view of coherent twin plane. (c) Stacking-fault interface in stainless steel observed in the transmission electron microscope. (d) Coherent twin boundary in stainless steel observed in the transmission electron microscope.

systematically or unsystematically unidirectionally shuffled or sheared array of $\{111\}$ planes as illustrated in Fig. 4.37. Such interfaces therefore can be observed to contain partial dislocations indicative of the termination of an overlapping stacking fault, or ledges composed of several partial dislocations. Because of the irregular structure of deformation twin and twin-fault interfaces, it is unlikely that such interfaces are energetically equivalent to a coherent twin boundary. It is more likely that the energy of such an interface exceeds that of a perfectly coherent interface.

Figure 4.47 illustrates some of the features of overlapping stacking faults and the resulting structure possible in a deformation twin interface. Note particularly the ledge-dislocation contrast in the deformation twin interfaces shown in Fig. 4.47(d). The interfaces in Fig. 4.47(d) could qualify as a special case of the ledge-dislocation–coincidence boundary.[126]

FIGURE 4.47   Stacking-fault, twin-fault, deformation twin interfacial structure in fcc alloys. (a) Overlapping (intrinsic) stacking faults in deformed 304 stainless steel. Contrast in the electron microscope results from 120° phase shifts for each intrinsic stacking fault. Two overlapping faults produce a 240° shift, three faults, a 360° (or zero) shift. Two overlapping faults on successive $\{111\}$ planes produce an extrinsic fault. Fault interfaces arise by splitting of dislocations into partials. (b) Partial dislocations emitted from grain boundary ledges in stainless steel to produce intrinsic stacking faults. (c) Numerous overlapping stacking faults produce n-layer twins with irregular boundary in shock-loaded 304 stainless steel. (d) Deformation twins showing complex boundary structure in fcc Inconel 600 after shock deformation.

It should also be noted with reference to Fig. 4.47 that twin faults or deformation twins could also contain hexagonal phase regions either in the interior or at the interface as a result of systematic faulting (movement of Shockley $a/6 <112>$ partial dislocations) on every other $\{111\}$ plane. This feature is illustrated schematically in Fig. 4.48. In Fig. 4.48(b) the fcc-hcp phase boundary is essentially similar in structure to that of a stacking fault. The energy of the boundary is not known with any accuracy, but it is less than that for the coherent twin boundary in fcc metals and alloys.

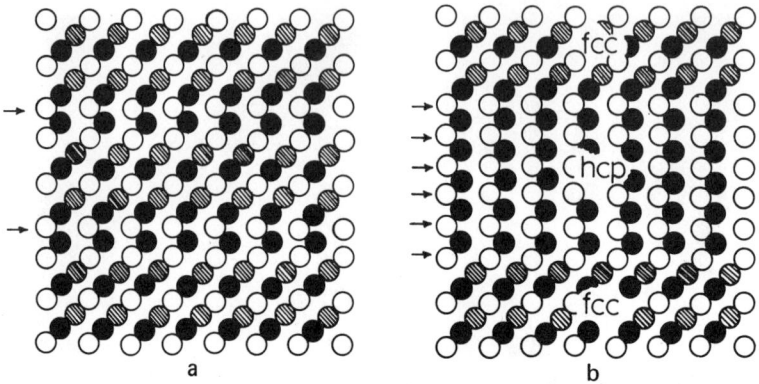

FIGURE 4.48 Stacking faults in fcc forming a hexagonal phase separated by a $\{111\}$ coherent phase boundary. (a) Two widely separated intrinsic stacking faults. (b) Intrinsic stacking faults on every other $\{111\}$ plane in fcc producing an hcp (ϵ) phase. Arrows denote stacking faults.

Noncoherent Twin Boundary Structure. Since a coherent twin in fcc metals and alloys results from the propagation of $a/6$ <112> Shockley partials or from $a/3$ <111> Frank partial dislocations,[148] a noncoherent step in a twin boundary or in the terminus of a twin band in an fcc matrix—that is, the noncoherent twin boundary—can be thought of in terms of an array of such partial dislocations which systematically terminate in an array (coincidence plane) as illustrated in Fig. 4.49(a). Deviations from a coincidence array gives rise to more complex structure as illustrated in Fig. 4.49(b).

Oblak and Kear[149] and Phillips[150] have demonstrated that, as illustrated in Fig. 4.49(a), noncoherent twin boundaries in fcc and diamond cubic (Si) lattices can be regular arrays of Frank or Shockley partial dislocations on consecutive planes. Such closely spaced partial dislocation arrays constitute "high-angle" dislocation boundary arrays (Fig. 4.25) and are therefore characteristically high-angle boundaries. This feature is evident when one compares coherent and noncoherent twin boundary free energies (Table 3.8). It is not difficult to realize that, because of the nature of noncoherent interfaces, the structure may, on deviation from regular dislocation and coincidence arrays, become complex. As a consequence, the range of structural units previously described for high-angle grain boundaries generally applies in the case of noncoherent twin boundaries in fcc. These features are illustrated in Fig. 4.50.

While we have specifically treated noncoherent fcc twin boundary dislocation structure in terms of partial dislocations, more complex arrays are possible (Fig. 4.50). Although partial dislocation structure of grain boundaries was not explicitly

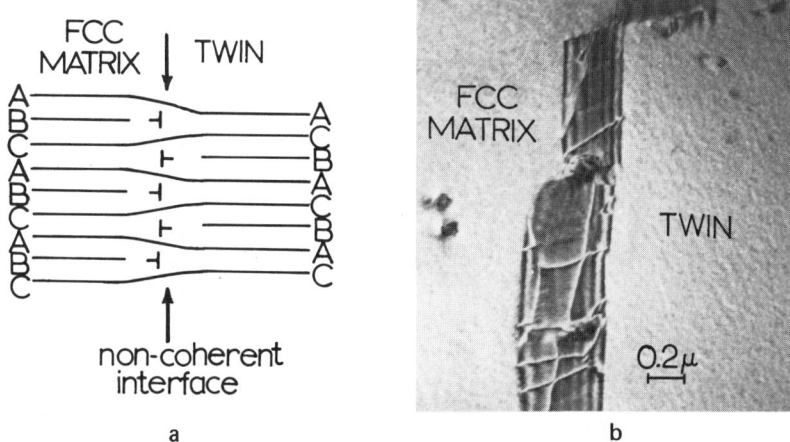

FIGURE 4.49 Noncoherent twin boundary structure in fcc metals and alloys. (a) Idealized structure characterized by an array of Frank partial dislocations of opposite sign. (b) Transmission electron microscope image of noncoherent twin boundary in Inconel 600 showing ledge structure. Note the deviation from coincidence (as evidenced by the boundary curvature).

discussed in Section 4.5.2, several grain boundary models have been proposed in which the boundary region is considered to be composed of a planar array of partial dislocations.[151,152] It is conceptually possible, in fact, to visualize the formation of grain boundary "faults" by the dissociation of GBD's. However, such a concept may not be advantageous when compared with other structural features already discussed in this chapter.

Regular (coincidence) arrays of noncoherent twin boundaries such as steps in coherent twin boundaries in cubic metals [Figs. 4.29(a) and 4.50(a) and (b)] have been studied by Whitwham and Lacombe,[153] Sargent,[154] Murr,[155] and others,[150,156] and found to be coincident with $\{112\}$, $\{551\}$, $\{711\}$, $\{110\}$, and $\{601\}$, consistent with expectations based on coincidence site lattice considerations.[97,157]

The equilibrium geometry and associated interfacial energetics of twin and twin boundary formation have been described in Chapter 2, based on the original proposals of Fullman and Fisher,[158] involving stacking-fault nucleation at grain corners. The formation of new $\{111\}$ planes can also be envisioned by the shuffling (Fig. 4.37) of lattice planes during grain boundary migration, to be discussed in detail in Chapter 5. Such a process can also be phenomenologically described by two-dimensional nucleation and growth of new planes on existing $\{111\}$ planes in a process analogous to the nucleation of new lattice planes on the surface of a perfect crystal growing from the vapor.[159-161] The point to be made is

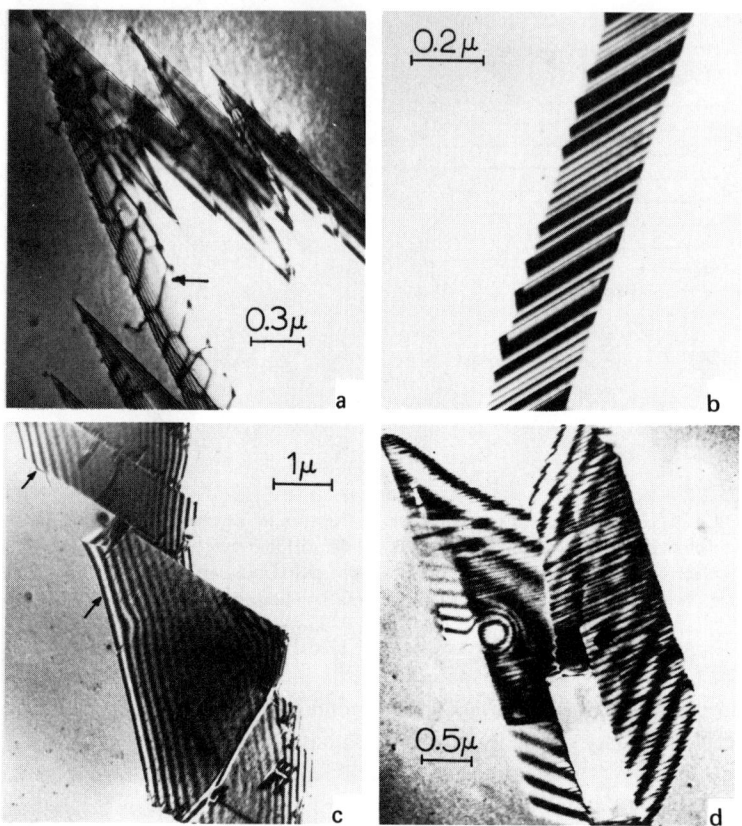

FIGURE 4.50 Noncoherent twin boundary structure in fcc metals and alloys. (a) Coherent/noncoherent boundary portions in nickel. Arrow indicates dislocation network resulting from slight deviation from $\{111\}$ coincidence. (b) Coincidence-related noncoherent steps in coherent twin boundary in 304 stainless steel. (c) Coherent/noncoherent twin boundary portions in stainless steel. Fine defect lines are shown by arrows. (d) Complex boundary structure on noncoherent twin boundaries at twin band terminating in a 304 stainless steel matrix.

that solid-solid interfacial structure and attendant phenomenology bear a similarity to solid-vapor (free surface) phenomenology.

It should also be pointed out that, although an attempt has been made to illustrate stacking-fault and twin boundary structure with reference to fcc metals and alloys, these interfacial features are not unique to fcc nor, for that matter, to cubic crystals.[148,162] Their occurrence is, however, as was pointed out in Chapters 2 and 3, sensitive to their associated interfacial energy, which, in the case of alloys, is composition-sensitive.

4.6 INTERPHASE INTERFACE STRUCTURE

Except in the case of the formation of an hcp phase in fcc alloys by the systematic formation of stacking faults as illustrated in Fig. 4.48, the descriptions of interfacial structure have been confined to those interfaces separating regions of identical composition and crystal structure, but of different crystallography. There are, obviously, many instances in metals and alloys where an interface separates regions of different composition or crystal structure, or both; the more prominent cases are illustrated in Fig. 1.11. However, Fig. 1.11 does not necessarily depict all possible interfacial circumstances. Additional situations include the boundary separating ordered and disordered regions in the same alloy composition, and the systematic oxidation and recrystallization of a pure metal resulting in the formation of oxide grains in a metal matrix or metal grains in an oxide matrix. In either case the situation is distinct from an inclusion or a precipitate, and the associated interfaces separate regions of different crystal structure and composition.

4.6.1 Domain Boundary Structure

There are several ordering processes which can occur in metals and alloys or other materials such as metal oxides. The simplest ordering phenomena involve the alignment of electron spins or polarization vectors which give rise to magnetic or electric domains, respectively. The boundaries that separate such magnetic or electric domains give rise to transition regions which require an energy of formation referred to as the domain boundary energy.[163] Because magnetic and electric domain boundaries do not require a structural change associated with shifts in the positions of lattice atoms, such interfaces arise by systematic electronic adjustment rather than atomic adjustment, and they are usually characterized by polarization directions at 90° or 180° with reference to the domain boundary interface or domain wall.[163]

Magnetic (ferromagnetic, ferrimagnetic) and electric (ferroelectric, ferrielectric) domain boundaries are observable in the transmission electron microscope as a result of the systematic deflection or phase shift of the electron beam on either side of the interface which produces contrast at the boundary.[163-167] In most cases domain boundaries are coincident with a lattice plane. This is particularly true of ferroelectric and ferrielectric domain boundaries which are analogous to a twin boundary. Figure 4.51 illustrates the conceptual features of interfaces resulting from systematic lattice polarizations.

Superlattices and Antiphase Boundary Structure. Many solid-solution alloys which are normally disordered become ordered at low temperatures. This process involves the regular arrangement of atoms of a particular species in an otherwise statistically random crystal structure, but the crystal structure is maintained. This feature is illustrated in Fig. 4.52 for an fcc structure of composition $A_3 B$, where A and B are unique atom species—for example, Cu and Au, Ni and Fe, Pt and Sn—and a tetragonal (fct) structure of composition AB, where A and B are Cu and Au, Pt and Fe, Bi and Na, etc.[162]

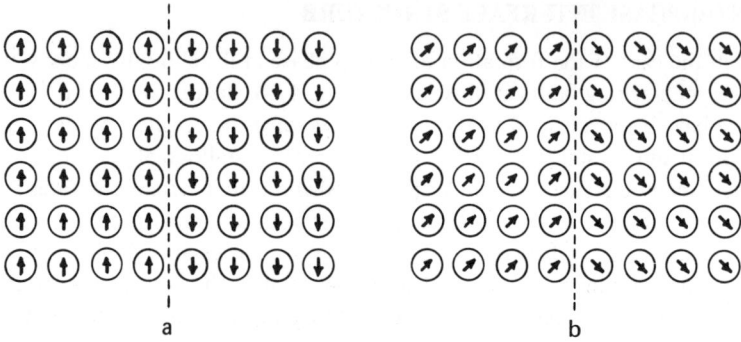

FIGURE 4.51 Idealized magnetic and electric domains resulting from systematic polarizations. (a) 180° boundary. (b) 90° boundary. Note that boundaries in (a) and (b) are coincident with <001>.

The ordering that occurs in a solid-solution alloy can appear over large areas or volumes of the lattice, or over small areas or volumes; that is, the ordering may be long range or short range. Such ordered regions, while having the same crystal

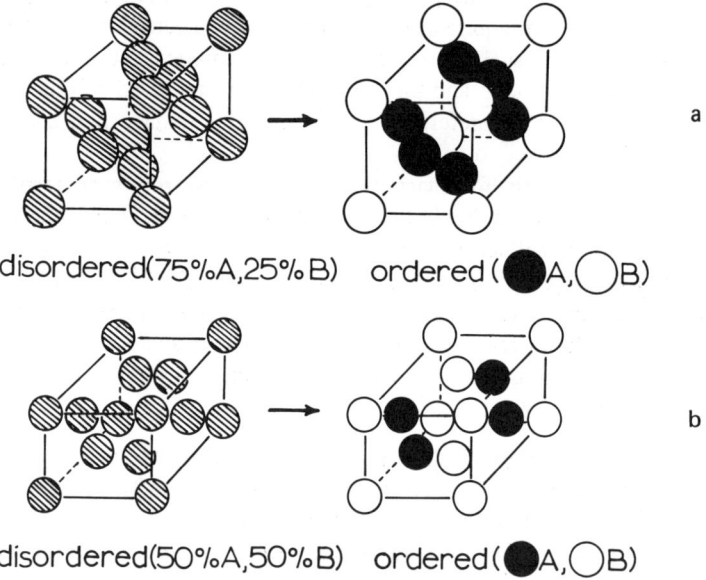

FIGURE 4.52 Order–disorder phenomena in solid-solution alloys. (a) A_3B—fcc lattice. (b) AB—fct lattice.

STRUCTURE OF INTERFACES

structure as the total matrix, take on a superlattice arrangement because of the periodic occupation of the lattice site by specific alloying atoms. In a disordered solid solution, crystallographically equivalent atomic planes are statistically identical, whereas in the ordered solid solution alternate planes may be A-rich or B-rich. This feature is illustrated in Fig. 4.53, which shows systematic variations in the regular atomic site occupation which results in domains of different order separated by a transitional interface or out-of-step boundary called an antiphase boundary (APB).

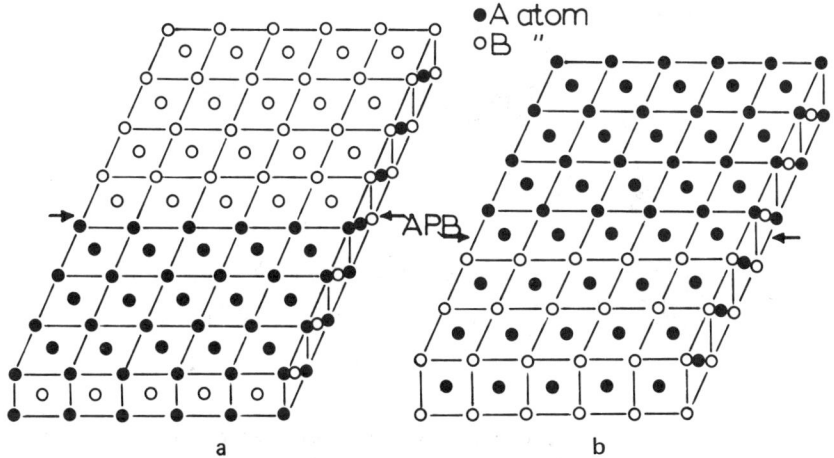

FIGURE 4.53 Schematic representation of antiphase (domain) boundaries (APB) in ordered solid-solution alloys. (a) AB type. (b) A_3B type.

As a general rule, an ordered alloy has a lower symmetry than the disordered alloy matrix, and it can form in a number of ways called variants which involve a translation or a combination of translation and lattice orientation changes. Both variants involve a phase transformation, the former characterized by an APB, the second by a twin-type interface.

In addition to the so-called antiphase boundaries in ordered alloys, interfaces having a similar structural nature are formed in compounds showing ordering of interstitials or vacancies (superlattice arrays). Crystals undergoing a systematic phase transformation such as the ε-phase boundary depicted previously in Fig. 4.48(b), and so-called shear-structure boundaries which can result from deviations from stoichiometry forming superstructures form anti-phase-like interfaces. In addition precipitation along certain crystallographic planes can produce crystallographic shear planes[168,169] which are analogous to antiphase boundaries. Figure

4.54 shows an example of shear planes in TiO_2. The shear planes (APB's) in Fig. 4.54 make up for the oxygen deficiency in the slightly nonstoichiometric TiO_2. The interface actually separates regions of the crystal where different rows of octahedral sites are occupied within a continuous hcp framework of oxygen ions.

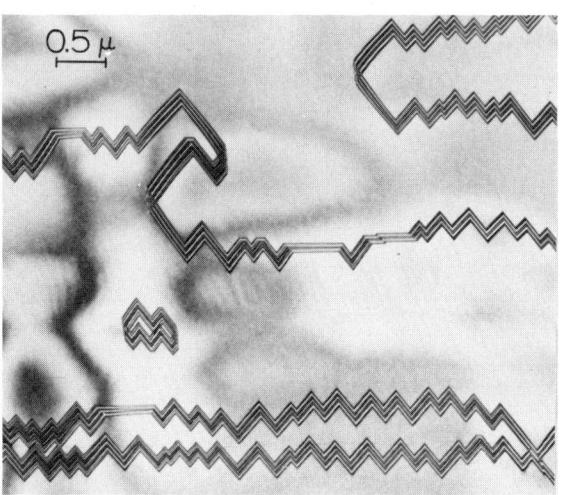

FIGURE 4.54   Isolated shear planes or antiphase boundaries in TiO_2 (rutile) observed in the transmission electron microscope. (Courtesy of Dr. J. VanLanduyt.)

It should be apparent that, because domain boundaries (magnetic, electric, or antiphase) are crystallographic, and involve an orderly transition, their structural features are characteristic of low-energy interfaces such as coherent twin boundaries; and although the energies are unknown in most cases, they are considerably less than those of associated grain boundaries in the same material. Many domain boundaries or boundary portions are twin-related and resemble coherent-noncoherent interfaces [Fig. 4.29(a)]. Also, although domain boundaries are coincident with crystallographic planes, they are distinct from coincidence site lattice boundaries described in Section 4.5.[170]

4.6.2 Structure of Phase Boundaries

In the perfectly general case of a phase boundary, like a grain boundary, there is generally no systematic crystallographic habit or orientation characterizing the interface. Phase boundaries are distinct from grain boundaries, however, because they separate regions of different crystal structure or composition, or both. Despite these differences, phase boundaries possess structural characteristics that are conceptually identical to those described in Section 4.5. These include disloca-

STRUCTURE OF INTERFACES   231

tions, ledges, and coincidence units or predominant interfacial portions which lie along particular sets of $\{hkl\}$ planes defined in the matrix or a reference phase. In many cases we can describe $\{h_1k_1l_1\}_A \mid\mid \{h_2k_2l_2\}_B$, where A and B represent the two phases separated by the interface in question, which is coincident with $\{h_1k_1l_1\}$ and $\{h_2k_2l_2\}$.

Two-phase interfaces such as those in duplex structures (austenite/ferrite, for example) are perhaps typical of the phase boundaries separating regions of different crystal structure and composition. Figure 4.55 illustrates the appearance of grain and phase boundaries in a duplex stainless steel (Uniloy 326: 26% Cr, 6.5% Ni,

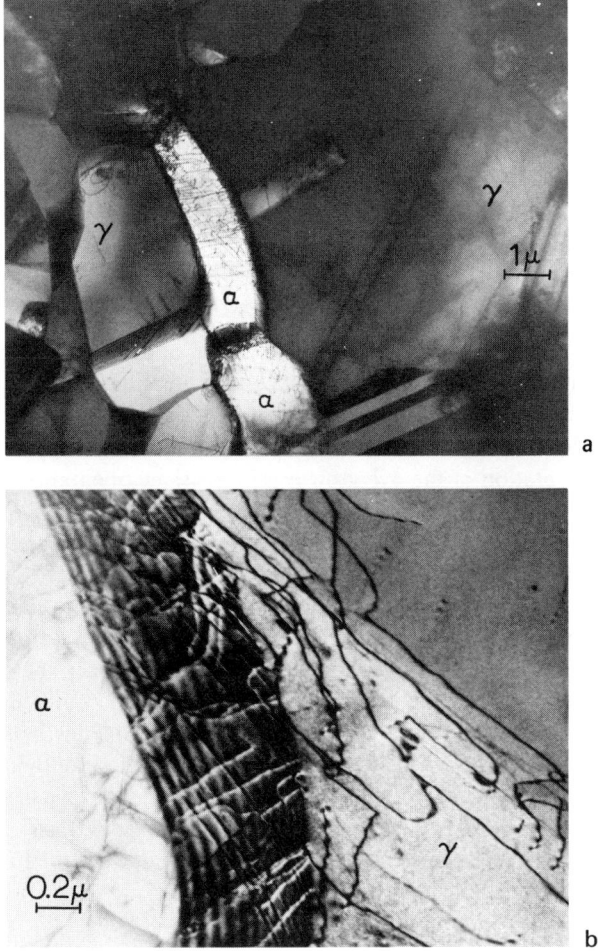

FIGURE 4.55 Grain and phase boundary structure in Uniloy 326. (a) Grain and phase boundary structure. (b) Phase boundary structure showing dislocations and ledges.

0.2% Ti, 0.05% C, balance Fe) which consists statistically of 30% ferrite in an austenite matrix. In Fig. 4.55(a) it is not possible to readily distinguish the grain boundaries from the fcc/bcc phase boundaries, and Fig. 4.55(b) is typical of the structure of grain boundaries as shown, for example, in Figs. 4.30 and 4.32. Similar structural features have been observed in the nearly coherent fcc/bcc boundaries in a duplex Cu–Cr alloy.[171]

The structure of many interphase boundaries can be described in terms of plane-matching phenomena where perfect matching of $\{h_1 k_1 l_1\}$ in grain A with $\{h_2 k_2 l_2\}$ in grain B can be achieved by the formation of periodic arrays of misfit dislocations as illustrated in Fig. 4.56. Bäro and Gleiter[172] recently showed such patterns of parallel lines in α/β brass interfaces to be identical to those observed as periodic misfit lines in grain boundaries as seen in Fig. 4.39(b). It is certainly possible to achieve the same plane matching by the incorporation of misfit dislocations in single-phase interfaces (grain boundaries), as discussed previously. Obviously the periodicity and therefore the interfacial structure will depend on the misfit and the associated lattice strain that can be accommodated. Accommodation of lattice misfit can be accomplished coherently, in which case the associated strain at the interface becomes appreciable. We shall describe this phenomenon in more detail in subsequent discussions of coherent and noncoherent interfaces at precipitates and related lattice inclusions which constitute two-phase systems having different compositions and crystal structures.

Structure of Epitaxial Interfaces. When a crystal of one structure and/or composition is grown on a different crystalline substrate, an epitaxial crystal interface usually results. Such growth usually involves nucleation from the vapor phase or ions in solution onto the solid substrate. Epitaxy is ideally the phenomenon in which the orientation of the overgrowth is in the same orientation as the solid substrate.[173-175]

The attempt to accommodate the overgrowth on the substrate is measured by the misfit between lattice parameters as a percentage difference. The crystallographic mismatch at the interface is accommodated, as was outlined above, by the creation of misfit dislocations conceptually illustrated in Fig. 4.56. When the misfit at the interface and the thickness of the overgrowth are small, the overgrowth is usually coherent because the misfit in such cases can be accommodated by homogeneous strain in the thin overgrowth. However, when the misfit and the film thickness are large, a homogeneous strain exists in addition to the misfit dislocations. As the strain increases in a coherent film whose interface lacks dislocations, misfit dislocations develop analogous to the loss of coherency at precipitates, for example. Misfit dislocations created at an epitaxial interface therefore form a boundary structure as illustrated typically in Fig. 4.57.

In Fig. 4.57 the interface is normal to the electron optical (viewing) axis. The structure is therefore analogous to that observed at bicrystal interfaces[113] and at many grain boundaries, as illustrated, for example, in Fig. 4.38. For fully coherent epitaxial interfaces, the misfit is fully accommodated by homogeneous strain, and

STRUCTURE OF INTERFACES 233

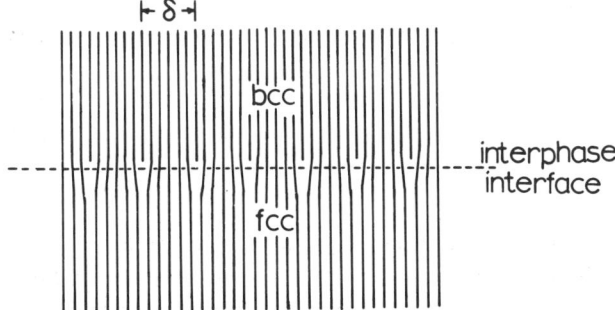

FIGURE 4.56 Misfit dislocations in plane-matched fcc/bcc phase boundary. The dislocation spacing in the phase boundary is denoted by δ. (After Bäro and Gleiter.[172])

FIGURE 4.57 Observation of misfit dislocations in [001] partially alloyed Ag-Pd bicrystal. (Courtesy of Dr. J. W. Matthews.)

the overgrowth is said to be pseudomorphic.[14,174] In addition, because of the accommodation features of epitaxial interfaces, minimum energy interfaces can occur at particular values of misalignment and misfit.[128,129,176,177] Such minima are observed when, as a result of epitaxial growth, some atoms of one crystal (overgrowth) lie precisely in atom sites in the substrate surface. The resulting interfacial structure is therefore analogous to, and an extension of, the coincidence-site lattice. Reiss[178] has suggested, however, that epitaxial interfaces possess low energy for orientations which are in addition to those predicted by the coincidence model.* These features and the observation of moiré fringes at epitaxial interfaces have been investigated with regard to misfit dislocations and the alignment of epitaxial islands by Matthews.[179]

Structure of Multicomponent Interphase Boundaries. Multicomponent phase interfaces can be characteristically a general interphase boundary possessing coherent as well as noncoherent or coincidence features. Included among the possible interfaces are epitaxial and/or pseudomorphic growths as outlined above, directionally solidified eutectic systems,[180,181] precipitation and other inclusions,[182,183] and decomposition-producing phase boundaries—for example, spinodal alloys[184] (Fig. 1.11). The resulting structures of such boundaries will be, in most cases, indistinguishable from those of other grain boundaries already discussed, but some differences in the complexity of structural units can occur as a result of differences in both crystal structure across the interface, and the attendant differences in the atom sizes and binding characteristics of the lattices joined at the interface. Figure 4.58 illustrates these features conceptually for a simple orienta-

FIGURE 4.58 Conceptual view of ideal multiphase boundary structure (interface separating regions of different crystal structure and composition).

---

*Reiss assumed a rigid monolayer—that is, one without relaxations to include interface dislocations. Such a relaxation could alter not only the number of minima but also the energy levels of them.

tion of two phases. The boundary structure as shown in Fig. 4.58 may be further complicated by the addition of phases within the two grains, terminating at random spacings along the interface. An example of this feature would be a regular grain boundary separating spinodal structures as shown in Fig. 4.59.

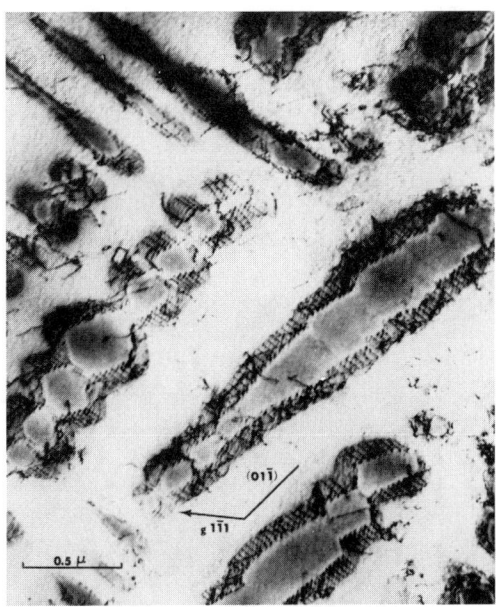

FIGURE 4.59   Transmission electron micrograph of spinodal phase boundary structure in 64Cu-27Ni-9Fe alloy showing dislocations, steps, and coincidence units tending to lie along $\{110\}$ planes in the fcc matrix. (Courtesy of Professor G. Thomas.[184])

Transformation Boundary Structure. The interface that results when a solid solution becomes unstable and partly decomposes into another crystalline phase can also be coherent, partially coherent, or noncoherent, and, like other phase boundaries, possess structural units similar to those already illustrated for various types of grain boundaries. Such structures, distinct from precipitation phenomena to be discussed below, can include spinodal, polymorphic, and ordering transformations discussed above. Such interfaces are also descriptive of the martensitic transformations which are characteristically crystallographic.[21,162] Although there is a degree of coincidence associated with many such boundaries, deviations from coincidence and the necessity to accommodate two different crystal structures at the interface give rise to structural features typical of higher-energy boundaries. Many transformation structures, in addition, undergo secondary transformations or twinning which forms serrations at their intersections along the phase boundary. This results in modifications of structural units along the interface, as well as the production of dislocations and/or ledges. This structural feature is illustrated in Fig. 4.60.

FIGURE 4.60   Coherent (shear) interface structure in a U–6.3 w/o Nb alloy. (Courtesy of A. O. Ehlinger.)

A prominent feature of many coherent or semicoherent transformation interfaces is the obvious shear movements as illustrated in Fig. 4.60. Massive transformations can take place, however, where changes of crystal structure occur, but the process involves the random transfer of atoms across the interface. Such transformations are therefore controlled mainly by interface diffusion resulting in displacement of noncoherent boundaries. Massive transformation boundaries are therefore generally higher-energy boundaries than the shear-type polymorphic interfaces. In addition, one might expect to find a higher probability of dislocation ledge structures and boundary spiral structures at massive transformation boundaries because, unlike the more coherent shear-type interfaces, the massive transformation interfaces result from nucleation and growth phenomena. This may be particularly true of ferrous eutectoid structures—for example, the ferrite/ cementite interface. In this regard it should be pointed out that ledge structures, either linear (planar) or spiral, not only compensate for lattice misfit at the interface, but also serve as growth ledges in a manner phenomenologically identical to growth upon a free surface or at a solid-liquid interface.[185] Dislocations are also associated rather copiously with nearly all transformation interfaces because of the interruptions in systematic shear, misfit, deviations from coherency and a crystallographic habit, or orientation changes along the interface. Mourey and Dabosi[186] have recently treated the crystallography (and consequently the interface) of the martensitic transformation in terms of the O-lattice theory discussed previously.

Interfacial Structure at Precipitates and Crystal Inclusions. In the discussion of phase boundaries above, the meanings of coherency and noncoherency have been twofold. In one instance coherency implies the satisfaction of bonds by "elastic stretching" to accommodate the lattice mismatch across the interface; in the other, coherency implies coincidence of the interface with a crystallographic plane or the matching of orientations such that along the interface $\{h_1 k_1 l_1\}_A \parallel \{h_2 k_2 l_2\}_B$, where A and B denote two phases separated by the interface. Deviations from coherency in both cases will necessitate the formation of dislocations to accommodate the misfit as shown ideally in Fig. 4.56. On the other hand, noncoherent interfaces are those where perfect plane matching is lacking or nonexistent, or where no systematic coincidence or orientation relationship exists.

When precipitates or inclusions occur in a matrix, the concept of coherency and noncoherency applies in both cases to the description of the interface. However, coherency normally implies the elastic accommodation of misfit, and thereby implies an associated elastic strain field, whereas noncoherency implies the lack of systematic accommodation and the absence of a detectable strain field. However, in the context of the coincidence connotation of coherency, noncoherent inclusions and precipitates can and do possess in many cases a crystallographic or coincidence relationship. Consequently, the particle morphology can appear faceted as in the case of single crystals discussed in Chapter 1 and the initial portions of this chapter.

Figure 4.61 illustrates ideally some of the features discussed above, and shows an example of both a coherent and a noncoherent inclusion. In many cases a coherent precipitate which nucleates in solid solution will appear as a small zone of segregated atoms forming an included phase of several atoms' thickness coincident with certain crystal planes of the matrix. In the case of Guinier-Preston (G.P.) zones,[187,188] these coincident phase plates consist of atoms of a single species (for example, copper atoms at G.P. zones in Al-Cu alloys) and are therefore a thin metal phase within an alloy matrix. The energy of the associated interface is presumed to be low, and in the range of coherent twin or stacking-fault free energies. However, as the precipitate phase forms, the associated phase boundary free energy will be larger and can approximate that associated with general phase or grain boundaries. This is also true of noncoherent inclusion boundaries (particle-matrix interfacial free energies). Thus, the range of structures and structural units associated with

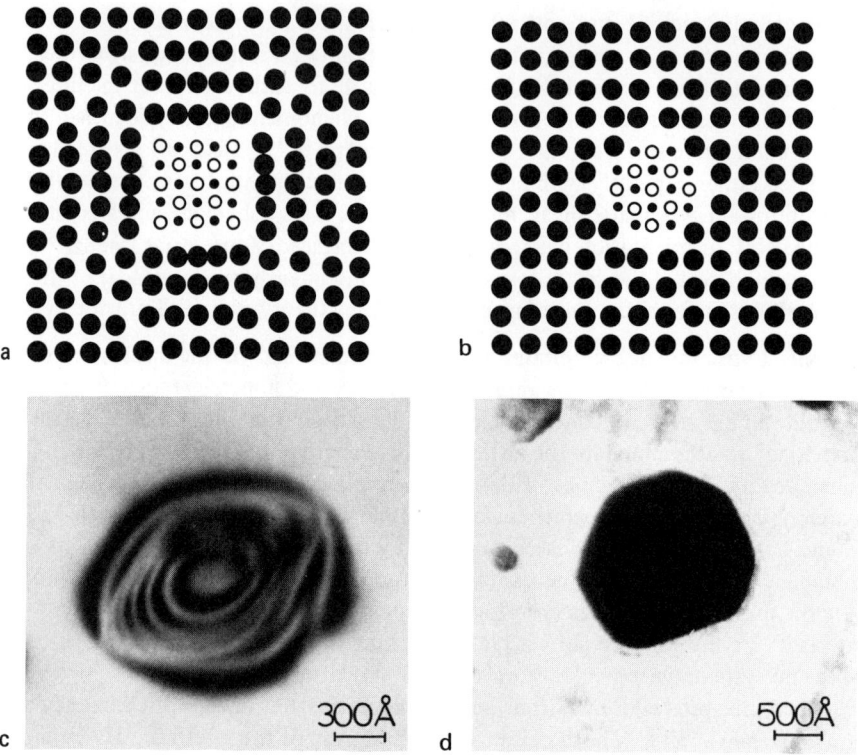

FIGURE 4.61 Coherent and noncoherent particle inclusions in a metal or alloy matrix. (a) Idealized view of coherent precipitate. Elastic adjustment to misfit produces radial strain field. (b) Idealized view of noncoherent precipitate or dispersoid. (c) Electron microscope image of tiny carbide precipitate in Inconel 600 alloy matrix. Image prominence is due to isostrain field contrast. (d) Noncoherent-faceted ThO_2 dispersoid in nickel matrix observed by transmission electron microscopy.

precipitates is expected to be varied, and the associated range of energies will, as a consequence, be large (see Table 3.10).

As was indicated above for massive transformation interfaces and associated boundaries at solid-state nucleated phases, the boundary structure is expected to be characterized by a high degree of dislocation and ledge structure. This is also true of precipitates (Fig. 4.62) which are a solid-state transformation product, as well as other inclusions—for example, dispersed particles. It can also be observed from Fig. 4.61(b) that in many cases the dispersed or noncoherent phase can appear as an ordered domain within the matrix, and the associated interfacial structure is characteristic of an antiphase boundary (APB).

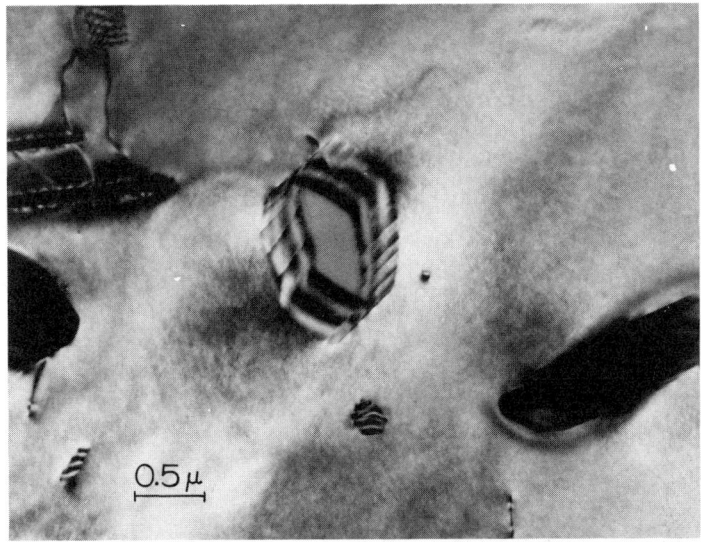

FIGURE 4.62   Dislocation ledge structure at a θ precipitate plate in Al–4 w/o Cu. The largest ledge heights are about 100 Å. The coincidence nature of ledges is also apparent. (Courtesy of Dr. G. C. Weatherly.[189])

Ledges and facets have been observed to form prominent structural units at precipitate and inclusion interfaces [Figs. 4.61(d) and 4.62; references 189-193], and to function in a manner similar to ledges and facets (terraces) at solid-vapor (free) surfaces with regard to nucleation and growth and dissolution. Both the latter growth features and the ledge structure of the precipitate–matrix interface, as well as the incorporation of coincidence units into the interface structure, are readily apparent in Fig. 4.63. We shall observe in Chapter 5 that ledge structure can be shown to account for grain boundary sliding as well as nucleation, growth, and dissolution phenomena at solid–solid interfaces. A detailed, systematic study of precipitate–matrix interfaces by field-ion microscopy has also shown ledge structure to dominate.[194] It will be instructive to compare the structural features of precipitates in Fig. 4.63(a) with those at a solid–vapor interface in Fig. 4.15(b). We shall also see in Chapter 5 that a precipitation reaction proceeds only when the reduction in the volume free energy exceeds the sum of the interfacial free energy plus the strain energy. Consequently, for small particles the interfacial free energy may dominate. In addition, the structure of the particle–matrix interface will be governed to a large extent by the minimization of the interfacial energy in a manner similar to that outlined in Chapter 1 with regard to γ-plots. That is, the Wulff theorem will have a conceptual if not physical effect on the precipitate or dispersoid morphology as implicit, for example, in Fig. 4.61(d).

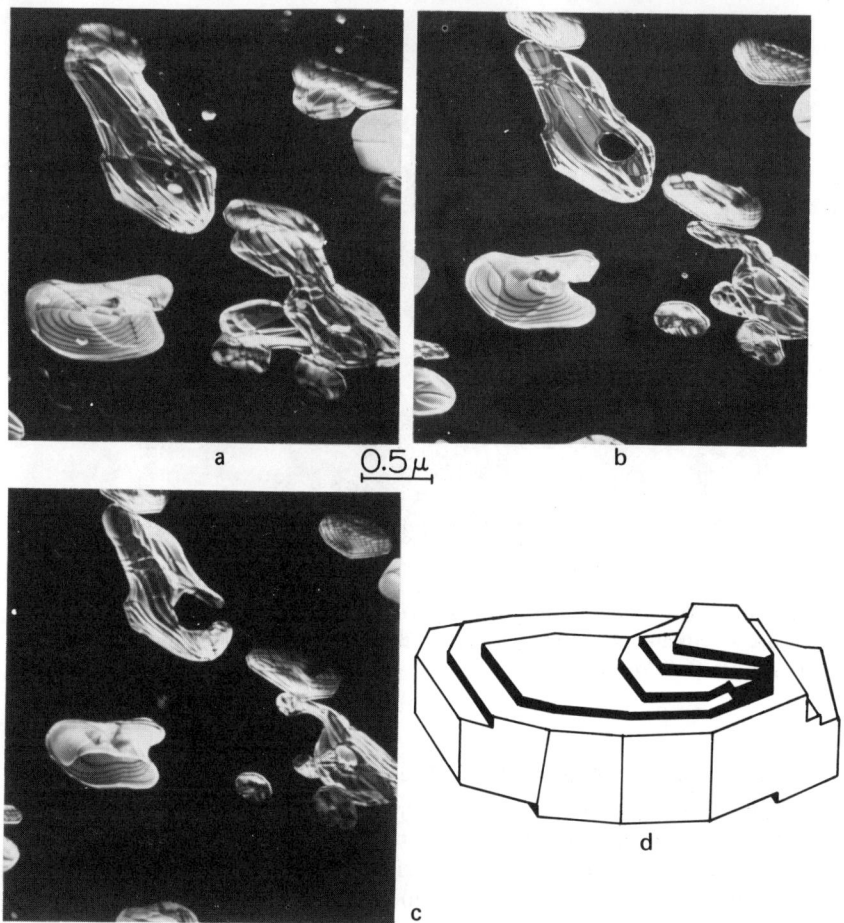

FIGURE 4.63   Ledge structure of precipitate plates. (a)–(c) Dark-field electron microscope images showing ledge structure and dissolution of θ' precipitates in Al–4 w/o Cu: (a) 15 minutes at 320°C. (b) 30 minutes at 320°C. (c) 45 minutes at 320°C. Ledge mobility during dissolution is apparent. (Courtesy of Dr. G. C. Weatherly.[189]) (d) Schematic view of ledge structure of precipitates showing coincidence interfaces (facets).

4.7   ADSORPTION, ABSORPTION, SEGREGATION, AND PRECIPITATION AT SOLID–SOLID INTERFACES

Under certain conditions, vacancies, interstitials, and other solute atoms can be attracted to regions of internal stress in a metal or alloy matrix.[148] Such regions include the strain field at an individual dislocation or the strain field associated with more extended defects such as stacking faults, and solid–solid interfaces. In dealing

STRUCTURE OF INTERFACES

with vacancies or other atoms attracted to an interface, one deals specifically with adsorption; however, if the attracted atoms diffuse into the interface itself and become incorporated in some way in the boundary, the phenomenology must be treated specifically as absorption. The accumulation of large numbers of solute atoms or vacancies either at or in the vicinity of a solid-solid interface is therefore treated as segregation, while the formation of a new phase within the interface or upon the interface is referred to structurally as interfacial precipitation, and is phenomenologically and thermodynamically identical to bulk phase transformations characteristic of precipitation.

While, in the specific case of segregation, it is possible to distinguish phenomenologically between nonequilibrium and equilibrium segregation, the latter is due to diffusion-controlled migration of atoms to the interface as compared with interface movement. McLean[71] has proposed that the driving force for equilibrium segregation involves the relief of lattice strain which occurs when a solute atom diffuses from the lattice interior to a solid-solid interface or a free surface (solid-vapor interface). The process of segregation also involves the lowering of interfacial free energies as discussed in Chapters 1 and 3, and this energy minimization is also important in the tendency toward segregation. Furthermore, as we noted in Chapters 1 and 3, equilibrium segregation occurs ideally over an interfacial layer having a width of only a few atom diameters, whereas nonequilibrium segregation and precipitation can extend over much greater distances.

4.7.1 Interfacial Adsorption and Absorption

It is perhaps a matter of common knowledge to most metallurgists and materials scientists that the structure and properties of interfacial adsorbates, segregates, and precipitates are essentially inseparable. However, we have attempted in this chapter to present only a structural view of interfacial phenomena; and this philosophy will be extended in the case of interfacial adsorption, absorption, and segregation. Ideally, interfacial adsorption and absorption of vacancies and/or solute atoms can be depicted as illustrated in Fig. 4.64. Absorption phenomena as shown in Fig. 4.64(b) can be envisioned as a nucleus for precipitation, particularly where solute complexes are formed which represent the stoichiometry of the precipitated phase. One should also realize that the fit of atoms in a boundary is an important feature determining the character and concentration of adsorption and absorption of solute atoms.[195]

In Chapter 3 we discussed several techniques for detecting solute or impurity adsorption or absorption; we shall defer until Chapter 5 the details of the properties exhibited as a result of such grain boundary or solid-solid interfacial phenomena. Techniques are nearly developed[196] whereby it will be possible to readily observe the distribution of an impurity atom or solute with respect to a solid-solid interface utilizing the basic techniques of field-ion microscopy—that is, the atom-probe FIM. Not only are impurity or solute atoms made visible, but all such atoms are simultaneously observable in their relationship to the matrix and the associated grain boundary.

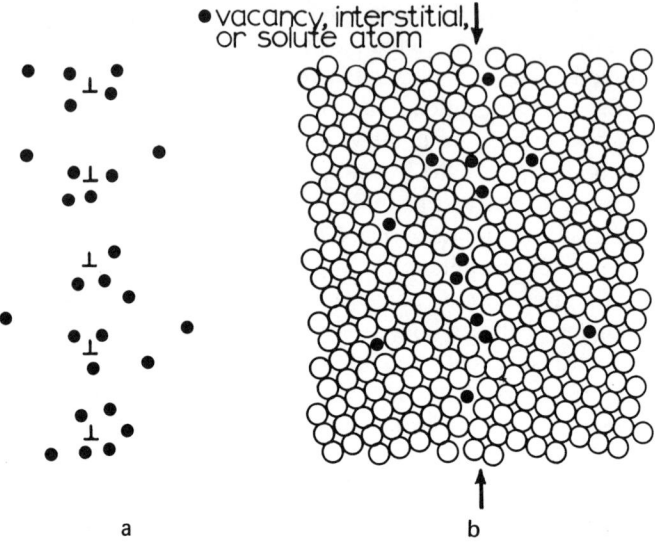

FIGURE 4.64 Idealized view of solute or vacancy adsorption and absorption at a grain boundary or phase boundary. (a) Adsorption at associated interfacial dislocation strain fields. (b) Absorption at grain or phase-boundary dislocations and ledges.

4.7.2 Solute Segregation and Precipitation at Interfaces

In Chapter 1 the concept of equilibrium segregation (Gibbs adsorption) was developed and discussed in terms of concentration profiles. The phenomenological effects of solute segregation on the interfacial free energy were outlined in Chapters 2 and 3; and Chapter 2 dealt specifically with interfacial equilibria at grain and phase boundary precipitates. Structurally speaking, one would expect equilibrium segregation to be represented by a random or regular distribution of solute atoms on the interface plane. In practice, this is usually not observed. However, observation (resolution) tends to be the limiting feature in discussing segregation structure at an interface. Direct observations of oxygen segregation to grain boundaries has been claimed by Fortes and Ralph,[197] but their results have been questionable not only because of their assumptions on oxygen resolution, but also because the distribution was found to vary in concentration across the interface as shown conceptually in Fig. 1.6(a).

In the equilibrium state, impurity segregation would be expected only at the distortion characterizing the interfacial phase, or at a thickness of 2 to 3 atom diameters. Nonequilibrium segregation therefore represents any deviation from this behavior. Obviously, dislocations near a grain boundary or vacancies attracted to the interface or the area near an interface will influence not only adsorption but also desorption phenomena. Interfacial strain fields which may develop for certain boundaries or boundary structural units may also influence the segregation kinetics because grain boundary segregation is affected by interaction forces between the solute and the boundary structure. Consequently, segregation behavior depends on the characteristics of the solute or impurity as well as the interfacial structure. The residual interaction forces which determine the segregation characteristics can also be identified with electronic and elastic interaction forces. Because segregation of a solute will lower the energy of the interface, it would be expected that the structure would also be altered. This feature has been observed for precipitate-ledge structure.*

We observed earlier in this chapter that grain boundary structure is dependent to a large degree on boundary misorientation. As a consequence, it would be expected that segregation and precipitation behavior would be strongly influenced by misorientation, and that inhomogeneities would be expected along an interface composed of varying structural units or characterized by systematic or nonsystematic changes in misorientation (as in the case of coherent–noncoherent units along an annealing twin boundary). Because of these features, grain boundary migration, corrosion, and related properties can be expected to be altered as boundary structure, misorientation, and consequently segregation, adsorption, desorption, or precipitation characteristics are altered. As we shall see in Chapter 5, grain boundaries deviating from or associated with coincidence lattice orientations possess higher mobilities and differ markedly in corrosion resistance and other properties when compared with random high-angle grain boundaries, for example,[198-200] as a result of the fact that less solute segregation seems to occur at coincidence-site lattice boundaries when compared with other (random) boundaries.[201] This is structurally simple to envision on considering the prospects of fitting a solute or impurity atom at the boundary in Fig. 4.27(a) as compared with Fig. 4.46(b). However, Pumphrey[202] has recently argued that in many instances it is the interfacial plane rather than the boundary defect structure that is particularly important for precipitation, and this view is consistent with nucleation theory.†

The driving force for precipitation is proportional to the reduction in the Gibbs free energy, ΔG, as expressed by

$$\Delta G = \Delta G_s + \Delta G_\epsilon + \Delta G_\phi \tag{4.33}$$

where ΔG_s is the surface free energy term, ΔG_ϵ is the strain energy term, and ΔG_ϕ is the chemical free energy change. Solute segregation to an interface will, as we noted above, reduce the total free energy of the matrix. This causes the chemical free energy term to dominate, resulting in heterogeneous boundary nucleation. However, if the precipitation reaction is dominated by the surface and strain energy terms, then nucleation at grain boundary structure such as dislocations or other

---

*See R. Sankaran and C. Laird, *Mater. Sci. Eng.*, **14**, 271 (1974).
†See W. C. Johnson et al., *Met. Trans.*, **6A**, 911 (1975).

structural units causing strain fields will be favored. Since a precipitate nucleated in a grain boundary or other interface generates a new interface between the matrix and the precipitate phase, this structure may actually dominate nucleation and growth of the precipitate, because, as Pumphrey[202] pointed out, the reduction in the surface and strain energy terms resulting from the creation of this new interface may be more important than reductions resulting from variations in boundary structure. This is readily apparent when the precipitate becomes coherent or semi-coherent with one of the matrix grains when nucleated in a random (noncoherent) interface.

Precipitation, like segregation at an interface, is influenced by boundary structure and coincidence.[203] Unwin et al.[204] have also shown that all boundaries except low-angle boundaries ($\Theta \leqslant 2°$) act as sinks for vacancies, and Vaughn[205] has argued that precipitation at coincidence boundaries will reduce the free energy needed for precipitation. As a result, many precipitates form low-energy (coincidence) interfaces with both grains at a boundary.

Figure 4.65 gives several examples of the structural and structurally associated characteristics of segregation and precipitation described above. It should be noted that the interfacial structures of precipitates in the boundary are the same as those described previously for a precipitate in the matrix. That is, they can consist of coincidence–dislocation–ledge arrays or structural units characterized by these features. The precipitation-free zone observed in Fig. 4.65(b) is explained in terms of matrix diffusion-controlled loss of excess vacancies at the grain boundary combined with a solute concentration profile across the interface. The width and size of the zone depend on the structure and misorientation of the interface at any location where a solute concentration profile exists.[204]

Although it can be generally argued that low-energy or coherent interfaces are less likely sites for segregation and precipitation, there are numerous examples of such activity at twin and stacking-fault interfaces. Murr et al.[206] have argued that the positive slope for twin and stacking-fault temperature coefficients of interfacial free energy is due to equilibrium segregation, and nonequilibrium segregation and precipitation at such interfaces have also been discussed by Vaughn.[207] Aluminum segregation to coherent twin boundaries in Ni–Al solid solutions has also been observed.[208] Since the elastic strain fields associated with such interfaces are small, electronic interactions appear to be responsible.

4.8 CALCULATION OF SOLID INTERFACIAL FREE ENERGIES

Because many solid interfaces can be structurally characterized with varying degrees of accuracy, more quantitative evaluations of solid interfacial energy are possible than for free surface (solid–vapor), solid–liquid, or liquid–vapor interfacial free energies as described in Section 4.4. As the translational symmetry improves, the calculational accuracy increases. In the majority of contemporary attempts to calculate the solid interfacial free energy, pairwise interactions have been used, and instead of choosing the translation lattice a priori, the lattice has been determined

STRUCTURE OF INTERFACES   245

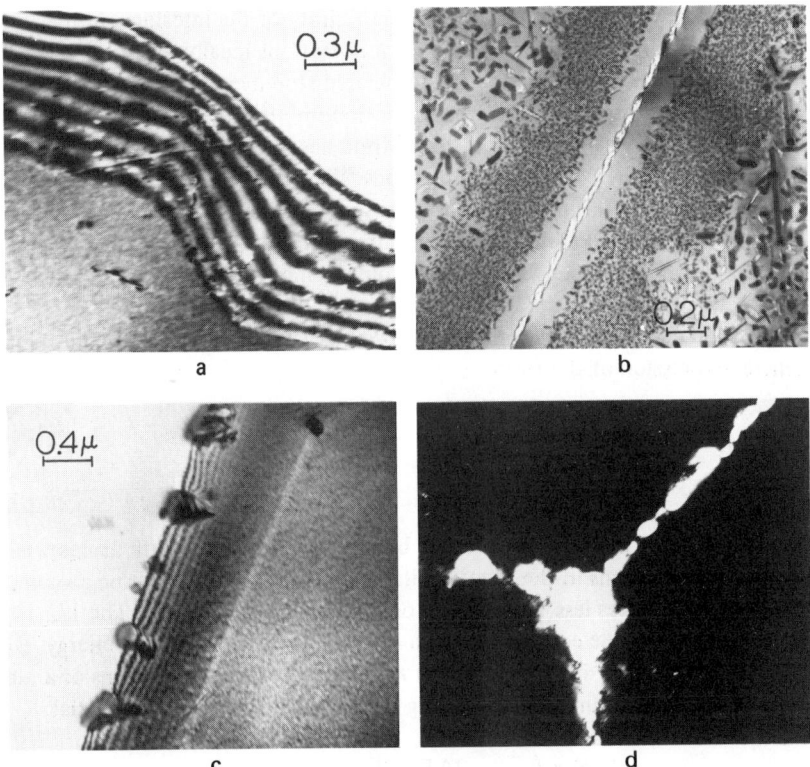

FIGURE 4.65   Precipitation and segregation at grain boundaries in metals and alloys.
(a) Vacancy and impurity segregation to grain boundary in aluminum. (b) Precipitate-free zone
in Al–Zn–Mg alloy. (Courtesy of Dr. G. W. Lorimer.[204]) (c) Carbide precipitates in Inconel
600 grain boundary. (d) $M_{23}C_6$ precipitates in 304 stainless steel cause systematic depletion of
chromium in vicinity of the interface resulting in enhanced grain boundary etching. Note
systematic profile around the triple junction.

by minimizing the total energy of the associated bicrystal, and the calculations of
the energies of the atomic configurations at the boundary based upon the assumption of interatomic forces derived from some appropriate potential function[209-211]
or pseudopotential.[56]

4.8.1 Theoretical Calculation of Grain Boundary Free Energies

The quantitative evaluation of grain boundary free energy is, as we have pointed
out in preceding sections, and as originally demonstrated by Read and Shockley,[81]
totally dependent on the concomitant interfacial structure. Such evaluations
generally begin with the construction of a translation lattice by changing the atomic

arrangement of the adjacent matrices in the vicinity of the interface, primarily on the basis of geometrical assumptions. Such operations ideally produce an elastic deformation of both crystals, and the associated stored energy has been calculated by considering the crystals as an elastic continuum[81,212] or as a regular crystalline array.[213] More recently, however, the method has involved the minimization of the total bicrystal energy by relaxation methods.[209,211] In the calculations of Weins,[211] a rigid translation determining the relative translations of the crystals (grains) composing the interface that resulted in minimum energy was followed by atomistic movement which allowed the atoms to move from their lattice sites to sites of lower energy.

The total energy of the atomic array at a bicrystal interface can be described by a general expression of the form

$$E = \frac{1}{2} \sum_i \sum_j \psi(r_{ij}) \tag{4.34}$$

where $\psi(r_{ij})$ is the interaction potential between the ith and the jth atoms; i is summed over all atoms in the array, while j is summed only over those atoms separated by distances less than second or third nearest neighbors. The 1/2 corrects for the double counting of such interactions. The potential energy, $\psi(r_{ij})$, between atoms i and j at a separation of r_{ij} has been expressed in terms of a number of pairwise potential functions including the Lennard-Jones 6-12 potential:

$$\psi(r_{ij}) = \frac{-4\Delta F_S r_0^6}{C_6 r_{ij}^6} + \frac{2\Delta F_S r_0^{12}}{C_{12} r_{ij}^{12}} \tag{4.35}$$

where F_S is the sublimation energy, r_0 is the equilibrium separation, and C_6 and C_{12} are crystal constants, with C_6(fcc) = 14.46 and C_{12}(fcc) = 12.13; the so-called 4-7 potential:

$$\psi(r_{ij}) = \frac{-4\Delta F_S r_0^4}{C_4 r_{ij}^4} + \frac{2\Delta F_S r_0^7}{C_7 r_{ij}^7} \tag{4.36}$$

where C_4(fcc) = 25.34 and C_7(fcc) = 13.36; and the more successful Morse potential:

$$\psi(r_{ij}) = D \exp[-2\alpha(r_{ij} - r_0)] - 2 \exp[-\alpha(r_{ij} - r_0)] \tag{4.37}$$

where D (eV) is the dissociation energy of the bond, r_0 is the equilibrium separation, r_{ij} is the distance between atoms i and j, and α is a positive constant having dimensions of $\overset{\circ}{A}{}^{-1}$. Values for D, α, and r_0 have been given for aluminum, copper, and gold.[211,214]

STRUCTURE OF INTERFACES   247

Figure 4.66 gives several examples of the calculation of the grain boundary free energies of a number of symmetrical tilt boundaries in aluminum and copper utilizing the Morse potential of Eq. (4.37). On comparing Fig. 4.66 with Fig. 4.22, we see that the experimental difference between the <100> and <110> tilt boundary energy variations with misorientation, Θ, is also characteristically demonstrated by the calculated data. In addition, the mean grain boundary energies depicted in the calculated curves of Fig. 4.66, although generally larger by a factor of 2 than the corresponding experimental values at high temperature, vary for aluminum and copper by the same proportion: $\gamma_{gb}(Cu) \cong 2\gamma_{gb}(Al)$ (Table 3.6). However, the values calculated in Fig. 4.66 refer to grain boundaries at $0°K$. Consequently, when the experimental values in Table 3.6 are adjusted to $0°K$ by considering the corresponding temperature coefficients for the interfacial free energies, the difference between the average experimental and theoretical (calculated) grain boundary free energies is decreased. Figure 4.66 illustrates that, although very close agreement between experimental (Fig. 4.22) and calculated grain boundary energies

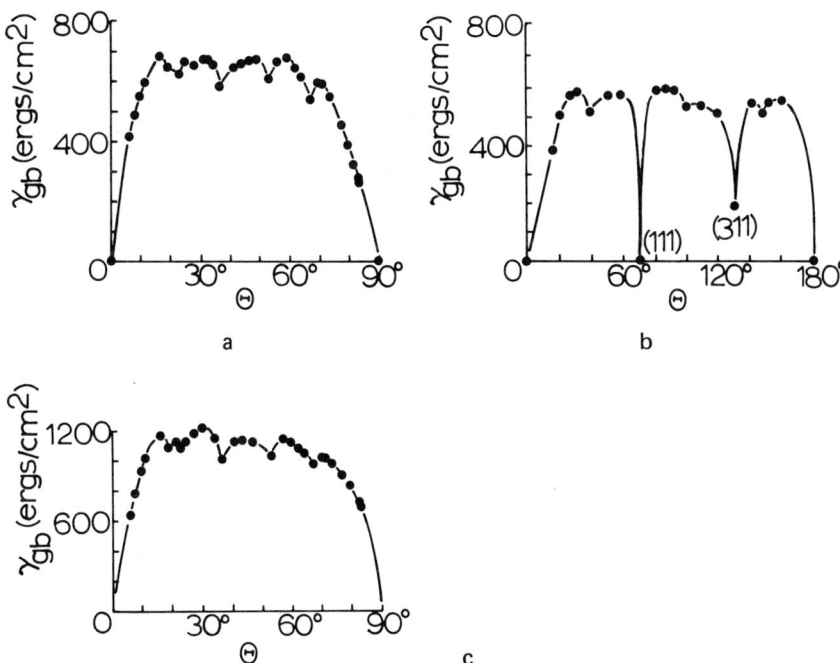

FIGURE 4.66 Energy of symmetrical tilt boundaries (calculated). (a) Aluminum <100> tilt boundaries, (b) Aluminum <110> tilt boundaries, (c) Copper <100> tilt boundaries. (After Hasson et al.[209,210,215])

might be lacking, the experimental variations with grain boundary misorientation for various rotation axes seem to be similarly depicted on comparison of the calculated variations.

Calculation of Grain Boundary Entropy. An upper limit on the entropy of a coincidence-related grain boundary has been determined by a statistical thermodynamic approach by Ewing,[216] and the Einstein model of a crystal, in conjunction with the Morse potential [Eq. (4.37)], has also been adapted to the computer calculation of grain boundary entropy.[217] For a pure metal, such calculations represent the temperature coefficient for the grain boundary free energy.

In the Einstein crystal model, each atom is treated as a triharmonic oscillator, and the entropy per atom is given by the expression

$$S = 3k[1 + \ln(Tk/h\nu_c)] \tag{4.38}$$

where k is Boltzmann's constant, T is the absolute temperature, h is Planck's constant, and ν_c is the Einstein frequency. The total interfacial entropy, however, is composed not only of vibrational entropy (or frequency-originated entropy), but also configurational and anharmonic entropy terms:

$$S = S_\nu + S_c + S_\lambda \tag{4.39}$$

where S_ν, S_c, and S_λ represent the frequency-originated, configurational, and anharmonic entropy contributions, respectively. Evaluating, we obtain

$$\left. \begin{array}{l} S_\nu = 3K \ln(\nu_c/\nu_b) \\ S_c = k \ln \Omega \\ S_\lambda = \dfrac{15}{16} \dfrac{h^2 TQ^2}{\pi^2 m \theta_E^2} \left(\dfrac{\epsilon_0}{\epsilon_1} \right) \end{array} \right\} \tag{4.40}$$

where ν_c and ν_b refer to matrix (crystal) and boundary atom vibrational frequencies, respectively, Ω is the number of distinguishable atomic configurations, Q is an anharmonic factor, m is the atomic mass, θ_E is the Einstein temperature, and ϵ_0 and ϵ_1 are the bond energies of an interior (matrix) and a boundary atom, respectively.

The entropy of symmetric tilt boundaries in pure gold has been determined from Eqs. (4.39) and (4.40) to be in the range of 0.5 to 1.0 erg/cm² °C.[216,217] This range of values appears to be considerably higher than the experimentally determined value of 0.10 erg/cm² °C (Table 3.6). On the other hand, a solid-vapor surface entropy of 0.50 erg/cm² °C, calculated from Eq. (4.39) for gold,[216] agrees fairly well with the experimental value of 0.43 erg/cm² °C (Table 3.5).

4.8.2 Theoretical Calculation of Twin Boundary and Stacking-Fault Free Energies in FCC Metals and Alloys

Hirth and Lothe[148] have argued on strictly geometrical grounds that the stacking-fault and coherent twin boundary free energies at $0°K$ can be represented by the number of pairs of separation, N, which are not in the proper fcc stacking sequence, multiplied by the distortional energy, ψ_N, per pair. As we indicated in considering the structure of fcc stacking-fault and coherent twin boundaries in Section 4.5.3, there is a systematic relationship between the intrinsic and extrinsic stacking faults and a coherent twin boundary. The coherent twin boundary has one pair of second-nearest-neighbor planes ($N = 2$) out of sequence, two third neighbors ($N = 3$), one fourth neighbor plane ($N = 4$), etc. From this near-neighbor plane consideration, and the relationship of the fault boundaries, the twin boundary and intrinsic and extrinsic stacking-fault free energies can be expressed by[148]

$$\gamma_{tb} = \sum_{N=1}^{\infty} [2N\psi_{3N} + N(\psi_{3N-1} + \psi_{3N+1})]$$
$$= \psi_2 + 2\psi_3 + \psi_4 + \ldots$$

$$\gamma_{SF}(\text{extrinsic}) = \sum_{N=1}^{\infty} [2\psi_{3N-1} + (3N+1)\psi_{3N} + 3N\psi_{3N+1}]$$
$$= 2\psi_2 + 4\psi_3 + 3\psi_4 + \ldots$$

$$\gamma_{SF}(\text{intrinsic}) = \sum_{N=1}^{\infty} [(3N-1)\psi_{3N-1} + 3N\psi_{3N}]$$
$$= 2\psi_2 + 3\psi_3 + (0)\psi_4 + \ldots \quad (4.41)$$

In a hard-sphere central force model, ψ_j represents the energy of the jth-nearest-neighbor atom pair separated by a vector, r_j, between atom centers. The associated shear displacements for the fault formations can be represented by $r_j \to r_j'$ and $r_j \to r_j''$. Hirth and Lothe[148] have shown the pair energies to be expressed in terms of the bond energies in the form

$$\left.\begin{array}{l} \psi_2 = \psi_3' - 3\psi_3 + 6\psi_3'' - 3\psi_4 - 6\psi_5 + 6\psi_4 \\ \psi_3 = -\psi_6 + 3\psi_6' \\ \psi_N = 0; \quad N > 3 \end{array}\right\} \quad (4.42)$$

Consequently, from Eq. (4.41), and considering only eight nearest-neighbor bonds, we obtain

$$\left.\begin{array}{l} \gamma_{tb} = \psi_3' - 3\psi_3 + 6\psi_3'' - 3\psi_4 - 6\psi_5 + 6\psi_4' \\ \gamma_{SF}(\text{extrinsic}) = 2\psi_3' - 6\psi_3 + 12\psi_3'' - 6\psi_4 - 12\psi_5 + 12\psi_4' \\ \gamma_{SF}(\text{intrinsic}) = 2\psi_3' - 6\psi_3 + 12\psi_3'' - 6\psi_4 - 12\psi_5 + 12\psi_4' \end{array}\right\} \quad (4.43)$$

It is observed in Eq. (4.43) that

$$\gamma_{SF}(\text{extrinsic}) = \gamma_{SF}(\text{intrinsic}) = 2\gamma_{tb} \qquad (4.44)$$

which is consistent with Eq. (3.62). By utilizing the same arguments, it can also be shown that the fcc–hcp interfacial (ϵ-phase boundary) energy in fcc metals and alloys is equal to that of a coherent twin boundary.

The calculation of stacking-fault energies as outlined in Eq. (4.43) is contingent upon the determination of the associated bonding or cohesive energies. Although experimental values for perfect-crystal cohesive energies are available,[218] accurate estimates of bonding energies at an interface are not known. Cohesive energies in transition metals, for example, have been shown to be related to the formation of the d-band. Therefore, electronic structure plays an important role,[219,220] and a precise calculation involves numerous difficulties. Ducastelle and Cyrot-Lackmann[219,220] have computed cohesive energies using an expansion of the density of states of the d-band in its moments, and, utilizing the plane-packing concepts outlined above, have calculated the stacking-fault energies for some of the transition metals—namely, nickel and cobalt. Their calculated values agree reasonably well with experimental values.

Cotterill and Doyama[214] adjusted the truncation radius for a Morse potential to give the proper stacking-fault energy for several pure metals, and Hodges[221] derived expressions for the stacking-fault energies in close-packed metals by carrying out peturbation calculations to second order in the pseudopotential.[56] However, a favorable comparison between calculated and experimental stacking-fault free energy values was observed only for aluminum.

Although accurate calculations of stacking-fault or twin boundary free energies have been lacking, the computations do verify the considerably lower value expected, as compared with grain boundaries, for example. In addition, because of the low interfacial energy associated with coherent twin and stacking-fault interfaces, the effects of interfacial entropy, which have been thus far ignored, would seem to be of considerable importance. Furthermore, as is suggested in the calculations of Ducastelle and Cyrot-Lackmann,[220] electronic contributions, although not appreciable at grain boundaries, can also become important at stacking-fault or twin boundaries, particularly in alloy systems where, as we can observe experimentally (Table 3.9), the stacking-fault energy varies with electron/atom ratio.

4.9 SUMMARY

It should now be clear that the distinct ranges of interfacial free energies recorded experimentally in Chapter 3 are, as suggested in Section 3.5, the result of (and a manifestation of) distinct structural features. These features have been described in terms of structural units involving specific geometric-crystallographic phenomena. It can be observed, retrospectively, that structural units involving dislocations, ledges, and coincidence units are the essential features of all metal and alloy interfacial structure, as is readily demonstrated on comparing Figs. 4.15 and 4.63, for example, and on considering the coincidence-ledge features of solid-solid interfaces, solid-liquid interfaces, and the free surface (solid-vapor interface).

It should be apparent that any metal or alloy interface can be described in terms of the characteristic structural units outlined in this chapter, and that the concomitant energy and associated properties will be the result of this structure. The complexity of interfacial structures has been shown to be the result of deviations in regular crystal structure and/or composition. Such deviations can give rise to superimposed structural units such as those characterizing lamellar interfaces in the Al–CuAl$_2$ eutectic alloy,[222] as outlined in detail in the range of interphase interface structures treated in Sections 4.6 and 4.7.

An attempt has been made in this chapter to present an integrated overview of the structure of interfaces in metals and alloys, and to demonstrate the remarkable similarities and characteristically identical structural units that essentially all interfaces possess. Some simple models for interfacial structure have been developed, and calculated interfacial free energies based on such models and structural units have been compared with experimentally determined values (Chapter 3). It was observed that calculated interfacial energies lose accuracy when the associated energies are small.

Having established many of the basic structural features of metal and alloy interfaces in this chapter, and having dealt in some detail with the thermodynamics and associated energetics of such interfaces in previous chapters, we have now essentially set the stage for the treatment of properties and applications of metals and alloys which depend in a most direct way on these interfacial phenomena. It will be shown in Chapter 5 that interfacial structure is a controlling feature of the flow characteristics, strength of metals and alloys, creep, and related phenomena which depend on grain growth and boundary migration, stacking-fault energy, precipitation, etc.

REFERENCES

1. R. C. Tolman, *J. Chem. Phys.*, **17**, 333 (1949).
2. J. G. Kirkwood and F. Buff, *J. Chem. Phys.*, **17**, 330 (1949).
3. R. Defay, I. Prigogine, A. Bellemans, and D. H. Everett, "Surface Tension and Adsorption," John Wiley & Sons, New York (1966).
4. K. P. Staudhammer and L. E. Murr, "Atlas of Binary Alloys: A Periodic Index," Marcel Dekker, New York (1973).
5. P. V. Geld and S. K. Chuchmarev, *Dokl. Akad. Nauk SSSR*, **83**, 877 (1952).
6. R. J. Good, W. G. Givens, and S. C. Tucek, *Advan. Chem. Ser.*, **43**, 211 (1963).
7. E. W. Müller and T. T. Tsong, "Field-Ion Microscopy," American Elsevier Publishing Co., New York (1969).
8. L. E. Murr, *Thin Solid Films*, **20**, 81 (1974).
9. M. C. Fleming, "Solidification Processing," McGraw Hill Book Co., New York (1974).
10. U. T. Son and J. J. Hren, *Surface Sci.*, **23**, 177 (1970).
11. T. T. Tsong and E. W. Muller, *J. Appl. Phys.*, **38**, 545 (1967).
12. R. W. Newman and J. J. Hren, *Phil. Mag.*, **16**, 211 (1967).

13. H. N. Southworth, *Surface Sci.*, **23**, 160 (1970).
14. L. E. Murr, O. T. Inal, and H. P. Singh, *Thin Solid Films*, **9**, 241 (1972).
15. F. C. Frank, *Discussions Faraday Soc.*, **5**, 48 (1949).
16. F. C. Frank, "Growth and Perfection of Crystals," John Wiley & Sons, New York, p. 411 (1958).
17. H. J. Pick, G. G. Storey, and T. B. Vaughan, *Electrochim. Acta*, **2**, 165 (1965).
18. W. K. Burton, N. Cabrera, and F. C. Frank, *Proc. Roy. Soc., Ser. A*, **243**, 299 (1951).
19. M. Volmer, "Kinetik de Phasenbildung," Steinkopff, Dresden and Leipzig (1939).
20. R. Becker and W. Doring, *Ann. Phys.*, **24**, 719 (1935).
21. J. W. Christian, "The Theory of Transformations in Metals and Alloys," Pergamon Press, Oxford and New York (1968).
22. T. Suzuki, *J. Cryst. Growth*, **20**, 202 (1973).
23. J. Frenkel, *Zh. Fiz. Khim.*, **19**, 392 (1945).
24. W. J. Dunning, "The Solid-Gas Interface," E. A. Flood (ed.), Vol. 1, Marcel Dekker, New York, p. 292 (1967).
25. D. P. Woodruff, "The Solid-Liquid Interface," Cambridge University Press, London (1973).
26. L. B. Lyon and G. A. Somorjai, *J. Chem. Phys.*, **46**, 2539 (1967).
27. P. W. Palmberg and T. N. Rodin, *J. Chem. Phys.*, **49**, 134 (1968).
28. R. S. Averback and D. N. Seidman, *Surface Sci.*, **40**, 249 (1973).
29. P. J. Estrup, "Modern Diffraction and Imaging Techniques in Material Science," S. Amelinckx *et al.* (eds.), American Elsevier Publishing Co., New York, p. 377 (1970).
30. E. A. Flood (ed.), "The Solid-Gas Interface," Vol. 1, Marcel Dekker, New York (1967).
31. G. A. Somorjai (ed.), "The Structure and Chemistry of Solid Surfaces," John Wiley & Sons, New York (1969).
32. R. L. Park and H. E. Farnsworth, *J. Chem. Phys.*, **43**, 2351 (1965).
33. T. Edmonds and R. C. Pitkethley, *Surface Sci.*, **15**, 137 (1969).
34. K. Baron, D. W. Blakely, and G. A. Somorjai, *Surface Sci.*, **41**, 45 (1974).
35. L. H. Germer, J. W. May, and R. J. Szostak, *Surface Sci.*, **7**, 430 (1965).
36. J. W. May and L. H. Germer, *Surface Sci.*, **11**, 443 (1968).
37. R. L. Park and H. E. Farnsworth, *J. Appl. Phys.*, **35**, 2220 (1964).
38. W. W. Mullins, "Metal Surfaces," ASM, Metals Park, Ohio (Joint ASM-AIME Seminar), p. 23 (1963).
39. R. Stratton, *Phil. Mag.*, **44**, 519 (1953).
40. A. S. Skapski, *Acta Met.*, **4**, 576 (1956).
41. S. N. Zadumkin, *Dokl. Akad. Nauk SSSR*, **101**, 507 (1955).
42. V. K. Semenchenko, "Surface Phenomena in Metals and Alloys," Pergamon Press, New York (1962).
43. Y. S. Avraamov and A. G. Gvozdev, *Fiz. Metal Metalloved.*, **23**, 405 (1967).
44. S. N. Zadumkin, Institute for Powder Metallurgy and Special Alloys, Izd. Akad. Nauk, Ukr. SSR, Kiev, p. 7 (1963).
45. A. Kh. Breger and A. A. Zhukhoritskii, *Zh. Fiz. Khim.*, **20**, 355 (1946).
46. S. N. Zadumkin, A. I. Temrokov, I. G. Shebzukhora, and I. M. Aliev, Institute for Materials Science Problems, Izd. Naukova Dumka, Kiev, p. 9 (1968).
47. S. N. Zadumkin, *Fiz. Metal. Metalloved.*, **11**, 331 (1961).
48. W. Missol, *Phys. Status Solidi (b)*, **58**, 767 (1973).
49. J. Frenkel, *Z. Phys.*, **51**, 232 (1928).
50. J. Bardeen, *Phys. Rev.*, **49**, 653 (1936).
51. P. Hohenberg and W. Kohn, *Phys. Rev.*, **136**, B864 (1964).
52. W. Kohn and L. J. Sham, *Phys. Rev.*, **140**, A1133 (1965).
53. J. R. Smith, *Phys. Rev.*, **181**, 522 (1969).
54. N. D. Lang and W. Kohn, *Phys. Rev.*, **1**, B4555 (1970).

55. M. Djafari Rouhani and R. Schattler, *Surface Sci.*, **38**, 499 (1973).
56. W. A. Harrison, "Pseudopotentials in the Theory of Metals," W. A. Benjamin, New York (1966).
57. B. V. Belogurov, Institute for Powder Metallurgy and Special Alloys, Izd. Akad. Nauk Ukr. SSR, Kiev, p. 19 (1963).
58. N. D. Lang, *Solid State Commune*, **7**, 1047 (1969).
59. N. W. Ashcroft, *Phys. Lett.*, **23**, 48 (1966).
60. J. F. Nicholas, *Australian J. Phys.*, **21**, 21 (1968).
61. D. Turnbull, *J. Appl. Phys.*, **21**, 1022 (1950).
62. D. Turnbull, *J. Chem. Phys.*, **18**, 768 (1950).
63. J. W. Taylor, *Phil. Mag.*, **46**, 857 (1955).
64. M. Brillouin, *Ann. Chem. Phys.*, **13**, 77 (1898).
65. G. Quincke, *Proc. Roy. Soc., Ser. A*, **76**, 431 (1905).
66. W. Rosenhain and J. C. W. Humphrey, *J. Inst. Metals*, **10**, 125 (1913).
67. W. Rosenhain and D. Ewen, *J. Inst. Metals*, **10**, 125 (1913).
68. N. F. Mott, *Proc. Roy. Soc.*, **60**, 391 (1948).
69. R. King and B. Chalmers, *Progr. Metal Phys.*, **1**, 127 (1949).
70. K. T. Aust and B. Chalmers, "Metal Interfaces," ASM, Cleveland, p. 153 (1952).
71. D. McLean, "Grain Boundaries in Metals," Clarendon Press, Oxford (1957).
72. S. Amelinckx and W. Dekeyser, *Solid State Phys.*, **8**, 325 (1959).
73. F. Weinberg, *Progr. Mater. Sci.*, **8**, 105 (1959).
74. R. C. Gifkins, *Mater. Sci. Eng.*, **2**, 181 (1967).
75. K. T. Aust and B. Chalmers, *Met. Trans.*, **1**, 1095 (1970).
76. H. Gleiter, *Phys. Status Solidi (b)*, **45**, 9 (1971).
77. H. Gleiter and B. Chalmers, *Progr. Mater. Sci.*, **16** (1972).
78. F. Hargreaves and R. J. Hills, *J. Inst. Metals*, **41**, 257 (1929).
79. W. L. Bragg, *Proc. Roy. Soc.*, **52**, 54 (1940).
80. J. M. Burgers, *Proc. Roy. Soc.*, **52**, 52 (1940).
81. W. T. Read and W. Shockley, *Phys. Rev.*, **78**, 275 (1950).
82. F. L. Vogel, *Acta Met.*, **3**, 245 (1955).
83. J. J. Gilman and W. G. Johnson, in "Dislocations and Mechanical Properties of Crystals," J. C. Fisher *et al.* (eds.), John Wiley & Sons, New York, p. 116 (1957).
84. L. E. Murr, R. J. Horylev, and W. N. Lin, *Phil. Mag.*, **22**, 515 (1970).
85. A. H. Cottrell, "Dislocations and Plastic Flow in Crystals," Oxford University Press, Fairlawn, New Jersey (1953).
86. J. Weertman and J. R. Weertman, "Elementary Dislocation Theory," The Macmillan Co., New York (1964).
87. N. Gjostein and F. N. Rhines, *Acta Met.*, **7**, 319 (1959).
88. R. J. Wagner and B. Chalmers, *J. Appl. Phys.*, **31**, 581 (1960).
89. L. E. Murr, *Phys. Status Solidi (a)*, **3**, 447 (1970).
90. G. Hasson, J. Y. Boos, I. Herbeural, M. Biscondi, and C. Goux, *Surface Sci.*, **31**, 115 (1972).
91. J. C. M. Li, *J. Appl. Phys.*, **32**, 525 (1961).
92. W. Bollmann, *Phil. Mag.*, **7**, 1513 (1962).
93. W. Bollmann, *Discussions Faraday Soc.*, **38**, 26 (1964).
94. M. L. Kronberg and F. H. Wilson, *Trans. AIME*, **85**, 501 (1949).
95. F. C. Frank, "Conference on Defects in Crystalline Solids," The Physical Society, London, p. 150 (1955).
96. D. G. Brandon, B. Ralph, S. Ranganathan, and M. S. Wald, *Acta Met.*, **12**, 813 (1964).
97. D. G. Brandon, *Acta Met.*, **14**, 1479 (1966).
98. S. Ranganathan, *Acta Crystallogr.*, **21**, 197 (1966).
99. P. H. Pumphrey and K. M. Bowkett, *Scripta Met.*, **5**, 365 (1971).

100. Y. Ishida, *J. Mater. Sci.*, **7**, 72 (1972).
101. T. S. Ke, *J. Appl. Phys.*, **20**, 274 (1949).
102. G. Baro, H. Gleiter, and E. Hornbogen, *Mater. Sci. Eng.*, **3**, 92 (1968).
103. H. Gleiter and G. Baro, *Mater. Sci. Eng.*, **2**, 224 (1967).
104. M. J. Marcinkowski, "Conference on Fundamentals Aspects of Dislocation Theory," Gaithersburg, Maryland, April 21-25 (1969), NBS Special Publication No. 317 (Washington, D.C.),(1970).
105. M. J. Marcinkowski, in "Electron Microscopy and the Structure of Materials," G. Thomas, R. M. Fulrath and R. M. Fisher (eds.), University of California Press, Berkeley, California, p. 382 (1972).
106. H. F. Ryan and J. Suiter, *Acta Met.*, **14**, 847 (1966).
107. J. Levy and C. Goux, *Mem. Sci. Rev.*, **LXIV**, 663 (1967).
108. J. Levy, *Phys. Status Solidi*, **31**, 193 (1969).
109. Y. Ishida, T. Hasegawa, and F. Nagata, *J. Appl. Phys.*, **40**, 2182 (1969).
110. T. Schober and R. W. Balluffi, *Phys. Status Solidi (a)*, **44**, 115 (1971).
111. B. Loberg and H. Norden, *Ark. Fysik*, **40**, 413 (1970).
112. B. Loberg, H. Norden, and D. A. Smith, *Phil. Mag.*, **24**, 897 (1971).
113. R. W. Balluffi, Y. Komen, and T. Schober, *Surface Sci.*, **31**, 68 (1972).
114. Y. Ishida and M. McLean, *Phil. Mag.*, **27**, 1125 (1972).
115. I. M. Bernstein and B. B. Rath, *Surface Sci.*, **31**, 101 (1972).
116. L. E. Murr, *Met. Trans.*, **6A**, 505 (1975).
117. J. C. M. Li and Y. T. Chou, *Met. Trans.*, **1**, 1145 (1970).
118. A. Mascanzoni and G. Buzzichelli, *Phil. Mag.*, **22**, 857 (1970).
119. L. E. Murr, *Appl. Phys. Lett.*, **24**, 533 (1974).
120. H. F. Ryan and J. Suiter, *Phil. Mag.*, **10**, 727 (1964).
121. T. L. Lin and D. McLean, *Met. Sci. J.*, **2**, 108 (1968).
122. Y. Ishida and M. H. Brown, *Acta Met.*, **15**, 857 (1967).
123. R. E. Hook, *Met. Trans.*, **1**, 85 (1970).
124. R. J. Bayuzich and R. S. Goodrich, *Surface Sci.*, **23**, 225 (1970).
125. L. E. Murr, O. T. Inal, and G. I. Wong, in "Electron Microscopy and the Structure of Materials," G. Thomas, *et al.* (eds.), University of California Press, Berkeley, California, p. 415 (1972).
126. G. H. Bishop and B. Chalmers, *Scripta Met.*, **2**, 133 (1968).
127. K. Sadananda and M. J. Marcinkowski, EMG Report No. CXXII, University of Maryland, Center of Materials Research, College Park, Maryland, February (1973); *J. Appl. Phys.*, **45**, 1521, 1533 (1974).
128. W. Bollmann, *Phil. Mag.*, **16**, 363 (1967).
129. W. Bollmann, *Phil. Mag.*, **16**, 383 (1967).
130. W. Bollmann, "Crystal Defects and Crystalline Interfaces," Springer-Verlag, New York (1970).
131. P. H. Pumphrey, *Scripta Met.*, **6**, 107 (1972).
132. P. H. Pumphrey, *Scripta Met.*, **7**, 893 (1973).
133. A. R. Thölen, *Phys. Status Solidi (a)*, **2**, 537 (1970).
134. Y. Ishida and D. A. Smith, *Scripta Met.*, **8**, 293 (1974).
135. M. J. Marcinkowski, W. F. Tseng, and E. S. Dwarakadasa, *J. Mater. Sci.*, **9**, 29 (1974).
136. W. F. Tseng, M. J. Marcinkowski, and E. S. Dwarakadasa, *J. Mater. Sci.*, **9**, 41 (1974).
137. M. J. Marcinkowski and E. S. P. Das, *Phil. Mag.*, **26**, 1281 (1972).
138. M. J. Marcinkowski, K. Sadananda, and W. F. Tseng, *Phys. Status Solidi (a)*, **17**, 423 (1973).
139. M. J. Marcinkowski and K. Sadananda, *Phys. Status Solidi (a)*, **18**, 361 (1973).
140. M. J. Marcinkowski and E. S. Dwarakadasa, *Phys. Status Solidi (a)*, **19**, 597 (1973).
141. F. R. N. Nabarro, "Theory of Dislocations," Oxford University Press, Oxford, p. 120 (1967).

142. R. DeWitt, *J. Appl. Phys.*, **42**, 3304 (1971).
143. J. C. M. Li, *Surface Sci.*, **31**, 12 (1972).
144. J. C. M. Li, in "The Nature and Behavior of Grain Boundaries," H. Hu (ed.), Plenum Press, New York, p. 71 (1972).
145. H. B. Aaron and G. F. Bolling, *Surface Sci.*, **31**, 27 (1972).
146. A. Seeger and P. Haasen, *Phil. Mag.*, **3**, 470 (1958).
147. H. Gleiter, *Acta Met.*, **17**, 565 (1969).
148. J. P. Hirth and J. Lothe, "Theory of Dislocations," McGraw-Hill Book Company, New York (1968).
149. J. M. Oblak and B. H. Kear, "Proceedings of the Electron Microscopy Society of America," C. J. Arceneaux (ed.), Claitor's Publishing Division, Baton Rouge, Louisiana, p. 432 (1970).
150. V. A. Phillips, *Acta Met.*, **20**, 1143 (1972).
151. J. H. Van der Merwe, *Proc. Roy. Soc., Ser. A*, **63**, 616 (1950).
152. J. C. M. Li and B. Chalmers, *Acta Met.*, **11**, 243 (1963).
153. D. Whitwham and P. Lacombe, "Properties of Grain Boundaries," 4th Colloquium of Metallurgy, Saclay, 1960, Presses University de France, p. 149 (1961).
154. C. M. Sargent, *Trans. AIME*, **242**, 1183 (1968).
155. L. E. Murr, *Acta Met.*, **21**, 791 (1973).
156. K. T. Aust, *Trans. AIME*, **221**, 758 (1961).
157. D. H. Warrington and P. Bufalini, *Scripta Met.*, **5**, 771 (1971).
158. R. L. Fullman and J. C. Fisher, *J. Appl. Phys.*, **22**, 1350 (1951).
159. H. Gleiter, *Acta Met.*, **17**, 142 (1969).
160. M. Volmer and I. Estermann, *Z. Phys. Chem.*, **102**, 267 (1922).
161. W. Kossel, *Chemie*, **56**, 33 (1943).
162. C. Barrett and T. B. Massalski, "Structure of Metals," 3rd ed., McGraw-Hill Book Company, New York (1966).
163. L. E. Murr and A. Sosin, "Solid-State Electronics: A Text," Marcel Dekker, New York, in press.
164. H. W. Fuller and M. E. Hale, *J. Appl. Phys.*, **31**, 1699 (1960).
165. L. C. De Jonghe, "Proceedings of the Electron Microscopy Society of America," C. J. Arceneaux (ed.), Claitor's Publishing Division, Baton Rouge, Louisiana, p. 26 (1973).
166. S. Amelinckx *et al.* (eds.), "Modern Diffraction and Imaging Techniques in Materials Science," North-Holland Publishing Company, Amsterdam (1971).
167. S. Amelinckx, *Surface Sci.*, **31**, 296 (1972).
168. J. S. Anderson and B. G. Hyde, *J. Phys. Chem.*, **28**, 1393 (1967).
169. P. Delavignette and S. Amelinckx, *Phil. Mag.*, **6**, 661 (1961).
170. G. Thomas and J. Washburn (eds.), "Electron Microscopy and the Strength of Crystals," Wiley-Interscience, New York (1963).
171. M. G. Hall, H. I. Aaronson, and K. R. Kinsman, *Surface Sci.*, **31**, 257 (1972).
172. G. Bäro and H. Gleiter, *Acta Met.*, **21**, 1405 (1973).
173. D. W. Pashley, *Advan. Phys.*, **14**, 327 (1967).
174. K. L. Chopra, "Thin Film Phenomena," McGraw-Hill Book Company, New York (1969).
175. J. H. Van der Merwe, *Surface Sci.*, **31**, 198 (1972).
176. C. W. Tucker, *J. Appl. Phys.*, **37**, 3013 (1966).
177. J. C. de Plessis and J. H. Van der Merwe, *Phil. Mag.*, **11**, 43 (1965).
178. H. Reiss, *J. Appl. Phys.*, **39**, 5045 (1968).
179. J. W. Matthews, *Surface Sci.*, **31**, 241 (1972).
180. J. L. Walter, H. E. Cline, and E. F. Koch, *Trans. AIME*, **245**, 2076 (1969).
181. H. E. Cline, J. L. Walter, E. F. Koch, and L. M. Osika, *Acta Met.*, **19**, 405 (1971).
182. H. I. Aaronson, C. Laird, and K. R. Kinsman, "Phase Transformations," ASM, Cleveland, Ohio, p. 313 (1970).
183. K. R. Kinsman, H. I. Aaronson, and E. Eichen, *Met. Trans.*, **2**, 1041 (1971).

184. M. Bouchard, R. J. Livak, and G. Thomas, *Surface Sci.*, **31,** 275 (1972).
185. C. Laird and H. I. Aaronson, *Acta Met.*, **17,** 505 (1969).
186. M. Mourey and F. Dabosi, *Mater. Res. Bull.*, **9,** 379 (1974).
187. A. Guinier, *Nature,* **142,** 569 (1938).
188. G. D. Preston, *Nature,* **142,** 570 (1938).
189. G. C. Weatherly, *Acta Met.*, **19,** 181 (1971).
190. C. Laird and H. I. Aaronson, *Acta Met.*, **15,** 73 (1967).
191. H. I. Aaronson, "Decomposition of Austenite by Diffusional Processes," John Wiley & Sons, New York (1962).
192. H. I. Aaronson, K. R. Kinsman, and C. Laird, Mechanisms of Diffusional Growth of Precipitate Crystals, in "Phase Transformations," ASM, Metals Park, Ohio (1969).
193. C. Laird and H. I. Aaronson, *J. Inst. Metals,* **96,** 222 (1968).
194. R. G. Faulkner and B. Ralph, *Acta Met.*, **20,** 703 (1972).
195. H. Gleiter, *Acta Met.*, **18,** 117 (1970).
196. J. A. Panitz, *Rev. Sci. Instrum.*, **44,** 1034 (1973).
197. M. A. Fortes and B. Ralph, *Acta Met.*, **15,** 707 (1967).
198. J. W. Rutter and K. T. Aust, *Acta Met.*, **13,** 181 (1965).
199. J. C. M. Li, "Recovery and Recrystallization of Metals," Interscience, New York, p. 160 (1963).
200. C. Crois and O. Dimitrov, *C. R. Acad. Sci. (Paris),* **252,** 1465 (1961).
201. H. Gleiter, *Acta Met.*, **18,** 117 (1970).
202. P. H. Pumphrey, *Scripta Met.*, **7,** 1043 (1973).
203. P. N. T. Unwin and R. B. Nicholson, *Acta Met.*, **17,** 1379 (1969).
204. P. N. T. Unwin, G. W. Lorimer, and R. B. Nicholson, *Acta Met.*, **17,** 1363 (1969).
205. D. Vaughn, *Acta Met.*, **18,** 183 (1970).
206. L. E. Murr, G. I. Wong, and R. J. Horylev, *Acta Met.*, **21,** 595 (1973).
207. D. Vaughn, *Phil. Mag.*, **25,** 281 (1972).
208. H. Gleiter, *Acta Met.*, **16,** 1167 (1968).
209. G. C. Hasson, J. B. Guillot, B. Baroux, and C. Goux, *Phys. Status Solidi (a),* **2,** 551 (1970).
210. G. Hasson, J.-Y. Boos, I. Herbeuval, M. Biscondi, and C. Goux, *Surface Sci.*, **31,** 115 (1972).
211. M. J. Weins, *Surface Sci.*, **31,** 138 (1972).
212. M. Biscondi, B. Baroux, and C. Goux, *C. R. Acad. Sci. (Paris),* **264C,** 483 (1967).
213. B. Baroux, M. Biscondi, and C. Goux, *Phys. Status Solidi,* **38,** 415 (1970).
214. R. M. J. Cotterill and M. Doyama, *Phys. Rev.*, **145,** 465 (1966).
215. G. Hasson and C. Goux, *C. R. Acad. Sci. (Paris),* **627B,** 1314 (1971).
216. R. W. Ewing, *Acta Met.*, **19,** 1359 (1971).
217. R. W. Ewing and B. Chalmers, *Surface Sci.*, **31,** 161 (1972).
218. K. A. Gschneider, *Solid State Phys.*, **16,** 275 (1964).
219. F. Ducastelle and F. Cyrot-Lackmann, *J. Phys. Chem. Solids,* **31,** 1295 (1970).
220. F. Ducastelle and F. Cyrot-Lackmann, *J. Phys. Chem. Solids,* **32,** 285 (1971).
221. C. H. Hodges, *Phil. Mag.*, **15,** 371 (1967).
222. I. G. Davies and A. Hellawell, *Phil. Mag.*, **19,** 1285 (1969).

PROBLEMS

(4.1) Draw a crystal unit cell for an fcc metal indicating the (102) plane. Make a ball-model sketch of the structure of a (102) surface along the lines depicted in Fig. 4.7. Show the location of the (102) surface in Fig. 4.6(*d*).

STRUCTURE OF INTERFACES 257

(4.2) Calculate the difference in surface free energy (F_S) for nickel and silver, and compare the calculated values with measured values extrapolated to $0°K$ ($-273°C$). Does the difference occur in the expected direction? Discuss any disagreement between measured and calculated values. Calculate the solid surface free energies for these metals from liquid metal data at the melting point, and compare with the measured solid-state values at temperature.

(4.3) Calculate the spacing of dislocations in the low-angle grain boundaries of Fig. 4.20(a) and (b), using the values of misorientation which can be measured from the micrographs, if $\phi = 90°$. Compare the calculated spacings with those you can actually measure from the micrographs of Fig. 4.20(a) and (b). Note that the Burgers vectors for dislocations in LiF and stainless steel are of the type $a[011]$ and $(a/2)[011]$, respectively.

(4.4) For a misorientation of (110) grains in pure nickel of $15°$, what would be the expected dihedral groove angle for a grain boundary intersecting the surface following an equilibrium anneal at $1060°C$? What would the groove angle be for $\Theta = 60°$?

(4.5) Referring to Fig. 4.22(d), indicate the probable Σ values for coincidence boundaries which may account for the cusps in the curve. Explain the absence of other possible cusps not observed. Make a sketch, similar to Fig. 4.24, illustrating the variation of grain boundary energy with misorientation for rotations about [210] in aluminum in the range $0 < \Theta < 180°$.

(4.6) Measure the average ledge height in Fig. 4.34(a). Since the misorientation was measured to be approximately $39°$, estimate the total number of *partial* dislocations which could be emitted from either of the two ledges in Fig. 4.34(a). (Note: Stainless steel possesses the fcc crystal structure.) What percentage of the source has already been emitted in Fig. 4.34(a)? If the grain in Fig. 4.34(a) had a mean diameter of 24 microns, and if all its boundaries could be characterized by the same misorientation, what could be the potential dislocation density in the grain when all grain boundary dislocations are emitted? [Hint: Consider only total dislocations, and refer to Fig. 4.33(b).]

(4.7) Make a sketch similar to Fig. 4.48 showing an extrinsic stacking fault in an fcc metal or alloy. If this is an alloy that forms an ordered structure of the form ABX_2, make a sketch of the fcc unit cell, and show the schematic representation of an antiphase (domain) boundary in such an ordered solid solution.

(4.8) A thin film of gold is electrodeposited onto a chromium substrate. In a large-grain [001] section of the substrate, the gold overgrowth is observed to grow epitaxially by systematically removing chromium to attain a thin Au/Cr section for transmission electron microscopy. While the growth is epitaxial, the interface is characterized by a regular network of dislocations spaced approximately 60 Å. Make a schematic depicting the interface. Calculate the misfit. List two other fcc metals which, when grown epitaxially on the same chromium (001) grain, would produce a smaller misfit. How would the dislocation spacing at the interface be expected to change?

(4.9) Assume the solid circles in Fig. 4.64(b) to represent solute atoms in a dilute solid solution alloy. Estimate the concentration. What is the grain boundary misorientation in Fig. 4.64(b)? Estimate the solute concentration in the grain boundary. If the misorientation is increased $10°$, would the solute concentration be expected to be altered? Why? Would it increase or decrease? Make a sketch similar to Fig. 4.64(b) to support your argument.

(4.10) Describe a simple bicrystal experiment to verify the theoretical difference between (311) coincidence boundaries and general grain boundaries in aluminum as depicted in Fig. 4.66(b). Use sketches to illustrate a simple numerical calculation.

5

PROPERTIES OF INTERFACES

5.1 ADSORPTION AND REACTION AT A FREE SURFACE

It was demonstrated in Chapter 1 that, when two immiscible phases are separated by an interface, the concentration of one phase is usually greater at the interface. The concentration which occurs—that is, surface excess—is the result of adsorption of the excess by the interface as a result of imbalance in the binding forces of the surface atoms which produces an attraction perpendicular to the interface plane. Adsorption therefore takes place with a decrease in surface free energy [$d\gamma_I$ in Eq. (1.25)].

5.1.1 Physisorption and Chemisorption

In an ideal gas–solid interface, two distinct binding characteristics are normally distinguishable. These involve a weak binding analogous to van der Waals' binding, which is primarily coulombic in nature and does not involve electronic bonds, and, conversely, a strong electronic binding, which resembles the binding involved in the formation of molecules. The presence of weakly bound excess gas atoms or molecules on a free surface is normally referred to as physical adsorption or physisorption, and the electronically bound excess gas atoms or molecules represent the phenomenon referred to as chemisorption. These features are illustrated schematically in Fig. 5.1 as a difference in the energy of adsorption.

Thermodynamically, a gas in contact with a solid surface can be phenomenologically treated as a two-phase single-component system. At equilibrium,

$$[-S\,dT + V\,dP]_{gas} = [-S\,dT + V\,dP]_{solid} \tag{5.1}$$

or

$$\frac{1}{P}\left(\frac{\partial P}{\partial T}\right)_n = \frac{\Delta H_{ads}}{RT^2} \tag{5.2}$$

where P is the pressure, T is the temperature, R is the gas constant, n refers to the number of moles of adsorbate, and ΔH_{ads} refers to a partial molar quantity termed

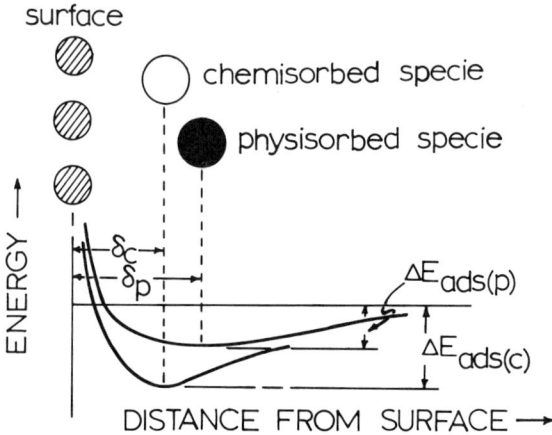

FIGURE 5.1 Idealized energy-surface displacement curves representing phenomenological aspects of physisorption and chemisorption. The terms ΔE_{ads} and δ represent the energy of adsorption and the equilibrium distance respectively.

the isosteric heat of adsorption, since it refers to a uniform gas adsorbate. Since the heat of adsorption and the energy of adsorption can be related by

$$\Delta H_{ads} \simeq \Delta E_{ads} + RT \tag{5.3}$$

it is observed from Fig. 5.1 that the heats of adsorption for physisorption and chemisorption will normally differ at constant temperature, and that $\Delta H_{ads(p)} < \Delta H_{ads(c)}$; and this is found experimentally to be perhaps the best single criterion for determining the characteristic adsorption type operating. Although there is considerable spread in values of heats of adsorption within each category, chemisorption is usually characterized by values of $\Delta H_{ads(c)} > 0.25$ eV (where 1 eV/molecule = 23.05 kcal/mole).[1,2]

For the adsorption of a single component the Gibbs adsorption equation [Eq. (1.25)] can be written

$$-d\gamma_I = \Gamma d\mu \tag{5.4}$$

where Γ is referred to a Gibbs dividing surface located such that the surface excess of the solid is zero. For the solid–gas interface

$$-\int d\gamma_I = F_S - F_{S(ads)} \tag{5.5}$$

where F_S is the surface free energy of the clean surface at some constant tempera-

PROPERTIES OF INTERFACES   261

ture, and $F_{S(\text{ads})}$ is the surface free energy when adsorption occurs. Substitution of Eq. (5.5) into Eq. (5.4) allows a two-dimensional (surface) equation of state to be written as

$$F_S - F_{S(\text{ads})} = RT \int_0^P d \ln P \tag{5.6}$$

where R is the gas constant, T is the absolute temperature, and P is the pressure of the adsorbate. The term on the left side of Eq. (5.6) represents a two-dimensional (surface) pressure exerted on the surface by the adsorbed monolayer. In effect, the surface pressure of a physisorbed monolayer is therefore numerically equal to the rate of change of surface free energy with the area of the monolayer.

Where temperature is not constant, the variation in surface free energy with adsorbant can be observed from an expanded form of Eq. (1.25) or (1.37) to be given by

$$dF_S/dT = -S_S + \Gamma S - (\Gamma d\mu/dT) \tag{5.7}$$

It is observed from Eq. (5.7) that, depending on the nature of adsorption, the surface free energy can either increase or decrease with temperature. This response is applicable to liquid metals as well as solid metals and is illustrated typically in Fig. 5.2, which also shows the variation of surface energy with the quantity of adsorbant—that is, surface excess. Similar variations in surface energy with surface excess are also apparent in Fig. 3.21.

Adsorption Isotherms. The Langmuir equation,[5]

$$P = K \frac{\theta}{1 - \theta} \tag{5.8}$$

where P is the gas pressure, θ is the monolayer coverage, and

$$\frac{1}{K} = a_0 e^{\Delta H_{\text{ads}}/RT} \tag{5.9}$$

(where $a_0 = 1/P$ when $\theta = 1$), has been widely used to describe the adsorption of gases on solids.[6-8] Since in many cases θ is unknown, θ may be written in terms of the volume, v, of gas adsorbed so that $\theta = v/v_m$. The quantity v_m is the volume of gas required for the adsorption of a complete monolayer (full coverage), and we observe from Eq. (5.9) that $a_0 = 1/P$ when $v = v_m$. Equation (5.8) is often written in the form

$$v = \frac{v_m bP}{1 + bP} \tag{5.10}$$

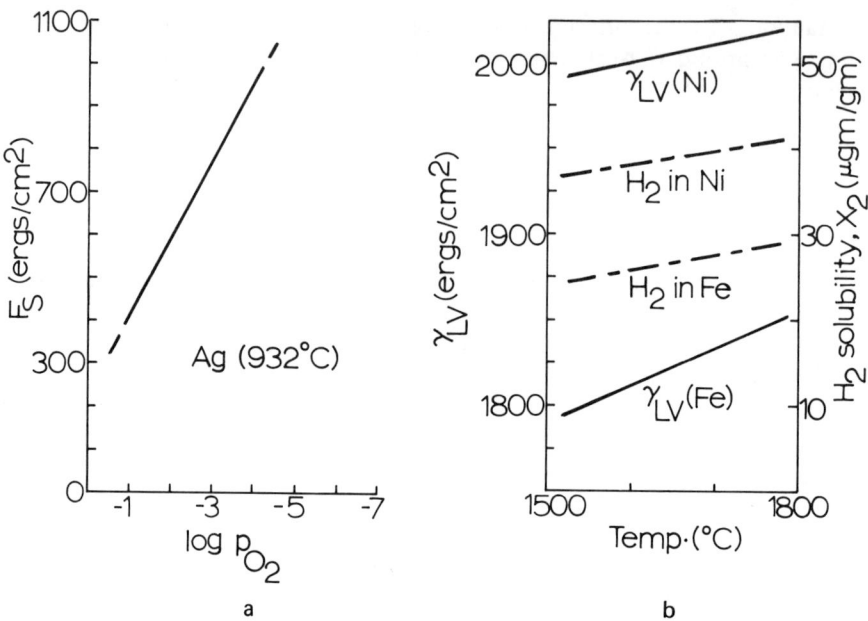

FIGURE 5.2 Variation of surface free energy with surface excess. (a) Surface energy variation of silver with oxygen adsorption.[3] (b) Surface tension variation of nickel and iron with hydrogen excess. (After calculations from ref. 4.) It should be noted that surface excess is related to bulk concentration, X_2, by combining Eqs. (1.39) and (1.41); that is, $\Gamma_{2(1)} = (X_2/kT)\, d\gamma_{LV}/dX_2$.

where $b = 1/K$ in Eq. (5.9). It is clear from Eq. (5.10) that at low pressures $v = v_m bP$, while at high pressures, where $bP \gg 1$, $v \simeq v_m$. Consequently, at high pressures the adsorption tends toward a limiting value (v_m) for which the monolayer is complete.

By converting Eq. (5.10) to a linear form such as

$$\frac{P}{v} = \frac{P}{v_m} + \frac{1}{v_m b} \quad (5.11)$$

a plot of P/v versus P should be a straight line from which the constants v_m and b may be obtained. In addition, in a Langmuir isotherm, ΔH_{ads} varies according to an equation of the form

$$\Delta H_{ads} = \Delta H_{ads(0)}(1 - A\theta) \quad (5.12)$$

where A is a constant and $\Delta H_{ads(0)}$ is the heat of adsorption at $\theta = 0$.

PROPERTIES OF INTERFACES

Figure 5.3 illustrates some examples of adsorption isotherms for pure metals as well as some examples of $\Delta H_{ads} - \theta$ curves. It is usually found that, in systems that exhibit chemisorption, θ values approach unity at low temperatures even at low gas pressures as a result of the generally higher values of ΔH_{ads}.

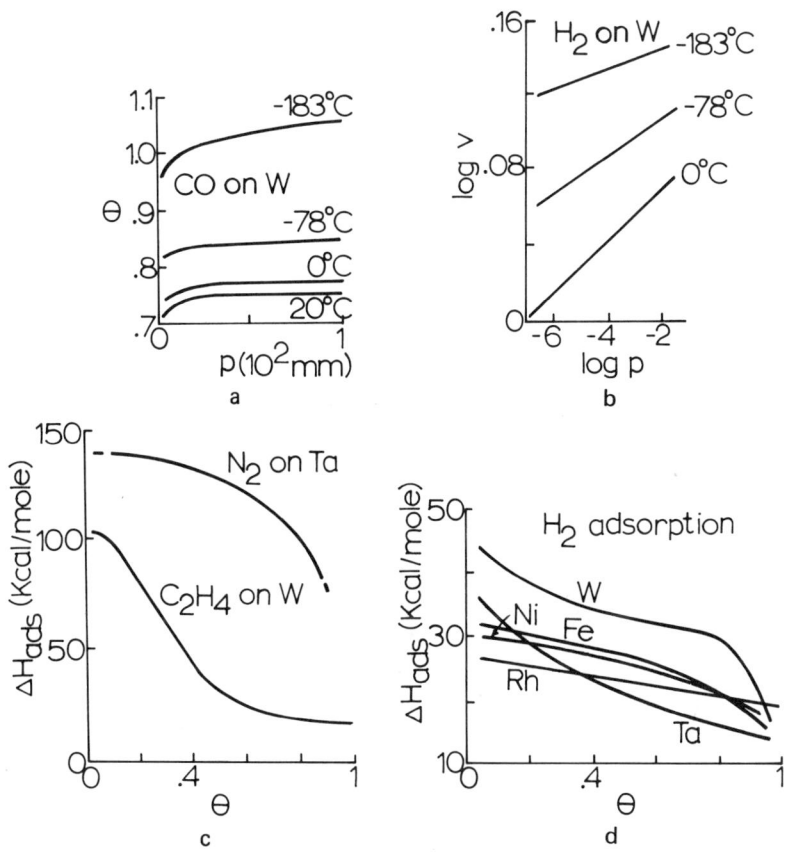

FIGURE 5.3 Adsorption kinetics. (a) Adsorption isotherm for CO on tungsten (after Rideal and Trapnell[9]). (b) Adsorption isotherm for H_2 on tungsten (after Trapnell[10]). (c) $\Delta H_{ads} - \theta$ curves for tantalum and tungsten (from data of Beeck et al.[11] and Beeck[12]). (d) $\Delta H_{ads} - \theta$ curves for hydrogen on different metals (after Beeck[12]).

Chemisorption and the Surface Molecule Concept. As indicated previously, chemisorption is distinguished from physisorption as a result of the chemical interaction of the adatom with the surface atoms. A useful picture of chemisorption on a metal surface is therefore one in which a localized molecule is formed between the adatom and the near-neighbor surface atoms. The theoretical situation has been outlined in several reviews[13,14] and recently treated in a rigorous manner by Thorpe[15] and by Gadzuk.[16] Various experimental techniques have also been developed which permit studies of the electronic nature of chemisorbed species to be performed.[17,18]

A simple concept of chemisorption involves a strong adatom interaction with a limited number of surface atoms to form a surface molecule. Such a molecule interacts only weakly with the other surface atoms. Desorption of CO from nickel, as well as observations of the catalytic properties of alloys,[19] to be discussed later, provides an indication of this possibility. Ideally, the surface molecules which characterize chemisorption arise by an interaction of the valence electrons of the adsorbate with a group orbital formed by some combination of surface atom orbitals; a simple picture of this phenomenon is schematically presented in Fig. 5.4.

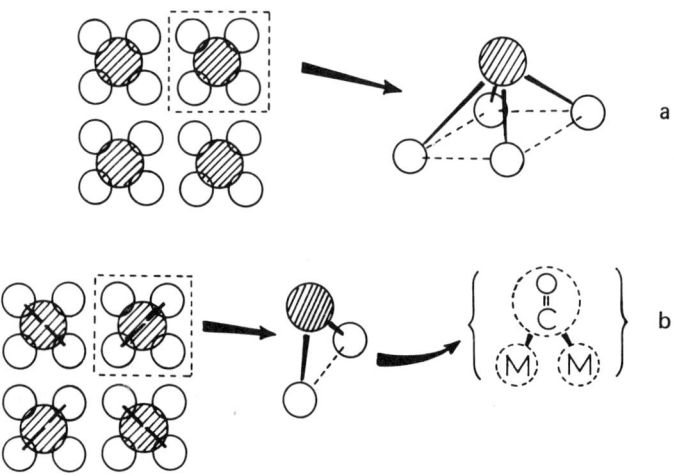

FIGURE 5.4   Schematic representation of surface molecules formed by chemisorption on metals. (a) $M_4 X$ pyramidal surface molecule on (100) bcc metal (M) surface; X denotes the adatom or adsorbate species. (b) $M_2 CO$ molecule on (100) bcc metal surface.

Because the adatom interactions are not confined to valence electrons (outer orbitals), the energetics of chemisorption would be expected to vary with coverage as a result of necessary variations in the type of surface molecule formed. This feature has been demonstrated experimentally in Fig. 5.3, and a simple schematic representation of such surface-structural variations with an increase in surface coverage is illustrated in Fig. 5.5 for (100) bcc metal surface chemisorption. In Fig. 5.5 we note that the terminal density shown need not be the only possible case

PROPERTIES OF INTERFACES   265

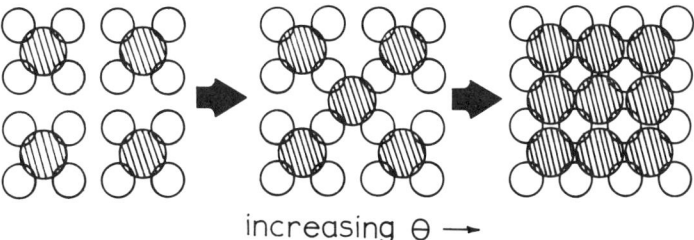

FIGURE 5.5   Systematic chemisorption on (100) bcc metal surface.

for $\theta = 1$; however, this is, in the particular case depicted, representative of the fully dense monolayer.

It should be apparent with reference to Figs. 5.4 and 5.5 that, since the characteristics of the surface molecule can vary for any particular surface crystallography as a result of electronic variations in adatom-surface atom binding, the formation of surface molecules will also be sensitive to the arrangement of the surface atoms (surface crystallography). Thus, if it is energetically more favorable to form an M_3X complex than an M_4X complex, chemisorption would be favored on $\{111\}$ planes as compared with chemisorption on $\{110\}$ planes of a surface, as illustrated in Fig. 5.6. As a consequence, the heat of adsorption should be less for $\{111\}$

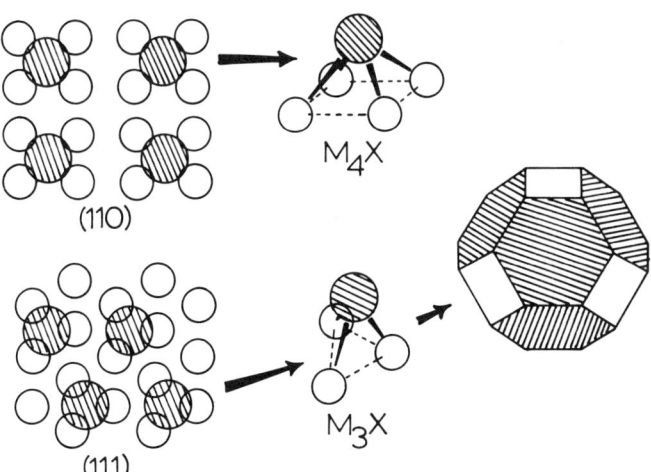

FIGURE 5.6   Variation of chemisorbed monolayer structure and energetics with metal surface orientations. The shading on the $\{111\}$ crystal faces of the tetrakaidecahedron suggests preferential chemisorption on a metal crystal.

chemisorption than for {110} or {100} chemisorption. This concept is supported experimentally, as illustrated, for example, in Fig. 5.7. The structure of the chemisorbed layer on the surface plane as depicted in Fig. 5.6 applies to gas adatoms or molecules such as O_2, H_2, N_2, and CO, as well as other adatoms. Structures similar to those depicted in Fig. 5.6 were demonstrated for the chemisorption of bismuth on copper, for example.[21]

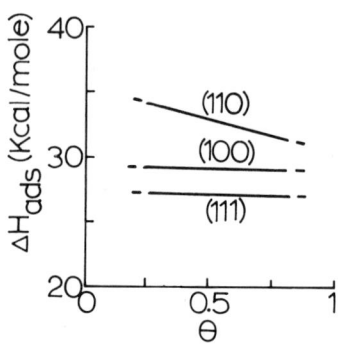

FIGURE 5.7 Heat of adsorption of sulfur on different crystallographic surface planes of silver. (From Bernard et al.[20])

5.1.2 Electronic Phenomena: Work Function Alteration

The work function of a metal can be defined as the difference between the electrostatic potential of an electron outside the surface, $-eV$, and the electrochemical potential of an electron inside the same metal, μ, or

$$E_W = -eV - \mu \tag{5.13}$$

or alternatively in terms of the Fermi energy, E_F:

$$E_W = E_B - E_F \tag{5.13a}$$

where E_B is the energy difference separating the top of the valence band from the bottom of the conduction band (the surface state). The emission of electrons from any metal surface can be viewed as a transmission through an atomic matrix. Such transmission is expected to be altered by the atomic arrangement for any particular crystal surface, and by the addition of adatoms to the surface which may alter the crystal structure or the electrostatic potential energy just outside the surface; this may serve to accelerate or suppress the emission of electrons.

PROPERTIES OF INTERFACES 267

The adsorption of a monolayer or more of positive or negative ions at a metal surface, as depicted schematically in Fig. 5.1, creates a modification in the electrostatic energy at the surface which can change the work function according to an expression of the form

$$E_W' = E_W - \frac{n_{A(+)} e^2 \delta}{\epsilon_0} \tag{5.14}$$

or

$$E_W' = E_W + \frac{n_{A(-)} e^2 \delta}{\epsilon_0} \tag{5.14a}$$

where $n_{A(\pm)}$ denotes the number of positive (+) or negative (−) ions of charge e per unit area of surface monolayer, ϵ_0 is the permittivity of free space, and δ is the equilibrium distance (δ_c or δ_p depicted in Fig. 5.1). In terms of the surface coverage, θ, the change in work function can also be written

$$\Delta E_W = (E_W' - E_W) = 4\pi n_{A(\pm)} \theta \mu' \tag{5.15}$$

where μ' is an effective dipole moment associated with each adatom or adsorbed molecule.

It is observed that, not only is the work function of a metal surface sensitive to surface orientation, but it will also vary with the geometry (crystallography) of the adsorbed layer or layers, and it can increase or decrease, depending on the nature of the adsorbate. These features are typically illustrated in Fig. 5.8.

5.1.3 Oxidation and Corrosion

In a general sense, whenever a metal surface undergoes chemisorption—that is, the metal surface passes from an elementary to a combined condition—corrosion of the surface occurs. Phenomenologically then, the formation of an oxide layer represents a form of corrosion because a portion of the metal surface is consumed in the formation of the surface compound. While oxidation characterizes one form of corrosion, a phenomenologically identical surface reaction can occur as a result of numerous other chemisorption processes—for example, sulfide reactions. There are many other forms of corrosion in which surface reactions are assisted or controlled by electrochemical phenomena or mechanical stress (strain energy). Because of the preference for energetically favorable reaction sites, corrosion in many instances is more common at certain surface structures, including defects. The etching at emergence sites of dislocation lines on a surface [Fig. 4.20(a)] and along grain boundaries or other interfaces which intersect a surface [Fig. 4.64(d)] are but a few typical examples of site preference. We shall defer until the following section a discussion of the effects of surface structure on reaction kinetics and catalysis.

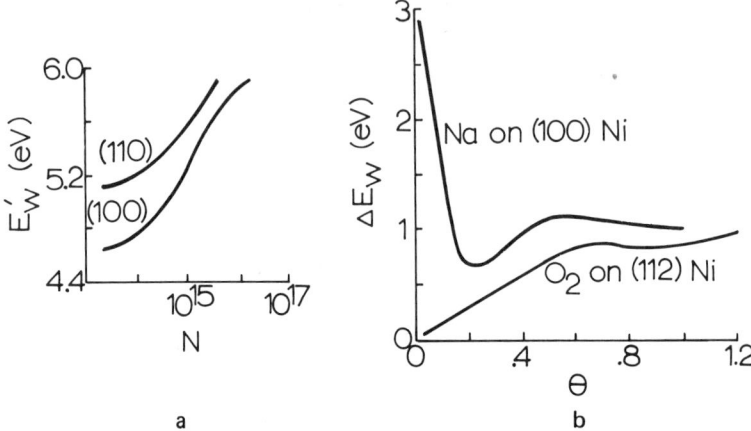

FIGURE 5.8 Change in work function with adsorption. (a) Oxygen on tungsten. N refers to the number of O_2 molecules colliding with each square centimeter of surface (from Hopkins and Pender[22]). (b) O_2 and sodium on nickel (from Blakely[23]).

Oxidation of Metals. Oxidation, as was pointed out above, begins with the chemisorption of oxygen on a metal or alloy surface. The surface coverage is usually not a monolayer, and it begins at nuclei which may have a preference for certain crystallographic surface orientations as well as surface structural features such as dislocations, steps, or ledges. The conditions for oxide nucleation may tend to vary with temperature, as surface structure has a tendency to vary with temperature. In addition, the energy required for chemisorption on specific crystallographic planes may decrease for certain surface-molecule formations, a feature which is illustrated conceptually in Figs. 5.4 and 5.6.

For an oxide layer to grow in excess of a monolayer or unit oxide layer also requires diffusion of either the oxygen or the metal through the oxide layer, or both. This will depend on the ease of diffusion of one or more of the reacting species through the oxide. Diffusion rates are temperature-dependent and in general follow an equation of the Arrhenius form:

$$D = D_0 \exp(-Q/RT) \qquad (5.16)$$

where D_0 is a constant for any particular oxidation process, and Q is the activation energy for the process. Studies of oxidation kinetics have indicated a variety of rate laws as a result of variations in the rates of chemisorption and continued oxide growth which, for various metals or alloys, will depend on the initial surface structure and crystallography, temperature, oxygen pressure, and diffusion rates.[24,25] These features are illustrated in the experimental oxidation data reproduced in Fig. 5.9.

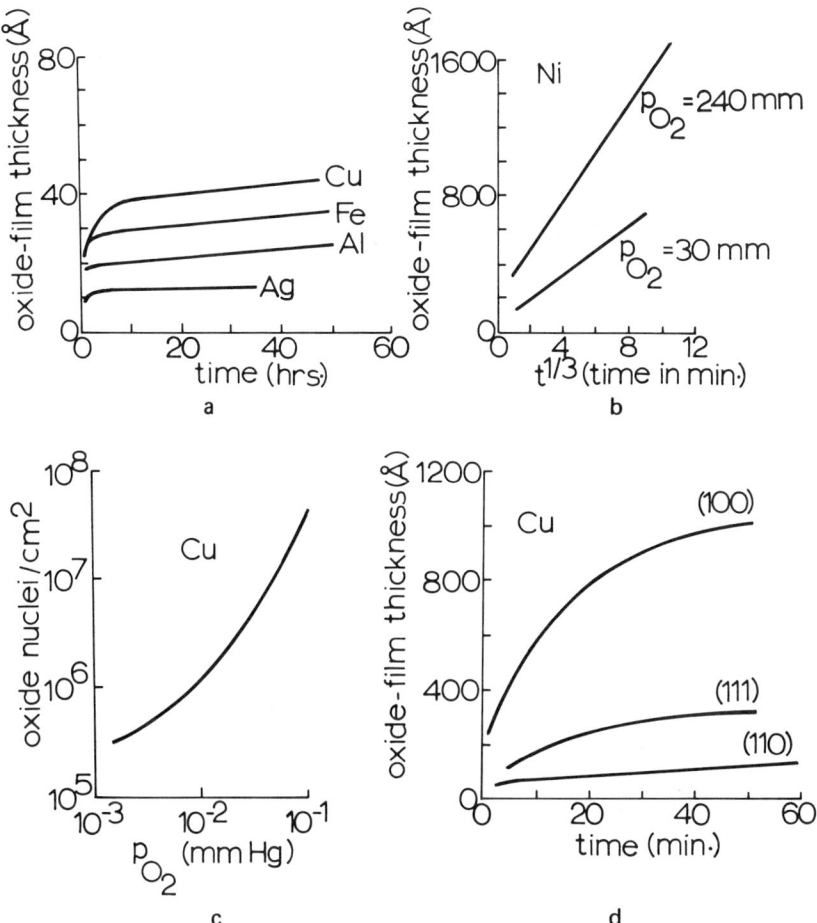

FIGURE 5.9 Examples of oxidation kinetics for a number of metals. (a) Thickness–time relationships for oxide films formed at room temperature.[24] (b) Thickness–time relationships for oxidation of nickel at 400°C.[24] (c) Oxide nuclei versus oxidation pressure for copper at 500°C.[26] (d) Oxidation rates on various crystallographic planes of copper at 178°C.[27]

There is considerable evidence that oxidation occurs primarily by the heterogeneous formation of oxide nuclei as a function of crystallographic orientation, oxygen pressure, and temperature,[28-32] and similar reactions—for example, surface metal sulfide formation—have been observed to follow a similar pattern.[25] In addition, Young[33] has indicated that there was a one-to-one correlation between oxide nuclei and dislocations in tellurium-doped copper crystals, whereas more recent work by Lawless[25] failed to show a correlation between oxide nuclei and dislocations in the metal. Phillips,[34] on the other hand, has shown that nuclei of Ag_2S

form at the emergence of stacking faults in thin single-crystal silver films. It is probably the variation in surface structure—that is, defects, polycrystalline versus single-crystal surfaces, etc.—which are responsible for the wide variations in experimental results of oxidation kinetics.

The morphology of oxide growth can vary considerably, depending on pressure, temperature, and crystallography; and oxide growth from a metal can be in the form of a continuous overgrowth or whiskers growing from some surface nucleus. The growth from the surface can be epitaxial or endotaxic, or the oxide structure can be crystallographically different from that of the metal. Since certain crystallographic orientations are preferred for chemisorption and oxide growth, oxidation rates may vary with temperature to the extent that recrystallization at high temperatures may reorient metal grains for preferred oxidation. Additionally, at high temperatures, the oxide itself may recrystallize. Several of these features are illustrated in Fig. 5.10, which shows a stage in the oxidation of erbium to Er_2O_3 near the melting point of erbium ($\sim 1300°C$).[35]

In the case of many oxide layers, the oxide formation is limited to only a few atom layers in thickness as a result of reduced diffusion and inability to perpetuate the reaction. This condition can act as a retardant to other corrosive reactions at the metal surface. Perhaps the best-known example is the oxidation of aluminum.

5.1.4 Reaction Kinetics and Catalysis

For a reaction to occur on a surface, it is necessary for the reactant molecules to diffuse to the surface. The reactants are then adsorbed, after which the surface reaction commences; this is followed by desorption of the reaction products and the diffusion of these products away from the surface. Heterogeneous catalytic reactions therefore consist in at least three single steps: adsorption, surface reaction, and desorption. The surface provides a reaction substrate for a particular reaction, and thereby promotes or catalyzes the reaction; that is, the reaction rate is changed. Since adsorption is sensitive to surface crystallography and surface structure such as the presence of defects, catalysis on the surface of metals is also likely to involve localized defects.

Following the notation of Clark,[36] the adsorption and desorption process may be expressed by

$$A + \Sigma \leftrightarrows (A - \Sigma) \tag{5.17}$$

where A represents an adatom or molecule adsorbed on a vacant site Σ. The rate of adsorption can be given in the form

$$r_a = [\sigma P_A/(2\pi mkT)^{1/2}] f(\theta) \exp(-E_a/kT) \tag{5.18}$$

where σ represents the fraction of molecules striking available surface sites with energies in excess of the activation energy, E_a, $P_A/(2\pi mkT)$ is the number of molecules of A striking unit surface area per second, P_A is the gas-phase pressure of

PROPERTIES OF INTERFACES

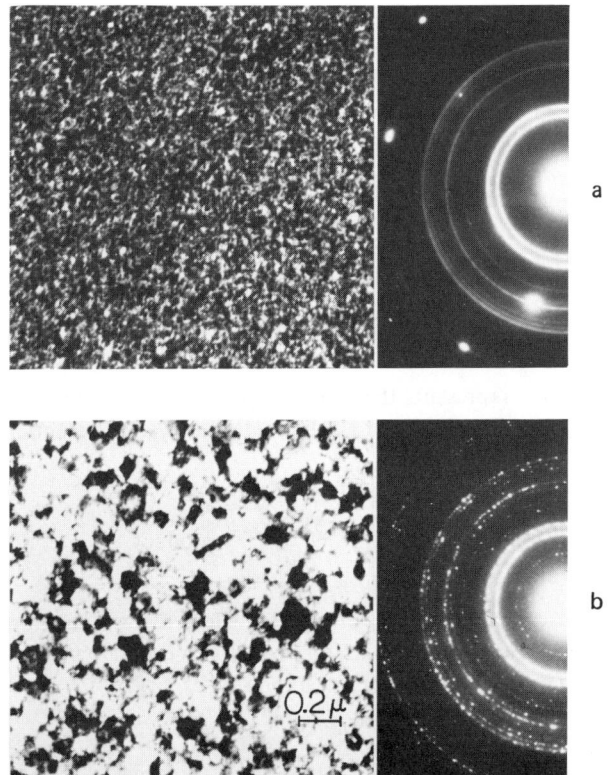

FIGURE 5.10 Erbium thin film oxidation in the electron microscope. (a) Pure erbium film, ~800 Å thick before oxidation. (b) Partially oxidized erbium film. Note grain growth and recrystallization which may have been conducive to oxidation or which may have occurred after oxidation. The associated selected-area electron diffraction patterns illustrate the change in the pure hcp erbium in (a) to a distribution of bcc Er_2O_3 grains in (b). Spots in (a) are from metastable ErO crystal in diffracting area.

A, m is the mass of A, and $f(\theta)$ represents the fraction of collisions of A taking place on available sites. If it is assumed that a molecule of A must collide directly with a vacant site to be adsorbed, then

$$f(\theta) = 1 - \theta \tag{5.19}$$

The rate of desorption can be expressed by an equation of the form

$$r_d = K_d \theta \exp(-E_d/kT) \tag{5.20}$$

where K_d is a desorption constant, and E_d is the activation energy of desorption. The activation energy of desorption can be expressed in terms of the activation energy of adsorption, E_a, and the heat of adsorption, ΔH_{ads}, in the form

$$E_d = E_a + \Delta H_{ads} \tag{5.21}$$

The effect of a catalyst on reaction rate may be illustrated by considering a simple bimolecular reaction A + B → product. For a homogeneous gas-phase reaction we can write

$$A + B \leftrightarrows (A \cdots B)^{\ddagger} \to product \tag{5.22}$$

where $(A \cdots B)^{\ddagger}$ represents the activated complex. If the reaction proceeds heterogeneously on the surface of a solid-metal catalyst, the reaction is expressed in the form

$$A + B + \Sigma_2 \leftrightarrows \begin{pmatrix} A \cdots B \\ | \quad | \\ -\Sigma - \Sigma - \end{pmatrix}^{\ddagger} \to product \tag{5.23}$$

where Σ_2 represents a vacant pair of adjacent sites. Clark[36] has shown that the ratio of the heterogeneous to the homogeneous reaction rate per unit volume of gas per unit area of catalyst surface can be expressed as

$$\left(\frac{r_{het}}{r_{hom}}\right) = \frac{C_{\Sigma_2}}{(q_{\ddagger}/V)} \exp\left(\frac{\Delta E}{RT}\right); \quad \Delta E = E_{hom} - E_{het} \tag{5.24}$$

where C_{Σ_2} is the concentration of vacant pairs of sites, $(q_{\ddagger}/V)$ is the partition function per unit volume of the activated complex, and E_{hom} and E_{het} are the activation energies of the homogeneous and heterogeneous reactions, respectively.

It can be shown from Eq. (5.24) that, in order that a significant advantage in rate be attained through the use of a catalyst, the activation energy of heterogeneous reaction must be considerably less than that of homogeneous reaction.[36] This is illustrated in the data for a number of metal catalysts reproduced in Table 5.1.

A typical example of a catalyzed reaction would be the Fischer-Tropsch synthesis,[37] which constitutes an important method for the production of gasoline and diesel fuels or the production of hydrocarbons in general from carbon monoxide and hydrogen using nickel or cobalt catalysts according to the reaction

$$nCO + (2n + 1)H_2 = nH_2O + C_nH_{2n+2} \tag{5.25}$$

Rideal[38] has suggested that the catalytic mechanism for this reaction is of the form illustrated in Fig. 5.11. This mechanism has been modified to illustrate oxygenated or alcoholic-type intermediate reactions which can also occur.[37,39]

PROPERTIES OF INTERFACES 273

TABLE 5.1
Activation Energies for Homogeneous and Heterogeneous Reactions[a]

Catalyst	Decomposition of	E_{hom} (kcal/mole)	E_{het} (kcal/mole)
Pt	HI	14.0	–
Au		25.0	44.0
Pt	N_2O	32.5	–
Au		29.0	58.5
Pt	CH_4	57.5	80.0

[a]Reproduced from Table 10.1 of Clark.[36]

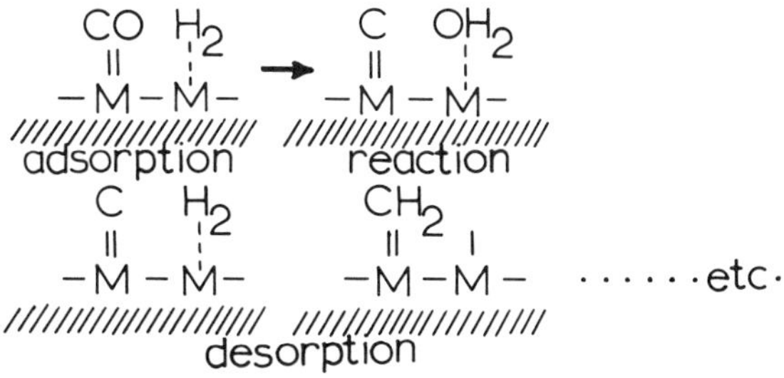

```
    CO  H₂           C   OH₂
    ‖   ‖       →    ‖    ‖
   -M - M-          -M - M-
   ///////////      ///////////
    adsorption       reaction

    C   H₂           CH₂
    ‖   ‖            ‖   |
   -M - M-          -M - M-    ······etc·
   ///////////      ///////////
          desorption
```

FIGURE 5.11 Catalytic mechanism for the Fischer-Tropsch synthesis.[38]

The importance of diffusion or related transport mechanisms in reaction kinetics and catalysis can be dramatically illustrated for tungsten single-crystal growth on tungsten filaments[40] in Fig. 5.12 by a reaction of the type

$$W\downarrow + 3H_2O\uparrow \leftrightarrows WO_3\uparrow + 3H_2\uparrow \qquad (5.26)$$

where the arrows, ↓↑, indicate a solid or vapor phase. The reaction proceeds by heating tungsten filaments in a gas mixture of 80 : 20 N_2 : H_2 in the presence of traces of H_2O vapor. At the hot regions of the filament (~2500°C) gaseous WO_3 is formed which diffuses through the gas to the slightly cooler tip (~2200°C) where it is reduced to metallic tungsten. The effect of temperature and H_2O vapor on the growth morphology and surface structure by reducing the surface mobility of W atoms and increasing their supply is shown in the sequence of scanning electron micrographs in Fig. 5.12.

Surface Structure and Catalysis. Since surface catalysis is initiated by adsorption, and since, as we attempted to illustrate in previous sections (Figs. 5.4 through 5.6), the surface-molecule complexes are controlled to a large extent by the atomic structure and crystallography of the surface, it is natural that catalysis itself would

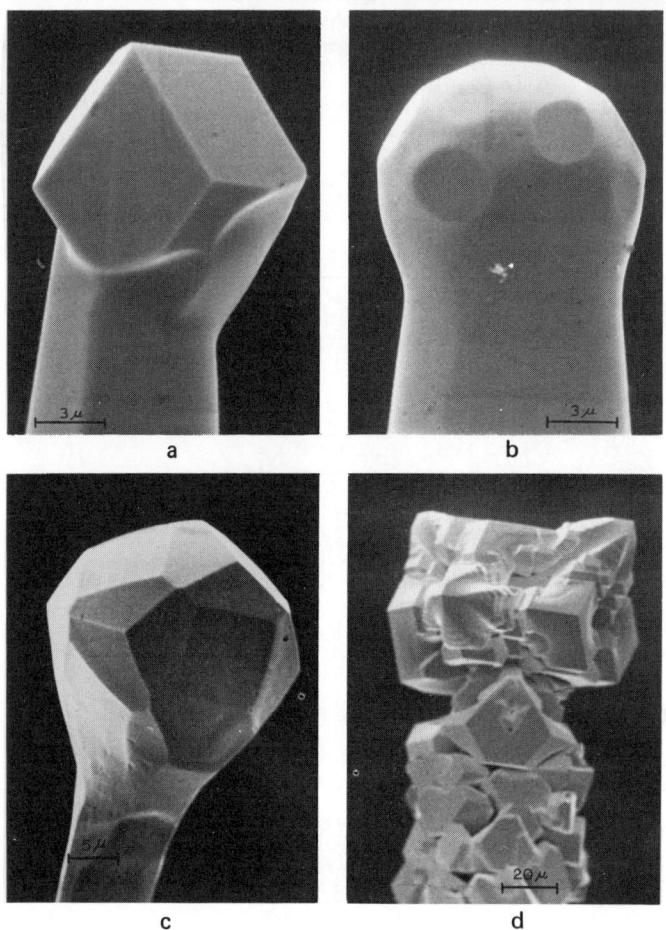

FIGURE 5.12 Reaction kinetics of tungsten single-crystal growth in a N_2-H_2 gas mixture. (a) Tungsten tip structure after 10 minutes in the pure gas mixture at 2600°C. (b) Same end form as (a) after the admission of H_2O vapor. (c) Tungsten end-form morphology after 7 minutes of exposure to the N_2-H_2 gas mixture and low H_2O vapor pressure at 1500°C. (d) End form of a tungsten filament exposed to a H_2O-rich N_2-H_2 gas mixture for 5 minutes at 2500°C. (Scanning electron micrographs courtesy of Dr. M. Prager.[40])

be affected by surface structure. A reaction might also be enhanced at ledge or kink sites as well as at vacant surface sites on particular crystallographic surface planes because of the energetically favorable coordination between the surface-adatom-reaction complex. These observations are supported by experimental

PROPERTIES OF INTERFACES   275

measurements of heats of adsorption and the calculation of activation energies for adsorption using various surface-lattice spacings.[12,41,42]

Dislocations, edge or screw, which emerge on the surface of metal catalysts might be expected to play some role in catalyzed reactions as a result of local adjustments in the coordination of surface complexes, or of adjustments in local adsorption energies. Sosnovsky[43] observed changes in catalytic activity of silver surfaces for formic acid decomposition when the surface was bombarded with positive argon ions; these changes were attributed to dislocation lines intersecting the surface. A quantitative correlation of dislocation density with rate of dehydrogenation of ethyl alcohol on LiF crystals has been given by Hall and Rase.[44] However, Bagg et al.[45] have claimed that the emergence of dislocations at the surface are unimportant for catalysis.

The establishment of the details of surface effects, particularly surface defects, in catalysis is a difficult problem because of the necessity to observe individual surface sites of atomic dimensions. No serious attempt has been made to resolve the problem of dislocation emergence as catalyst sites in the field-ion microscope where such resolution is available, and, as illustrated in Fig. 5.13, it is not always possible to associate dislocations in a metal with surface etch pits or other features that can be associated with preferential surface reaction sites. This is particularly true of electron transmission microscope studies in thin films as illustrated in Fig. 5.13 where dislocations may slip away from the site of an etch pit once formed or

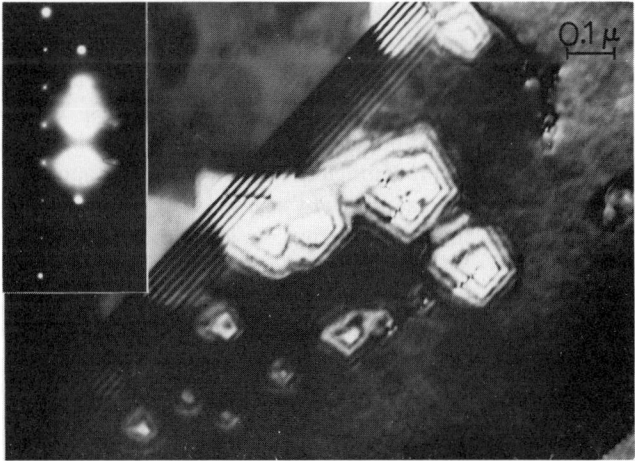

FIGURE 5.13   Etch pits associated with dislocations near a twin boundary in a chemically reacted nickel film. Note the asymmetry of the pits, which suggests preferential reaction along dislocation lines which nominally lie in slip planes inclined at ~55° to the foil (001) surface indicated by the selected-area electron diffraction pattern (insert).

in the state of formation. A possible means of direct observation would be the incorporation of a reaction stage directly within an electron microscope where the preferential reactions at dislocation lines (emergence sites on the surface or surfaces) in thin metal films could be observed at controlled temperature and reacting-gas pressure.

5.2 NUCLEATION AND GROWTH FROM THE VAPOR

Nucleation and growth from the vapor ultimately involve adsorption–desorption phenomena at a free surface as described above. Categorically, there are two types of vapor deposition and growth on a substrate: chemical and physical. In chemical vapor deposition, the process is essentially identical to a catalytic reaction and ideally involves the transport of the reactants to the substrate surface, reaction on the substrate (which may be either catalyzed or uncatalyzed), and the transport of products away from the deposition surface [Eq. (5.22)]. Since the diffusion rate of the reactants affects the diffusion rate (or transport) of the products, the transport of reactants to the substrate and the transport of products from the substrate are coupled by the reaction stoichiometry. Consequently, both diffusion and stoichiometry affect the reaction rate on the substrate.[46] The diffusion of reaction products over a substrate surface then forms a surface deposit which thickens as the process is continued. Coupled with the surface and/or gaseous reaction and continuous adsorption–desorption, the surface deposit is nucleated. Nuclei as well as individual reaction products can be involved in the surface diffusion processes. The residual growth of the deposit may be epitaxial, pseudomorphic, or endotaxic, and the surface morphology or growth characteristics—that is, surface structure of the overgrowth—may be controlled to a large extent by temperature, pressure, and the introduction of catalytic agents. These features are illustrated in the sequence of scanning electron micrographs in Fig. 5.12.

Physical vapor deposition is distinguished from chemical vapor deposition by the absence of any substrate reaction. Physical vapor deposition is thus characterized by adsorption–desorption phenomena coupled with diffusion and nucleation in a complex manner.

In the present treatment of vapor growth on a solid surface, we shall be concerned not only with the growth of a nonmetallic deposit upon a metal surface, but also with the growth of a metal upon a metal substrate, as well as the growth of a metal deposit upon a nonmetal substrate.

5.2.1 Homogeneous and Heterogeneous Nucleation

Homogeneous nucleation is characterized by a condition of critical supersaturation below which the nucleation rate is negligible, and above which the rate is large or infinite.[47-50] Heterogeneous nucleation, on the other hand, is characterized by a lack of critical supersaturation.[50,51]

Homogeneous Nucleation Theory. For homogeneous nucleation of a condensed vapor phase, the volume free energy change is given by

PROPERTIES OF INTERFACES

$$\Delta G_v = -(kT/v) \ln \Sigma \tag{5.27}$$

where k is Boltzmann's constant, T is the temperature, v is the molecular volume of the condensing adatoms, and Σ is the supersaturation ratio defined as the ratio of actual vapor pressure, P, to the equilibrium vapor pressure, P', in the system. If, for simplicity, a spherical nucleus is assumed,[52] the free energy of formation of a static critical nucleus of radius r^* is

$$\Delta G_v^* = 16\pi\gamma^3/3\Delta G_v^2 \tag{5.28}$$

where $\gamma = F_S$ or γ_{LV}, and F_S and γ_{LV} apply to the surface energies of the nucleus considered to be solid or liquid, respectively. If the concentration of critical-sized nuclei is then given by

$$N_{\text{hom}}^* = (PS'/kT) \exp(-\Delta G^*/kT) \tag{5.29}$$

where S' is an entropy factor described previously by Lothe and Pound,[53] then the steady-state nucleation rate can be regarded as the product of the diffusivity, D', and the concentration gradient, ZN_{hom}^*:

$$J_{\text{hom}} = D'ZN_{\text{hom}}^* \tag{5.30}$$

where Z is the Zeldovitch factor[54] expressed by

$$Z = \left[\frac{\Delta G^*}{3\pi kT} \left(\frac{4}{3} \frac{\pi r^{*3}}{v} \right)^2 \right]^{-1/2} \tag{5.31}$$

and for the impingement of single adatoms on any given nucleus,

$$D' = 4\pi r^{*2} P(2\pi mkT)^{-1/2} \tag{5.32}$$

On substituting for the parameters in Eq. (5.30), the rate of homogeneous nucleation can be expressed in terms of experimentally measurable terms in the form

$$J_{\text{hom}} = \frac{16\pi\gamma^2 P^2 ZS'}{\Delta G_v^2 kT(2\pi mkT)^{1/2}} \exp\left(\frac{-16\pi\gamma^3}{3\Delta G_v^2 kT} \right) \tag{5.30a}$$

Heterogeneous Nucleation Theory. Heterogeneous nucleation can be regarded as a geometrically modified case of homogeneous nucleation with regard to critical nucleus formation on the substrate as illustrated schematically in Fig. 5.14. As a consequence, the kinetics of heterogeneous nucleation can be described in terms of the same parameters that describe homogeneous nucleation, modified by a single additional parameter, Ω_C, in Fig. 5.14. It can be noted from Fig. 5.14 that the

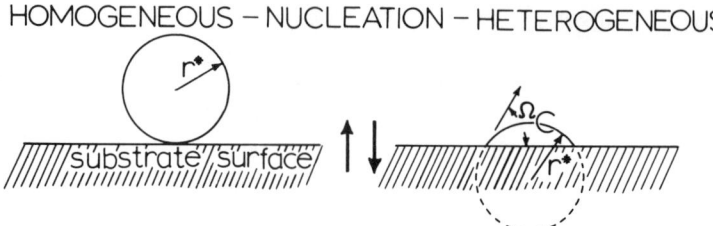

FIGURE 5.14 Critical radius and the concept of critical nuclei in homogeneous and heterogeneous nucleation. Arrows indicate movement of surface into the geometry of the spherical nucleus for deviations from homogeneous or heterogeneous nucleation—that is, for generating values of $\Omega_C \to 0$ to $180°$.

nucleus contact angle, Ω_C, is determined by interfacial equilibria as described in Chapter 2 [Eqs. (2.42) and (2.43)] (see Fig. 2.26).

The volume free energy change for heterogeneous nucleation is

$$\Delta G_v = -\frac{kT}{v} \ln \Sigma' = -\frac{kT}{v} \ln \left[\frac{J_{imp}(2\pi mkT)^{1/2}}{P'} \right] \quad (5.33)$$

where Σ' is the heterogeneous nucleation supersaturation ratio, expressed as the ratio of actual adatom concentration, n_a, to the equilibrium adatom concentration, $n_{a(e)}$, and J_{imp} is the impingement flux of atoms (of mass m), onto the surface. The free energy of formation of a critical nucleus of radius r^* (Fig. 5.14) is then

$$\Delta G^* = 16\pi\gamma^3 f(\Omega_C)/3\Delta G_v^2 \quad (5.34)$$

where $f(\Omega_C)$, the contact angle function, is given by

$$f(\Omega_C) = [(2 + \cos \Omega_C)(1 - \cos \Omega_C)^2]/4$$

$$= (2 - 3 \cos \Omega_C + \cos^3 \Omega_C)/4 \quad (5.35)$$

which represents the ratio of the volume of a spherical cap to the volume of the corresponding sphere in Fig. 5.14.

For heterogeneous nucleation, the concentration of critical nuclei is given by

$$N^*_{het} = [n_a n_{aw}/n_{a(e)}] \exp(-\Delta G^*/kT) \quad (5.36)$$

and the rate of addition of atoms to a nucleus is traditionally written

$$D' = W_c = n_{ads} 2\pi r^* \sin \Omega_C \, av \exp(-G_{sd}/kT) \quad (5.37)$$

where n_{ads} is the number of adsorption sites, a is the jump distance, ν is the surface vibration frequency, and G_{sd} is the activational energy for surface diffusion. If we again consider the nucleation rate to be the product of the diffusivity, W_c, and the concentration gradient, ZN^*_{het}, the rate of heterogeneous nucleation can be expressed in terms of experimentally measurable terms in the form

$$J_{het} = \frac{4\pi\gamma a n_a J_{imp} Z \sin \Omega_C}{-\Delta G_v} \exp\left[\frac{\Delta G_{des}}{kT} - \frac{\Delta G_{sd}}{kT}\right.$$

$$\left. - \frac{16\pi\gamma^3 f(\Omega_C)}{3kT \Delta G_v^2}\right] \tag{5.38}$$

where ΔG_{des} represents the activation energy for desorption. When the growth rate of nuclei is expressed by

$$D' = \omega_s = \pi a^2 J_{imp} \exp[(\Delta G_{des} - \Delta G_{sd})/kT] \tag{5.39}$$

Eq. (5.38) can be written in the form

$$J_{het} = \pi a^2 n_{ads} Z J_{imp} \exp\left[\frac{\Delta G_{des}}{kT} - \frac{\Delta G_{sd}}{kT} - \frac{16\pi\gamma^3 f(\Omega_C)}{3kT \Delta G_v^2}\right] \tag{5.38a}$$

where J_{het} represents the number of stable nuclei formed per square centimeter per second.

Although, as was indicated initially, homogeneous nucleation differs from heterogeneous nucleation because of the strong dependence of the former upon supersaturation, critical behavior has been predicted and observed for vapor-to-solid nucleation which is considered to be catalyzed by the free (solid) surface.[55,56] The dependence of heterogeneous nucleation on supersaturation can in fact be shown explicitly by differentiating Eq. (5.38) or (5.38a) to obtain

$$\frac{\partial \ln J_{het}}{\partial \ln \Sigma} = 1 - \frac{1}{\ln \Sigma} + \frac{32\pi F_S v^2 f(\Omega_C)}{3(k \ln \Sigma)^3}\left(\frac{F_S}{T}\right)^3 \tag{5.40}$$

The important parameters in Eq. (5.40) are the ratio F_S/T and $f(\Omega_C)$ (or Ω_C). Kenty[57] has made some computer-simulated heterogeneous nucleation experiments according to Eq. (5.38a) for silver on an unspecified substrate. Curves for a constant number of adsorption sites, $n_{ads} \simeq 10^{12}$ sites/cm^2,* for varying values of Ω_C and F_S are shown in Fig. 5.15. The results indicate that low values of contact angle and/or surface free energy are required for observable critical behavior. On

---

*Singh and Murr[58] have in fact measured the saturation density for indium and tin on (001) NaCl substrates to be 0.2 and 0.4 $\times 10^{12}$ cm^{-2}.

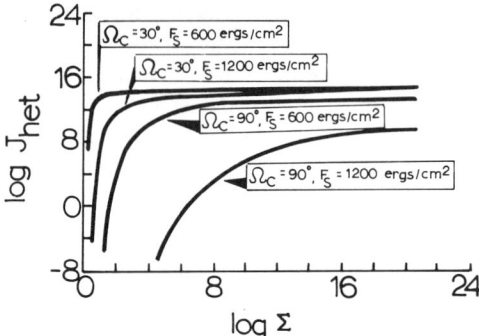

FIGURE 5.15 Heterogeneous nucleation rate versus supersaturation for solid silver with $n_{ads} = 10^{12}$ cm^{-2}. (From computer data of Kenty.[57])

the other hand, when both F_S and Ω_C are large, no critical behavior will occur because J_{het} varies slowly with supersaturation.

Venables[59] has given a rather complete theoretical treatment of the processes of thin film nucleation and growth on a defect-free substrate by taking into account the influence of random walk of the single atoms on the surface, which results in local variations in adatom concentration. In addition, Venables[59] has developed auxiliary diffusion equations which enable the cluster-cluster correlations and cluster size distributions[58,60] to be discussed.

5.2.2 Vacuum Vapor Deposition

Physical vapor deposition of metals onto metallic or nonmetallic substrates is normally carried out by the vaporization of the subject metal in a vacuum and its subsequent adsorption and condensation on the substrate surface, which may be polycrystalline or single crystalline.[46]

It can be observed from Eq. (5.38) that the nucleation of metal vapor upon a substrate will be governed by surface features similar to those described generally for adsorption phenomena above, and that nucleation rate is specifically controlled by the desorption energy, surface diffusion phenomena, temperature, and energetics associated with the formation of a metal-atom cluster upon the substrate. To a large extent, the nucleation phenomena will control the residual structure of the overgrowth, and therefore such features as the activation energy for adhesion [or desorption—Eq. (5.21)], the surface diffusion phenomena, cluster or overgrowth/substrate energetics, and the temperature of the substrate surface will be important parameters in the residual structure of the substrate-overgrowth interface as well as the residual structure of the overgrowth. As we have observed previously in Figs. 5.7 and 5.9(d), adatom adsorption will be noticeably influenced by the crystallographic orientation of the substrate. In addition, the structure of

PROPERTIES OF INTERFACES

the surface—that is, the presence of steps, kinks, ledges, etc.—will have a measurable influence on nucleation because of preferential sites occurring along such surface features as a result of a lower activation energy for adhesion, and the abrupt changes in the local values of ΔG_{sd}. Figure 5.16 illustrates these features for the nucleation of indium onto an irregular surface region of an (001) NaCl single-crystal surface. Similar variations in nucleation and residual film growth have also been observed following irradiation and other treatment of the substrate to induce surface-related defects.[61,62]

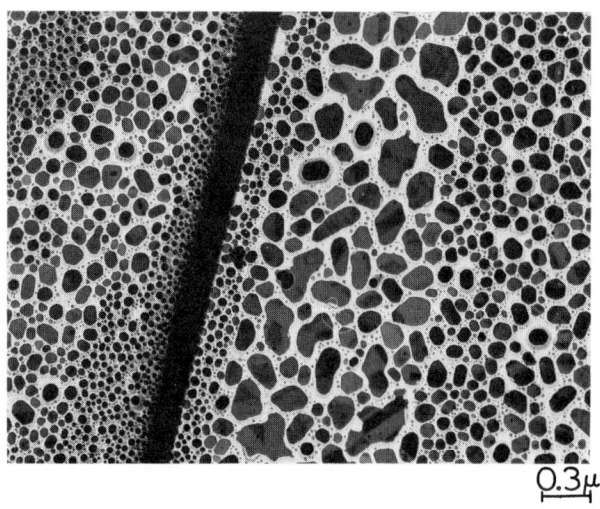

FIGURE 5.16 Variations in cluster density and size (nucleation rate) at surface steps and changes of crystallographic plane for indium vacuum vapor deposited onto (001) NaCl. The vacuum was 10^{-6} torr, and the substrate temperature was 100°C. Values of ΔG_{sd} and ΔG_{des} for indium on NaCl have been determined to be 0.13 and 0.36 eV, respectively.[58]

The growth from the vapor of a metal upon a metal will, if both substrate and overgrowth are sufficiently thick ($\gtrsim 10$ atom layers), be governed almost initially by the structural-energetic phenomena which describe the properties of solid–solid interfaces in Chapter 4. The misfit of different lattice atom sizes and crystallography will be accommodated by misfit dislocations as shown typically in Fig. 4.56.

Epitaxial Growth and Thin Film Structure. The growth of nuclei on a solid substrate surface can be considered categorically to occur in one of two modes: random orientation or epitaxial orientation. In general, an oriented overgrowth as a result of preferred nucleation might be expected on purely phenomenological grounds when one considers that, energetically, a coincidence-related interfacial adjustment might be favored. However, it must be realized that, for nuclei of the order of a few atom diameters in size, this criterion may be altered. The structure of the nuclei would also be expected to be crystalline, and to assume minimum energy configurations (see Fig. 4.4). Consequently, depending on the crystallographic orientation of the substrate, it would be expected that preferred orientations occur for metal nuclei, clusters, and subsequent overgrowths for which the interfacial energy would be smaller than that associated with random orientations.

H. P. Singh and L. E. Murr [*Met. Trans.*, **3**, 983 (1972)] have derived a ratio of epitaxial (oriented) nucleation rate to nucleation rate for randomly oriented nuclei on the same perfect substrate surface in the form

$$\frac{J_{\text{epi}}}{J_{\text{ran}}} = \exp\left\{\frac{16\pi F_S^3}{3kT\,\Delta G_v^2}\left[f(\Omega_C)_{\text{ran}} - f(\Omega_C)_{\text{epi}}\right]\right\} \quad (5.41)$$

where, $f(\Omega_C)_{\text{ran}}$ and $f(\Omega_C)_{\text{epi}}$ represent the geometrical factors for random and oriented (epitaxial) nuclei on the substrate. It is apparent from Eq. (5.41) that ΔG_v is a prominent parameter, and, if the substrate temperature is increased, the supersaturation will decrease, thereby increasing ΔG_v. Thus, epitaxy increases with increasing substrate temperature up to the point where desorption becomes controlling and vapor adatom adsorption declines. This is experimentally demonstrated in Fig. 5.17(*a*).

Even for an atomically perfect substrate surface, it is possible that the adsorption of foreign atoms or molecules as impurities would change the nucleation kinetics and the residual overgrowth structure because the local adsorption energy would increase, and simultaneously the values of ΔG_{sd} and ΔG_{des} would decrease. As a consequence, the epitaxial growth would vary with impurity concentration, decreasing with an increase in the number of impurities per unit area of substrate. Since there are two sources of impurities in a vacuum system—the background-vapor environment and the heated source during evaporation—the contamination of the substrate can be reduced in two associated ways: the evacuation of the system to high vacuum and the evaporation at rapid rates to eliminate entrapment of impurities within the vapor stream. As a consequence, epitaxy and thin film growth on a substrate would be expected to increase with a decrease in background pressure. This response is illustrated experimentally in Fig. 5.17(*b*). Figure 5.17(*a*) also depicts the variation in epitaxy with both pressure and evaporation rate.

It should be observed from Eqs. (5.38) and (5.41) that an increase in the flux rate at constant substrate temperature would normally decrease epitaxy because of the concomitant increase in the supersaturation for strictly heterogeneous nucleation. As a consequence, the experimental data depict a change in the mode of nucleation from homogeneous to heterogeneous, or the degree of nucleation character. In addition, the decrease in impurity impingement at higher evaporation rates also decreases the supersaturation until compensation as a result of strictly heterogeneous nucleation occurs.

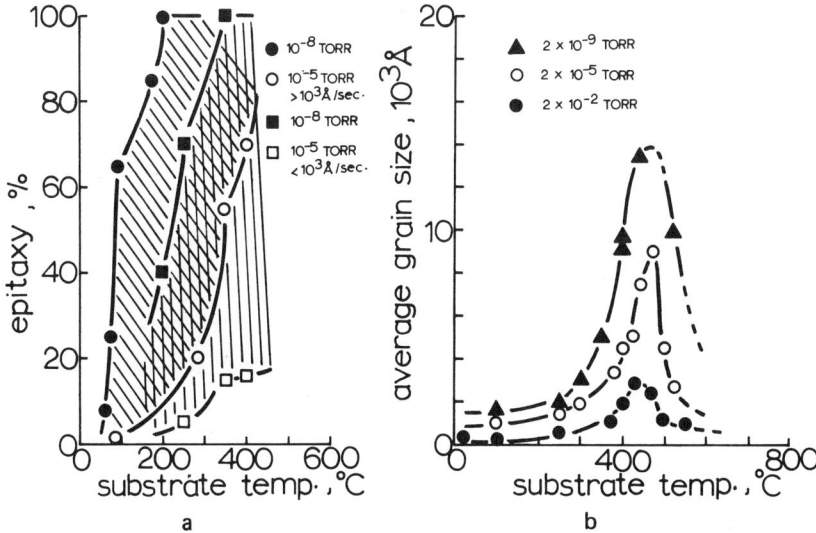

FIGURE 5.17 Effects of substrate temperature and background pressure on epitaxy and grain growth for metal vapor deposited onto (001) NaCl single-crystal substrates. (a) Epitaxial growth of palladium films having nominal thicknesses of 700 Å. The rate of metal evaporation was $\simeq 10^3$ or 3×10^3 Å/sec, indicated by $<10^3$ or $>10^3$ Å/sec, respectively. (From Murr.[63]) (b) Substrate temperature–pressure effects on nucleation and growth of pure aluminum films vapor-deposited at a rate of $\simeq 10^3$ Å/sec. (From Murr and Inman.[64])

Figure 5.18 illustrates somewhat typically the variation in metal overgrowth structure with pressure,[64] as well as the tendency toward epitaxy. There have also been experimental observations of overgrowths which accommodate atom for atom on the substrate, particularly at the very early stages of nucleation.[63,65] Figures 4.12 and 4.13 are examples of this type of pseudomorphic growth. Shifts in the overgrowth structure will occur with thickness from pseudomorphic to the normal epitaxial crystal structure with the creation of some residual strain. The continued vapor condensation, which for thick overgrowths must be considered as metal-metal adsorption, is accommodated by lattice defects in the overgrowth such as dislocations, stacking faults, or twins. Small microtwins and a high density of dislocations are observed in Fig. 5.18. Microtwins are particularly prominent in Fig. 5.18(b).*

*For the details of the effects of surface defect structure on epitaxial growth, the reader might consult the article by T. N. Rodin, P. W. Palmberg, and C. J. Todd in "Molecular Processes on Solid Surfaces," E. Drauglis, R. D. Gretz and R. I. Jaffee (eds.), McGraw-Hill Book Co., New York, p. 499 (1969). The effects of interfacial energy and related parameters on overgrowth recrystallization are also treated by J. L. Sacedon in *Thin Solid Films*, **22**, 165 (1974).

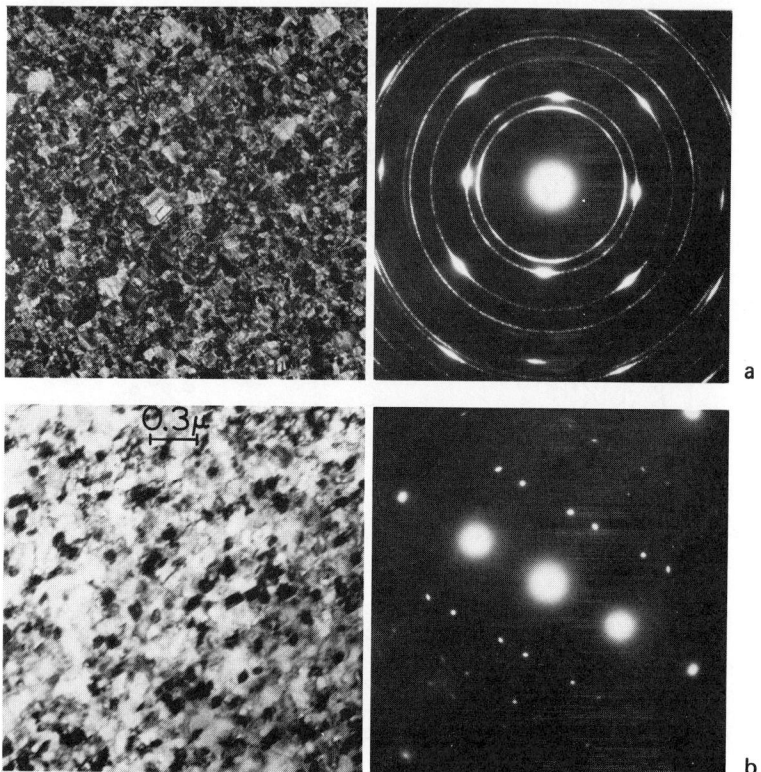

FIGURE 5.18 Structure and orientation of nickel films vapor-deposited onto (001) NaCl substrates at 275°C and a constant flux rate (evaporation rate). (a) Pressure of 10^{-5} torr; (b) pressure of 10^{-9} torr. Associated selected-area electron diffraction patterns illustrate relative orientation differences of clusters which evolve into residual grain structure. Part (b) illustrates perfect epitaxy [single crystal (001)]; oriented overgrowth.

5.3 METALLIC FRICTION AND ADHESION

Metallic surfaces in contact form a prominent and useful series of phenomena characteristic of surface interactions. These involve bearing surfaces, electrical contacts (static and dynamic), and a large variety of technological applications. Contacting surfaces which form a bond of any type are characterized by the associated energy of adhesion, while sliding contact characteristics are associated with surface friction.

To a large extent, the surface friction and adhesive properties of solid metal or alloy surfaces depend on the mechanical properties of the surface—for example, hardness. Rabinowicz[66] has shown that, on plotting the logarithm of liquid

PROPERTIES OF INTERFACES 285

surface free energies of metals as a function of the logarithm of their room-temperature hardness, a straight-line fit results with a slope of roughly 1/3. Thus, since the value of $\gamma_{LV} \propto F_S$ [Eq. (3.51)], we could write as a first approximation

$$H_S = KF_S^3 \qquad (5.42)$$

where H_S is the solid surface hardness (kilograms per square millimeter), and K is a constant of proportionality. It can also be shown that[66]

$$H_S = 3\sigma_y = 3\epsilon_y E_Y \qquad (5.43)$$

where σ_y is the yield strength, ϵ_y is the strain at the yield point on the stress–strain diagram, and E_Y is Young's modulus (see Fig. 3.19).

When two metal surfaces are brought together, the adhesive energy is characterized by the amount of work necessary to increase the separation of the surfaces from some equilibrium distance, of the order of a lattice spacing in one or the other component metals, to infinite distance. The energy of adhesion is given by

$$E_{Ad} = F_{S(a)} + F_{S(b)} - \gamma_{I(ab)} \qquad (5.44)$$

where $F_{S(a)}$ and $F_{S(b)}$ are the solid surface free energies of metals a and b, respectively, and $\gamma_{I(ab)}$ is the interfacial free energy at the contacting surfaces. The energy of adhesion, E_{Ad}, in Eq. (5.44) can be determined experimentally by utilizing Eq. (2.47). Note in Eq. (5.44) that when two identical metals are in contact, the adhesive energy becomes

$$E_{Ad} = 2F_S - \gamma_I \qquad (5.45)$$

If the lattices of each surface tend to match, the interface can be regarded as a low-energy coincidence boundary, where, for perfect registry of surface atoms, $\gamma_I = 0$. Ferrante and Smith[67] have calculated the variations in binding energy for perfectly coincident and noncoincident surface lattices.

The concept of adhesive energy can also be an important one in characterizing the adhesion of a solid metal and an immiscible, included phase such as a dispersoid particle in a metal lattice as shown in Fig. 4.61(d). In addition, adhesive properties are important for liquid metals on a solid metal or nonmetal surface, as this criterion is an indication of wettability—a feature of importance in solders and electrical contact applications. Table 5.2 lists several examples of experimentally determined adhesive energies.

5.3.1 Surface Structure and Friction

When two solid surfaces in contact are moved relative to one another, the force, F_f, opposing the movement is called the frictional force. If the applied force normal to the sliding surfaces in contact is denoted W, then we can define a coefficient of

TABLE 5.2
Experimental Values of Energy of Adhesion in Metal/Metal and Metal/Nonmetal Systems

System	Temperature (°C)	Contacting Phases	E_{Ad} (ergs/cm^2)	Ref.
Au/Al$_2$O$_3$	1000	Solid/solid	530	68[a]
Ag/Al$_2$O$_3$	700	Solid/solid	435	68[a]
Cu/Al$_2$O$_3$	850	Solid/solid	475	68[a]
Ni/Al$_2$O$_3$	1000	Solid/solid	645	68[a]
γ-Fe/Al$_2$O$_3$	1000	Solid/solid	800	68[a]
Stainless steel/Al$_2$O$_3$	1380	Solid/solid	1440	69
Ni/ThO$_2$	1200	Solid/solid	1100	69, 70
NiCr/ThO$_2$	1200	Solid/solid	640	70
Hg/Al	25	Liquid/solid	64	69
Hg/Pd	25	Liquid/solid	720	69
Hg/glass	25	Liquid/solid	92	69

[a]Calculated from data of Pilliar and Nutting.[68]

friction as

$$\mu = F_f/W \tag{5.46}$$

Amonton's law states that the coefficient of friction, μ, is independent of the apparent area of contact. Furthermore, the friction force, F, is independent of the sliding velocity, or the rate of displacement of one surface relative to the other.

The nature of the interaction between two metal surfaces or between any solid surfaces is determined by the real area of contact, A_r. Even in the case of two atomically smooth surfaces, the value of A_r will vary slightly with the movement of one lattice arrangement over the other. In practice, however, the surfaces are considered to be rough even on an atomic level owing to the presence of steps and ledges as depicted in Figs. 4.10 and 4.16. In addition, the roughening is enhanced by the presence of grain boundaries in a polycrystalline metal or alloy by the accommodation of the groove angle at the intersection of the grain boundary with the specimen surface as depicted schematically in Fig. 5.19. If the surface of Fig. 5.19 is inverted upon itself, the real area of contact is seen to be dependent upon the contacting force, W, and the surface hardness, and it will attain a value

$$A_r = W/H_S = W/KF_S^3 \tag{5.47}$$

when plastic deformation corresponding to the level of loading, W, has occurred.

In the case of rolling friction, we may also define a coefficient of rolling friction as

$$\mu_R = F_R/W \tag{5.47a}$$

PROPERTIES OF INTERFACES   287

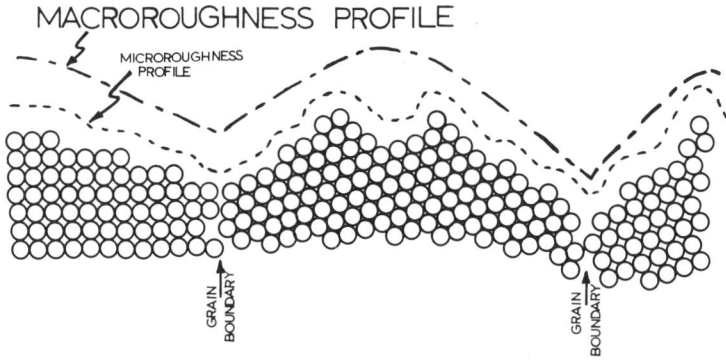

FIGURE 5.19   Enhanced surface roughening on the surface of polycrystalline metals and alloys.

where F_R is the force necessary to initiate rolling movement. If the specific case of a sphere rolling on a flat surface is considered, it must be observed that, since the surface of contact on the "flat" surface is elastically and plastically deformed, contact is made over an area determined by the deformation trough, and pure rolling action will generally not occur. The movement will be a combination of rolling and slipping; and both will influence the surface structure. An example of rolling-friction phenomena is shown in Fig. 5.20, which illustrates the residual surface and bulk deformation associated with a steel ball rolling over the surface of a thin stainless-steel film. For areas where plastic deformation has occurred as evidenced in the rolling path in Fig. 5.20, the friction force, F_R, varies as some power of the load, W, and inversely with the radius of curvature of the rolling elements.[66]

Wear. One of the practical examples of the applications and consequences of surface friction involves the wear of surfaces, particularly wear involving sliding or other abrasive contact, including surface-fatigue wear observed during repeated (or intermittent) sliding or rolling over a particular path or contact area. In the latter case, repeated loading and unloading (such as in an electrical contact) can cause the formation of surface or subsurface cracks which eventually result in the break-up of the surface, causing surface fragments and surface pits to form. It is this surface deterioration resulting in the formation of particles which characterizes surface wear.

The formation of surface matter as a result of wear can occur in two forms: the production of adherent or loose wear fragments. The latter particles are normally the result of sliding surfaces such as that shown in Fig. 5.19. Kerridge[72] has discussed the formation of wear particles as a two-stage process which involves the initial adhesion of an asperity on one surface to some small area of the other, followed by a shear within the asperity to form a fragment. Rabinowicz[66] has, on

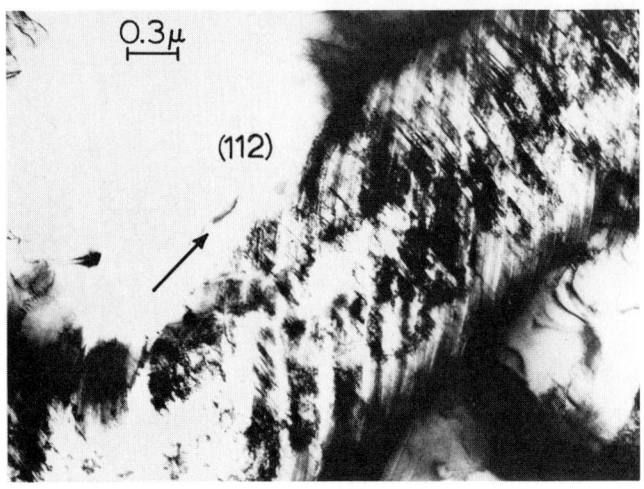

FIGURE 5.20   Residual deformation trough in a stainless-steel film resulting from the rolling of a carbon-steel ball over the surface as observed in the transmission electron microscope. Note the tendency of the surfaces to spin about the contact area as a result of frictional forces as evidenced by the bending of slip traces and deformation twins within the trough area. Arrow indicates rolling direction. (From Murr.[71])

assuming wear fragments to have a spherical shape, expressed the elastic-stored energy contained in such a particle as

$$E_e = (\nu^2 \sigma_y^2 / 2E_Y)(\pi d^3 / 12) \qquad (5.48)$$

where ν is Poisson's ratio, σ_y is the yield stress, E_Y is Young's modulus, and d is the average particle diameter. Similarly, the energy of adhesion at the point of contact can be written

$$E_a = E_{Ad}(\pi d^2 / 4) \qquad (5.49)$$

and, since the particle is removed as a free particle when $E_e \geqslant E_a$, we obtain

$$d \geqslant \frac{6E_Y E_{Ad}}{\nu^2 \sigma_y^2} = \frac{18}{\nu^2}\left(\frac{E_Y}{\sigma_y}\right)\left(\frac{E_{Ad}}{H_S}\right) \qquad (5.50)$$

It is observed in Eq. (5.50) that, for two contacting surfaces, abrasive fragmentation of the surface of the softer of the two surfaces will depend on the hardness (of the softer surface). It is also possible to write

PROPERTIES OF INTERFACES   289

$$d \doteq BF_S/H_S; \quad B = \text{constant} \tag{5.50a}$$

for the fragment size of the softer material, where F_S refers to the surface free energy of the softer surface (and consequently that of the fragments). As a consequence of the dependence of surface fragmentation (wear) on hardness, it would be expected that metallic surfaces with higher hardness and consequently higher surface free energy would be more wear-resistant. In addition, wear would also be proportional to the contact load. This feature is illustrated in the data of Kruschov and Babichev,[73] reproduced graphically in Fig. 5.21.

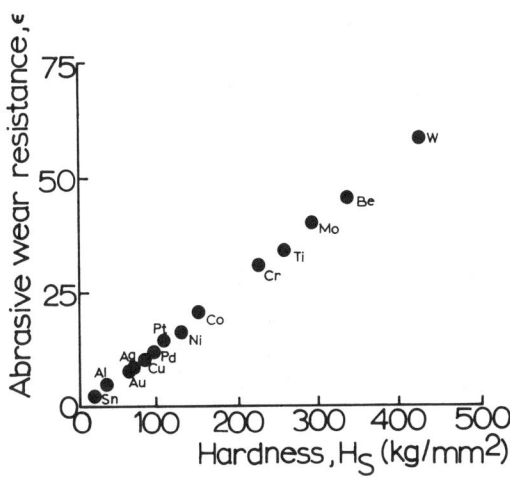

FIGURE 5.21   Wear resistance of pure metals as a function of surface hardness. The wear resistance is defined approximately by $\epsilon \propto H_S/W$, where W is the contact load. Consequently, for constant values of W, $\epsilon \propto H_S$. (After Kruschov and Babichev.[73])

Lubrication. The application of a film or monolayer adsorption on a surface can alter the frictional characteristics and wear phenomena as a result of variations induced in both surface hardness and surface free energy. A lubricating film is considered to be one that promotes a reduction in the contact or sliding friction, expressed as

$$\mu_l/\mu_u = S_l/S_u \tag{5.51}$$

where μ_l and μ_u represent the friction coefficients of the lubricated and unlubricated surfaces, and S_l and S_u represent the associated shear strengths; since $S_l < S_u$, $\mu_l < \mu_u$.

For a thin lubricating layer even on a "smooth" surface such as that shown in Fig. 5.19, areas of contact which penetrate the layer occur. If the fraction of solid-solid contact penetrating the lubricant is denoted α, and the remaining, lubricated area is denoted $(1 - \alpha)$ for a junction of area A_r, then the force required to shear the junction area, A_r, will be[74,75]

$$F_f = A_r[\alpha S_u + (1 - \alpha)S_l] \qquad (5.52)$$

Substituting for A_r from Eq. (5.47) we obtain

$$\mu = \frac{\alpha S_u}{H_S} + \frac{(1 - \alpha)S_l}{H_S} \qquad (5.53)$$

The variation in wear of lubricated surfaces can then be recognized by considering that the surface energies of the lubricated surfaces and the interfacial energy will be altered. Consequently, the energy of adhesion of the lubricated surfaces will be

$$E_{Ad(l)} = F_{S(al)} + F_{S(bl)} - \gamma_{I(abl)} \qquad (5.54)$$

where l and u designate the lubricated and unlubricated conditions. In the context of Eq. (5.52) we could also write as the effective energy of adhesion

$$E'_{Ad} = E_{Ad(u)} + (1 - \alpha)E_{Ad(l)} \qquad (5.55)$$

Consequently, the difference in abrasive wear as denoted by the difference in the size of surface fragments (fragment diameter) can be deduced from Eqs. (5.50) and (5.55) to be expressed by

$$\frac{d_l}{d_u} = \frac{E'_{Ad}}{E_{Ad(u)}} = \alpha + \frac{(1 - \alpha)E_{Ad(l)}}{E_{Ad(u)}} \qquad (5.56)$$

It should be recognized in Eqs. (5.51) through (5.56) that the subscript component, l, can also apply to an oxide layer on a metal or alloy surface as well as to any other reaction product or adsorbed layer.

The contact phenomena associated with friction and surface lubrication are difficult to study in detail by macroscopic surface measurements of friction forces or electrical resistance.[76] However, there have been several investigations of contact phenomena involving *in situ* experiments on whisker end forms in the field-ion microscope,[77-79] and this technique promises to contribute significantly to a detailed understanding of surface friction and contact phenomena, especially as these phenomena relate to surface structure and crystallography.

5.4 SURFACE MOBILITY AND TRANSPORT PHENOMENA

In discussing grain boundary grooving and interfacial equilibrium configurations at the intersection of grain boundaries with free (solid-vapor) surfaces in Chapter 2 (Figs. 2.10 and 2.11), the role of surface diffusion in the transport of matter to form the equilibrium groove morphology was indicated [Eq. (2.18a)]. In addition, surface diffusion was mentioned briefly in its role in adsorption and vapor condensation as discussed in Section 5.1. Changes in surface shape depend to a large extent on the transport of matter (surface atoms) from regions of steps, ledges, and kinks and the mobility of atoms on a surface; transport here refers to the relative movement of atoms with respect to their neighbors. The driving force for the transport of atoms on a surface usually involves temperature resolved as a temperature gradient or manifested in thermal vibration.

The transport of atoms upon a surface can be thought of as a movement of atoms from kinks and ledges in one region to another and constitutes a diffusion of atoms in a gradient of concentration. Atomic transport on a surface can also occur by a convective process as a result of evaporation or reevaporation from a surface. In addition, a type of mechanical "mixing" can occur when surfaces are mechanically contacting as in abrasive wear processes described in Section 5.3.1.

The mechanisms of matter transport in the formation of equilibrium grooves at grain boundary-surface intersections have been described in terms of associated surface and bulk diffusion processes which include evaporation-condensation (vapor transport) phenomena (surface transport), surface diffusion (diffusion along a solid-vapor or solid-liquid interface), volume diffusion in the solid, and volume diffusion in the vapor or liquid surrounding the groove.[80-82] It is important to note that, while surface diffusion and related surface transport phenomena contribute directly to changes in surface structure, the equilibrium morphology of a surface and the general surface mobility will also be influenced by, and under certain conditions dominated by, transport of matter from the solid, which can alternatively be viewed as the counterflow of vacancies.

5.4.1 Diffusion Phenomena

The transport of matter is represented ideally by the net flux, J, of particles per second per unit area of reference plane in opposite directions ($\pm x$-direction) in the presence of a concentration gradient, dc/dx, as given by the usual form of Fick's first law,[83]

$$J = -D(dc/dx) \tag{5.57}$$

where D is the diffusion coefficient as given previously in Eq. (5.16):

$$D = D_0 \exp(-Q/RT) \tag{5.16}$$

where Q is the activation energy for the particular transport process.

Fundamentally, the lattice vacancy can be considered to be the basic mechanism of all types of diffusion, of which four prominent ones can be recognized as illustrated schematically in Fig. 5.22. Each diffusion type, as shown in Fig. 5.22, requires an imperfect crystal, even if only vacancies (or in some cases interstitial atoms) are involved in the transport process. Each diffusion type, while fundamentally involving vacancy (or matter) transport, can alter the transport process by, in effect, causing the vacancy transport to be enhanced or retarded. That is, in the context of Eq. (5.16), the activation energy, Q, for each transport process will be different.* As a consequence, D, the diffusion coefficient, will vary

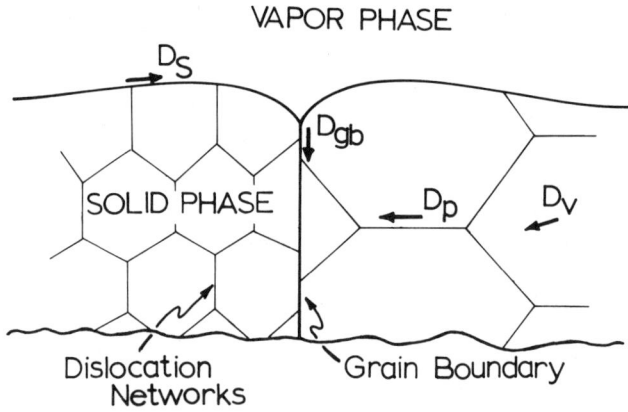

FIGURE 5.22 Schematic illustration of basic diffusion modes and associated diffusion coefficients in a polycrystalline solid.

such that at any temperature it is likely that the characteristic diffusion coefficients—D_S (surface transport), D_{gb} (grain boundary transport), D_p (transport by "pipe" diffusion through the region of a dislocation core), and D_v (volume diffusion or transport through the lattice)—will have different magnitudes. This feature is illustrated for a range of temperatures in the diffusion coefficient data for pure silver reproduced in Fig. 5.23.

In the case of an alloy, the transport process must be viewed as an interdiffusion of the component atoms. Thus, for the case of a simple binary alloy, the diffusion of the two components is expressed by

*As an indication of the values of the activation energies for diffusion processes, it might be pointed out that the activation energy for surface diffusion in molybdenum has been reported as roughly 60 kcal/mole, compared to approximately 100 kcal/mole for volume diffusion in the same metal.[82,84]

PROPERTIES OF INTERFACES   293

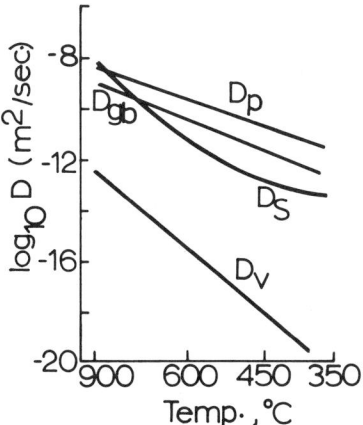

FIGURE 5.23 Variation of the diffusion coefficients with temperature in polycrystalline silver. (After Guy.[85])

$$J_1 = -D_{12}(dC_1/dx)$$
$$J_2 = -D_{12}(dC_2/dx)$$
(5.58)

since

$$dC_1/dx = -dC_2/dx \qquad (5.59)$$

and

$$D_{12} = C_1 D_2 + C_2 D_1 \simeq N_1 D_2 + N_2 D_1 \qquad (5.60)$$

where D_{12} is the interdiffusion coefficient, C_1 and C_2 are the concentrations of components 1 and 2, N_1 and N_2 denote the corresponding atomic fractions ($N_1 + N_2 = 1$), and D_1 and D_2 are the corresponding intrinsic diffusion coefficients for component atoms 1 and 2.

Figure 5.24 shows the variation of the intrinsic volume diffusion coefficients and the corresponding interdiffusion coefficient for the Ni–Cu system as a function of composition. The decrease in the diffusion coefficients with increasing nickel content is due to the decrease in vacancy concentration in the alloy which results in a corresponding increase in the activation energy for volume (lattice) diffusion.

Effect of Interfacial Structure on Diffusion. With reference to Fig. 4.10(*d*), it is observed that surface diffusion can be regarded as a combination of transport

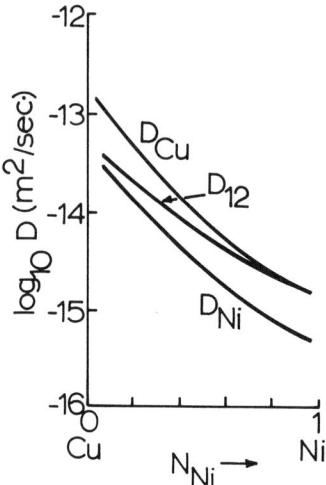

FIGURE 5.24 Variation of intrinsic diffusion and interdiffusion coefficients with composition in the Ni–Cu binary system. (After Guy.[85])

processes involving surface vacancy diffusion, adatom diffusion, and diffusion along monatomic steps. At low temperatures and for singular or vicinal surfaces, an approximate expression for the surface self-diffusion coefficient can be written in the form

$$D_S = \Sigma(n_i/M)D_i \qquad (5.61)$$

where n_i is the number of surface defects of type i, M is the number of sites per unit area of surface, and D_i represents the corresponding diffusion coefficients for the defects of type i. Obviously, i can vary systematically for certain crystallographic orientations, and n_i can be altered by surface treatment or temperature, or both. Unfortunately, accurate determinations of the mechanisms of surface diffusion have not been made for any particular surface. However, as we noted in Section 5.2 above, surface processes which depend to a large extent on diffusion can be measurably altered for crystallographic and structural variations of the surface, indicating that the surface transport processes are influenced by both surface crystallography and structure.

Since grain boundaries in general constitute a region of lattice distortion, diffusion along a grain boundary would be expected to be more favorable as this distortion increases. In addition, where the solid–solid interfacial structure varies from a regular array of dislocations (as in a low-angle grain boundary) to an arrangement of overlapping dislocation cores (as in the transitional structure for low-angle to high-

PROPERTIES OF INTERFACES   295

angle boundaries depicted in Fig. 4.25), the activation energy for self-diffusion or interdiffusion would be expected to decrease because of the increased transport facility. This latter feature is demonstrated experimentally in Fig. 5.25.

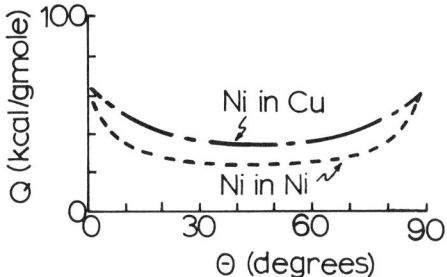

FIGURE 5.25   Activation energy for grain boundary diffusion of nickel into copper and the self-diffusion of nickel in nickel as a function of misorientation for [001] tilt boundaries. (From data of Yukawa and Sinnott[86] and Upthegrove and Sinnott.[87])

If one considers diffusion along a grain boundary such as that in Fig. 4.27(b), for example, it would be expected that the rate of diffusion as well as the energy required to activate the process would differ markedly in those regions where the grain boundary has a high degree of coincidence from those deviating from coincidence arrangements. That is, low-energy boundaries would generally be expected to possess different diffusion coefficients from high-energy (high-angle) grain boundaries at the same temperature. These expectations have in fact been observed experimentally as shown in the data reproduced in Fig. 5.26.

While the variation in grain boundary transport might be expected to result in large part because of variations in the effective thickness of the interface, Fig. 5.26 shows that the diffusion coefficient varies with grain boundary free energy, since the constant temperature curves in Fig. 5.26 are observed to closely parallel curves of grain boundary energy as a function of misorientation shown previously in Fig. 4.22(d) and especially Fig. 4.65(b).

In the case of coincidence boundaries, and in other special cases of interfacial structure, the grain boundary structure can be modified by a superimposed network of dislocations as discussed in Section 4.5. The diffusion associated with the interface may involve a transport enhancement as a result of the associated dislocation diffusion.

5.4.2   Sintering Mechanisms and Sintering Diagrams

One of the most important transport processes involving mechanical properties is in the area of powder metallurgy or the sintering of finely powdered materials—in-

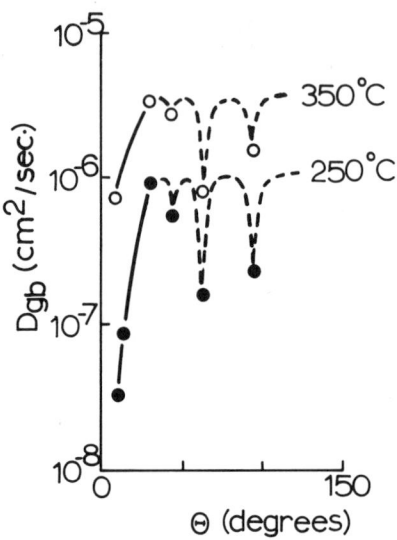

FIGURE 5.26 Variation of grain boundary diffusion coefficient with grain boundary misorientation in <110> aluminum tilt boundaries. A constant boundary thickness of 5 Å has been assumed. (After Jurisch.[88])

cluding metals, alloys, and nonmetallic materials. The process of sintering can be considered initially as one involving contacting surfaces. However, as soon as an interface forms as a result of the contact, further growth of the contact area becomes a more complex transport process involving diffusion along the interface and through the lattice, and surface diffusion, as well as the associated surface and interfacial free energies. The sintering process following contact is characterized by the growth of the neck formed between the contacting powder particles, and where an aggregate of particles is sintering, neck growth results in the densification of the aggregate.

When two particles are placed in contact as illustrated in Fig. 5.27(a), interatomic forces draw them together, deforming the contact areas of each particle elastically,[89] forming a neck of radius ρ, which increases with continued neck growth. The elastic distortion at the contact region during the initial sintering is evident in the contact (strain-field) contrast shown for two thoria (ThO_2) particles sintered together in a pure nickel matrix in Fig. 5.27(b). The size of the initial junction depends on the energy balance between the loss of surface due to contact, the formation of an interface (grain boundary), and the stored energy in the contacting particles. Since small changes in the contact geometry can increase adhesion, plastic deformation or other external stress can have a significant influence on the rate and character of powder-particle sintering. The energy of adhesion is given

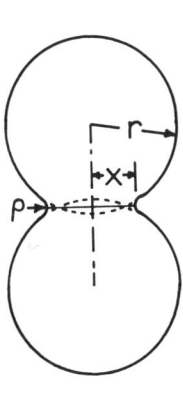 

a b

FIGURE 5.27   Two-sphere model and experimental example of contact and neck growth of two powder particles. (a) Schematic view of two spheres sintering. (b) Transmission electron micrograph showing two ThO_2 particles sintered together in a pure nickel matrix following initial consolidation, annealing at 1200°C for 250 hours in vacuum, shock deformation at 80 kilobars, and subsequent annealing for 10 hours at 800°C in vacuum.

generally by Eq. (5.44). However, for single-phase particles, $F_{S(a)} = F_{S(b)} = F_S$, and we can let $\gamma_{I(ab)} \equiv \gamma_{gb}$ so that Eq. (5.44) becomes

$$E_{Ad} = 2F_S - \gamma_{gb} \tag{5.44a}$$

and an upper limit for the radius of the neck [Fig. 5.27(a)] is[90]

$$x = (E_{Ad}/rG)^{1/3} r \tag{5.62}$$

where r is the particle radius and G is the shear modulus. It is observed from Eq. (5.62) that the neck is large when the particle size, r, is small. In addition, it should be noted that, if sintering is performed in an environment that permits or promotes adsorption of some impurity (for example, oxidation) at the surface, the energy of adhesion will be reduced from that of clean particles sintering together, since the surface energy will be less than the clean surface energy. In addition, in situations such as that illustrated in Fig. 5.27(b) where sintering occurs in a solid or liquid phase, the change in free energy when the two surfaces are brought together to form a grain boundary will be given by

$$E_{Ad} = 2\gamma_{P/M} - \gamma_{gb} \tag{5.63}$$

where $\gamma_{P/M}$ is the interfacial free energy between the particle and the matrix.

As indicated in Eq. (5.44a) above, the surface energy of the particles is the principal driving force for the sintering process. However, in terms of the sintering geometry, the surface energy can be resolved in terms of surface curvature in the same way that curvature is involved in contacting phases (Fig. 2.26) and changes of state—for example, melting [Eq. (3.65)], of the difference in surface curvature between sources and sinks of matter. As a consequence, the driving force will differ for different configurations and geometries of sintering particles.

Ashby[90] has pointed out that sintering, even in single-component systems in the absence of any applied stress, involves at least six distinguishable transport mechanisms which contribute to neck growth and concomitant densification by an aggregate of particles. The complexity of matter transport to the neck is further increased in systems of two or more components (alloys), and new mechanisms are introduced, or existing mechanisms altered (accelerated), by the application of an external stress or pressure. These mechanisms essentially involve the diffusion paths illustrated in Fig. 5.22, and contribute simultaneously to neck growth as originally pointed out by Wilson and Shewmon.[91] A complete description of the equations for neck growth rate (or sintering rate) and of the associated transport paths in a one-component system is given in Table 5.3. The net sintering rate during the initial stage of sintering (following adhesive contact) is given by

$$\dot{x}(\text{net}) = \sum_{i=1}^{6} x_i \tag{5.64}$$

As the neck grows, the curvature difference driving the mechanisms in Eq. (5.64) decreases until only two mechanisms are important: boundary diffusion from sources on the boundary, and lattice diffusion from sources on the boundary.[90] The net sintering rate during later stages of neck growth is then given by[90]*

$$\dot{x}(\text{net})' = \frac{1}{16} \frac{F_S \Omega K_3^3}{kT} (D_{gb}\delta_{gb} + D_v x) \left[\frac{1}{\log_e (x_f K_3/2) - \frac{3}{4}} \right] \tag{5.65}$$

where in the case of sintering of an aggregate of spheres as illustrated in Fig. 5.28,

$$x_f = (f/3)^{1/3} r + x \tag{5.66}$$

where f is the volume fraction of space (pore space) in the compact. The remaining parameters in Eq. (5.65) are defined in Table 5.3 with the exception of the curvature difference, K_3.

*The so-called dirty surface approximation is used where E_{Ad} is set equal to $F_S/10$.

TABLE 5.3
Sintering Mechanisms and Associated Neck Growth Rate Equations

Mechanism No.[a]	Transport Path (Fig. 5.22)	Associated Diffusion Coefficient	Source of Matter Transported to Neck	Associated Neck Growth Rate Equation
1	Surface diffusion	D_S	Surface	$\dot{x}_1 = 2 D_S \delta_S F K_1^3 \equiv \dfrac{dx_1}{dt}$ [b]
2	Lattice diffusion	D_v	Surface	$\dot{x}_2 = 2 D_v F K_1^2$ [b]
3	Vapor transport	None	Surface	$\dot{x}_3 = P_v F \left[\dfrac{\Omega}{2\pi \triangle_0 kT} \right]^{1/2} K_1$ [b]
4	Boundary diffusion	D_{gb}	Grain boundary	$\dot{x}_4 = \dfrac{4 D_{gb} \delta_{gb} F K_2^2}{x}$ [b]
5	Lattice diffusion	D_v	Grain boundary	$\dot{x}_5 = 4 D_v F K_2^2$ [b]
6	Lattice diffusion	D_p	Dislocations	$\dot{x}_6 = \dfrac{4}{9} K_2 N x^2 D_p F \left(K_2 - \dfrac{3}{2} \dfrac{Gx}{FS^t} \right)$ [b]

[a] Based on the data of Ashby.[90]

[b] δ_S and δ_{gb} refer to the effective surface and grain boundary thicknesses, respectively. Note that a grain boundary thickness of 5 Å has been assumed in the evaluation of diffusion data in Fig. 5.26. In the development of these equations a "dirty" surface approximation has been utilized with $E_{Ad} \simeq FS/10$; $F = F_S \Omega / kT$; where Ω is the atom or molecular volume, k is Boltzmann's constant, and T is the temperature (°K); $\triangle_0$ is the theoretical compact (aggregate) density; the K_i's represent the curvature differences—that is, the curvature driving mechanisms [see Eqs. (5.67)–(5.72)]; P_v in $\dot{x}_3$ is the vapor pressure $[P_v = P_0 \exp(-Q_{vap}/kT)]$; N in $\dot{x}_6$ is the dislocation density (cm^{-2}); and G is the shear modulus. The parameters r and x are defined in Figs. 5.27(a) and 5.28.

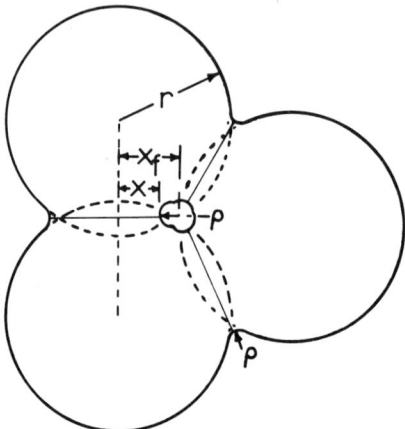

FIGURE 5.28   Sintering geometry schematic for a two-dimensional compact or aggregate of spheres or straight wires (cylinders).

The curvature differences, K's, in Table 5.3 and Eq. (5.65) can be deduced for a row of spheres as illustrated schematically in Fig. 5.27(a) to be

$$K_1(\text{row}) = \left(\frac{1}{\rho} - \frac{1}{x} + \frac{2}{r}\right)\left(1 - \frac{x}{r}\right) \tag{5.67}$$

and

$$K_2(\text{row}) = \left(\frac{1}{e} - \frac{1}{x}\right) \tag{5.68}$$

where by simple geometry

$$\rho = x^2/2(r - x) \tag{5.69}$$

Similarly, for an aggregate of spheres as shown schematically in Fig. 5.28 we can write[90]

$$K_1(\text{aggregate}) = \left(\frac{1}{\rho} - \frac{1}{x} + \frac{2}{r}\right)\left(1 - \frac{x}{x_f - (f/3)^{1/3}}\right) \tag{5.70}$$

$$K_2(\text{aggregate}) = \left(\frac{1}{\rho} - \frac{1}{x}\right) \tag{5.71}$$

PROPERTIES OF INTERFACES   301

$$K_3(\text{aggregate}) = 2/(x_f - x) \qquad (5.72)$$

Finally, Ashby[90] has noted that, for a two-dimensional compact of straight wires packed as shown in Fig. 5.28,

$$x_f = 0.55r \qquad (5.73)$$

while for a compact of spheres,

$$x_f = 0.74r \qquad (5.74)$$

Figure 5.29 illustrates the various modes and stages of sintering in systems involving multicomponent contact and other complex phenomena. The sintering of thoria particles in a metal matrix as demonstrated in Fig. 5.29(b) has also been concluded to be the mechanism of dispersoid growth in TD-Ni.[92]

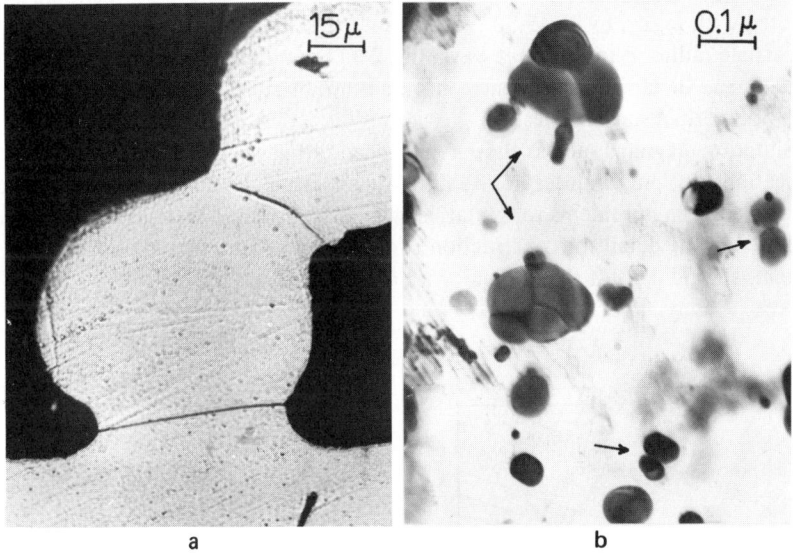

FIGURE 5.29   Experimental examples of sintering geometry and sintering stages. (a) Sintering of straight wires (cylinders) of nickel to one another and to a pure nickel plate after 200 hours at 1332°C in hydrogen (courtesy of U. M. Ahmad); (b) Various sintering stages of thoria (ThO_2) particle aggregates in a pure nickel matrix. Arrows denote fully dense clusters or clusters at various stages of sintering.

It is to be noted from Table 5.3 and Eqs. (5.64) and (5.65) that temperature affects each rate contribution to neck growth in an identical manner, and that the neck growth rate can vary depending on the dominance of any of the transport mechanisms. In addition, since neck growth and densification are diffusion-rate-controlled, the equations cited can be used to devise a time-temperature schedule which will lead to maximum density or maximum porosity.

Ashby[90] has recently devised a construction of diagrams called sintering diagrams which, utilizing the equations of Table 5.3 and Eqs. (5.64) through (5.74), show for a given temperature and neck size the dominant mechanism of sintering, and the net rate of neck growth or densification. Figure 5.30 illustrates one form of such sintering diagrams which describes the sintering of an aggregate of silver particles with two different radii. The axes are homologous temperature, T/T_m (where T_m is the melting point in °K), and the normalized neck radius, x/r [Figs. 5.27(a) and 5.28]. The space defined by these axes is divided into fields, within which a single sintering (transport) mechanism is dominant; that is, it contributes most to the transport of matter to the growing neck. The largest field in Fig. 5.30(a) and (b) is that corresponding to mechanism 4 of Table 5.3. Superimposed on the fields of Fig. 5.30(a) and (b) are contours of constant sintering time which show the neck growth after any given time for a given temperature. It should also be noted with regard to Fig. 5.30 that, because the sintering mechanisms depend on the particle radius, r, in different ways, the field boundaries move with a change in particle size. In addition, the time contours move because larger particles sinter more slowly than smaller ones.

Sintering diagrams such as those reproduced in Fig. 5.30 can have important applications not only in interpreting sintering experiments, but also in solving many practical sintering problems for metals as well as for ceramic systems. The reader interested in the details of construction of the various forms of such sintering diagrams should consult the original work of Ashby.[90]

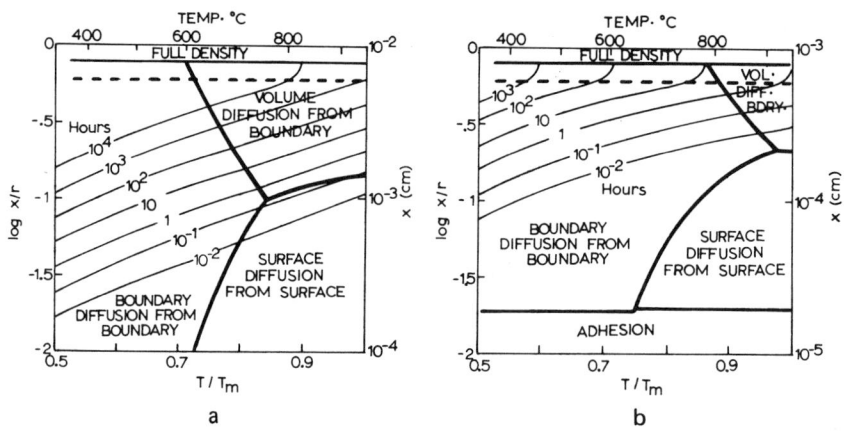

FIGURE 5.30   Sintering diagrams for the sintering of an aggregate of silver spheres with an initial relative density of 0.8. (a) r = 100 microns. (b) r = 10 microns. (After Ashby.[90])

5.5 FRACTURE OF METALS AND ALLOYS

Fracture in metals and alloys has been traditionally characterized by two modes: ductile and brittle. Ductile fracture, which normally characterizes metals and alloys, is associated with dislocation motion (slip) and residual necking, and there is no long-range association of microcracks with crystallographic or interfacial phenomena. In addition, ductile fracture is generally not initiated in the presence of a stress concentration. Cracks that do develop during ductile failure are characterized by plasticity effects which involve the movement and concentration of dislocations at the tip of the advancing crack.

Brittle fracture, on the other hand, is associated with elastic stress behavior—that is, a lack of readily recognizable dislocation motion. Microcracks develop along crystallographic directions and are associated with long-range interfacial phenomena. Brittle cracks advance primarily as a result of systematic decohesion—that is, cleavage. Thus, a typical ductile metal is soft but tough, since the stress required to move dislocations is small, whereas a brittle metal behaves like a covalently bonded material such as silicon; it is hard but fragile because the stress required to move dislocations is large. Fracture in metals and alloys can therefore be recognized as a matter of cohesion or cohesive strength. Whereas brittle elastic fracture is characterized by cleavage phenomena, brittle plastic fracture is characterized by features similar to ductile fracture—that is, hole growth on a microscopic level resulting in a dimpled fracture surface.

The ideal fracture strength, σ_{max}, of a crystalline solid can be defined as the tensile stress required to simultaneously rupture all the atomic bonds between two adjacent atomic planes at a temperature of absolute zero. The decohesion that characterizes the formation of two fracture surfaces is the reverse of the formation of an adhesive joint discussed in Section 5.3. The work of fracture, or the energy for decohesion, should equal the adhesive energy; as a consequence,

$$E_f = E_{Ad} = 2F_S - \gamma_I \tag{5.75}$$

However, for cleavage of a perfect single-crystal metal or alloy, $\gamma_I = 0$; therefore, $E_f \simeq 2F_S$—that is, the work of fracture is equal to the surface energy of the two new surfaces created. Alternatively, the ideal shear strength, τ_{max}, can also be defined as the shear stress required to simultaneously shear all atoms on two adjacent planes at a temperature of absolute zero.

Figure 5.31 illustrates a two-dimensional lattice which is sheared or separated along the line x-x'. In each case, the stress to separate or shear the perfect lattice will start from zero, reach a maximum, and then drop to zero, approximating a sinusoidal variation from which an estimate of σ_{max} and τ_{max} may be obtained in the form

$$\sigma_{max} = (E_Y F_S/a)^{1/2} \tag{5.76}$$

and

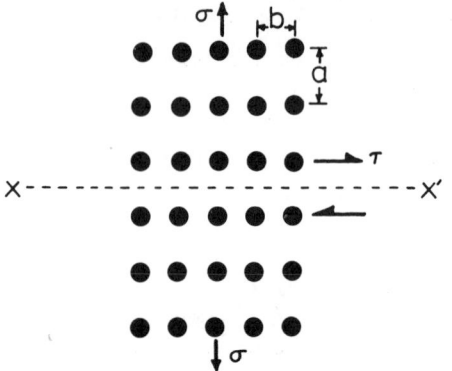

FIGURE 5.31   Ideal cleavage or shear of a perfect lattice.

$$\tau_{max} = Gb/2\pi a \qquad (5.77)$$

where E_Y is the elastic modulus (Young's modulus) and G is the shear modulus. In Eq. (5.77) the surface energy term, F_S, does not appear because new surface, except where the sheared material forms steps on a surface portion, is not created during shear. Note also that b can represent the Burgers vector (Fig. 5.31).

In crystals such as silicon or diamond, for example, the resistance to shear is much higher than for metals. As a consequence, τ_{max} is low for ductile metals and increases with the degree of brittleness. In addition, both E_Y and F_S in Eq. (5.76) are large for ductile metals and decrease for brittle metals; therefore, σ_{max}/τ_{max} will be large for ductile materials and generally small for brittle materials.

An alternative approach to the calculation of the fracture stress of a solid that presumably contains a crack or crack nuclei was originally proposed by Griffith.[93] In this approach, an existing crack will spontaneously increase in length when the decrease in strain energy associated with this increase is greater than the increase in the surface free energy required to produce the two new surfaces. For an elliptical crack as shown in Fig. 5.32, the strain energy reduction is given by

$$U = \pi C^2 \sigma^2 / 2E_Y \qquad (5.78)$$

where C is half the crack length, and σ is the applied tensile stress. When the crack lengthens by an amount dC, the strain energy is reduced further by an amount

$$dU = (\pi C \sigma^2 / E_Y) \, dC \qquad (5.79)$$

If the energy per unit area is E_u, then

PROPERTIES OF INTERFACES

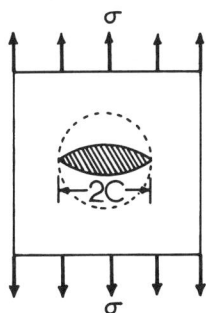

FIGURE 5.32   Griffith's model for crack propagation.

$$dF_S = 2E_u \, dC \tag{5.80}$$

Since the crack will grow spontaneously under a critical stress, σ_F, if the reduction in strain energy is sufficient to supply the new surface energy required, then

$$dU = (\pi C \sigma^2 \, dC/E_Y) = dF_S \tag{5.81}$$

and consequently the critical stress to initiate crack growth is

$$\sigma_F = (2E_Y F_S/\pi C)^{1/2} \tag{5.82}$$

Equation (5.82), as originally derived by Griffith,[93] applies most specifically to elastic cracks; for materials in which some plastic flow occurs during crack propagation, Orowan[94] and Irwin[95] have proposed that the Griffith equation [Eq. (5.82)] still applies, and one need only add the plastic work to the surface energy. However, Marcinkowski and Das[96] have studied the relationship between crack dislocations and crystal lattice dislocations in the elastic and plastic shear crack lamellae. In the elastic crack, it was shown that the elastic strain energy released by crack formation was equal to the sum of the self and interaction energies of the crack dislocations; if the crack dislocations were allowed to run out into the crystal against the lattice friction stress and be converted into crystal lattice dislocations, a plastic crack resulted. Such plastic cracks were found to have a lower energy than the corresponding elastic cracks. It is therefore incorrect to assume that plastic cracks add a plastic energy term, and that Eq. (5.82) would not apply.

Equation (5.76) represents a theoretical (ideal) strength, whereas Eq. (5.82) represents a more realistic strength. If we write

$$\sigma_F/\sigma_{max} = (2a/\pi C)^{1/2} \tag{5.83}$$

it is seen that the real fracture stress can be appreciably below the theoretical fracture stress, particularly where tiny microcracks occur in the lattice ($C > a$). It should also be pointed out that σ_{max} ideally characterizes the cohesive strength, since this is the stress required to cause lattice decohesion.

5.5.1 Surface Embrittlement by Segregation and Adsorption

It is now presumably well established that solute segregation to or adsorption of an impurity or impurities upon a surface will normally reduce the surface free energy (for example, Figs. 3.21 and 5.2) at a constant temperature. As a consequence of Eqs. (5.42) and (5.43), it is also likely that, as a result of surface excess, the mechanical strength (yield strength and tensile strength) will be reduced accordingly. This feature is illustrated for dilute Cu-Bi alloys in Fig. 5.33. It can be observed in Fig. 5.33 that neither the reduction in surface free energy nor the fracture stress can be taken as an indication of embrittlement, which is defined as induced brittle behavior. Hondros and McLean[97] have recently defined a cohesion

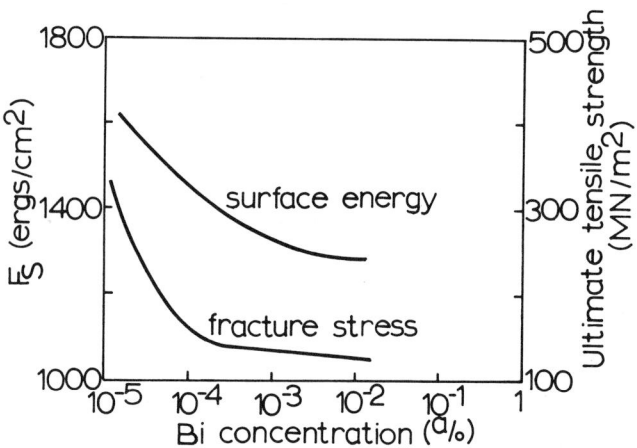

FIGURE 5.33 Variation of surface free energy and fracture stress with bismuth content in dilute Cu-Bi alloys. The surface energy measurements were made at 927°C, the tensile tests at 25°C. (From data of Hondros and McLean.[97])

margin as a quantitative expression of the percentage reduction in cohesion that would induce brittle behavior in any given case. Although some indication of the ductile-brittle transition is found in tensile (stress-strain) data, the two regimes are most effectively characterized by direct observation of the fracture mode (fracture surfaces) as comparatively illustrated in Fig. 5.34. In effect, such a ratio might be considered to be represented by Eq. (5.83). Alternatively, the requisite condition

PROPERTIES OF INTERFACES 307

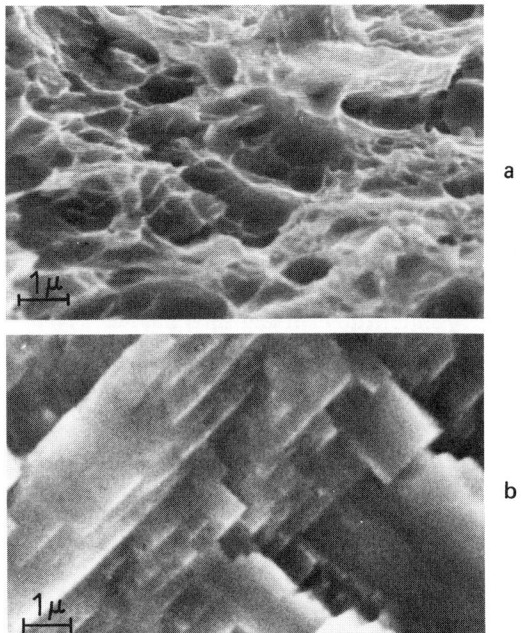

FIGURE 5.34 Comparison of fracture surface topographies for typical ductile (a) and brittle (b) fracture. (a) Dimple tensile fracture surface of carbon steel. (b) Cleavage-tensile fracture surface of single-crystal iridium.

for brittle behavior (embrittlement) might be defined as the point where the cohesive stress (σ_{max}) is reduced to a value equal to or close to the yield stress or flow stress (ensuring the absence of plastic flow). Such a cohesion margin must therefore be defined or determined experimentally for any particular metal or alloy. In addition, the cohesion margin may be governed by interfacial (grain boundary) decohesion; this is certainly true for polycrystalline metals and alloys because the effective energy of decohesion will be lower at a grain boundary than in the matrix.

Reduced matrix or surface decohesion would favor crack nucleation at any associated flaws such as surface grooves or valleys [as shown, for example, in Fig. 4.10(b)] where a stress concentration exists. This would also apply to grain boundary grooves on a free surface.

Liquid Metal Surface Embrittlement. As with segregation and adsorption, liquid metals can cause a reduction in cohesion by reducing the number of solid metal bonds in the surface. In addition, a liquid metal continuously migrating to a crack tip will maintain a sharp crack-tip radius which continues to propagate at reduced stress. Furthermore, the liquid/solid interfacial free energy may be considerably below the solid surface free energy (Table 2.2), and this interfacial energy may constitute the energy of cohesion, which is therefore drastically reduced, thereby causing surface embrittlement in the sense that elastic (cleavage) cracks are formed and failure is essentially of a brittle mode. It is widely accepted that the fundamental role of an embrittling liquid metal is to adsorb in the vicinity of strained atomic bonds and lower the binding energy, particularly at those bonds that define a crack tip.[98-100] A similar if not identical phenomenon occurs during stress corrosion cracking.

5.5.2 Grain Boundary Embrittlement and Intergranular Fracture

The energy of adhesion associated with a grain boundary is given by Eq. (5.44a). As a consequence, the cohesive stress at a grain boundary is

$$\sigma'_{max} = \left[\frac{E_Y(F_S - \gamma_{gb}/2)}{a}\right]^{1/2} \tag{5.84}$$

As a result, the grain boundary cohesive stress will be smaller than that for a single crystal or for the grain interior:

$$\frac{\sigma'_{max}}{\sigma_{max}} = \sqrt{\frac{2F_S - \gamma_{gb}}{2F_S}} \tag{5.85}$$

and fracture (crack nucleation and propagation) would be expected to be associated with, if not to give preference to, grain boundaries. Equation (5.85) illustrates that intergranular fracture is expected to be favored in polycrystalline metals and alloys, and that crack propagation might be expected to follow high-energy boundaries preferentially (as compared with twin boundaries or low angle boundaries, for example). However, if both F_S and γ_{gb} are lowered by segregation, some variations in crack nucleation and propagation can occur. Since, as was discussed in Chapter 4, segregation to grain boundaries can occur with some preference for the grain boundary structure, the propagation of an intergranular crack will be expected to be favored by interfaces structurally conducive to preferential segregation, which results in lowering of the cohesive strength or fracture stress.

It has already been demonstrated in Chapter 3 that the grain boundary energy as well as the surface free energy will be lowered by solute segregation to the interfaces. As a consequence, the reduction in grain boundary free energy by segregation or impurity adsorption will correlate with variations in fracture stress as shown in Fig. 5.33.

As in the case of a free surface, segregated atoms can lower the fracture strength of a grain boundary by weakening the grain boundary bonding. In addition, the structure itself, like the structure of a free surface, can nucleate cracks which, if they propagate along the interface, are aided by the cohesively weak interface.

PROPERTIES OF INTERFACES   309

Crack Nucleation at Grain Boundaries. It has been proposed that cracks are nucleated at grain boundaries by the stress concentrations associated with blocked shear bonds or dislocation pile-ups.[101-105] However, in many metals and alloys, particularly those that fail in a brittle, intergranular manner, as illustrated typically in Fig. 5.35, there is little evidence of dislocation pile-ups, and fracture appears to originate and propagate within the grain boundaries.

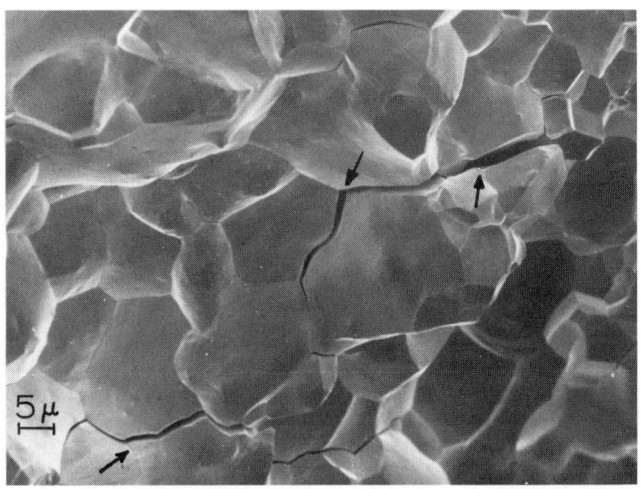

FIGURE 5.35   Intergranular brittle fracture of polycrystalline iridium at room temperature. Note the intergranular cracks indicated by the arrows which clearly suggest severe decohesion at the interface.

Although cracks can nucleate at precipitates or other grain boundary inclusions and propagate as a result of decreased cohesion caused by a monolayer or segregate or adsorbed impurity along the interface, there exist unique structural features of grain boundaries such as ledges which can effectively nucleate cracks at the interface. This is because the grain boundary ledges possess large strain fields[106] which, when subjected to deformation-induced stress, are relieved either by further slip or by microcrack initiation.[107] Although, as demonstrated in Fig. 4.31, ledges can occur as a result of GBD coalescence during grain growth (or related heat treatment), they can also be caused by plastic deformation of a polycrystalline material imposing a heterogeneous grain boundary shear. The latter is essentially a dislocation pile-up. However, it is not the pile-up itself that nucleates the crack but rather the residual ledge in the grain boundary. Figure 5.36 illustrates very simply the nucleation of either a transgranular or an intergranular microcrack at a grain boundary ledge. Preference of fracture mode—that is, microcrack nucleation mode

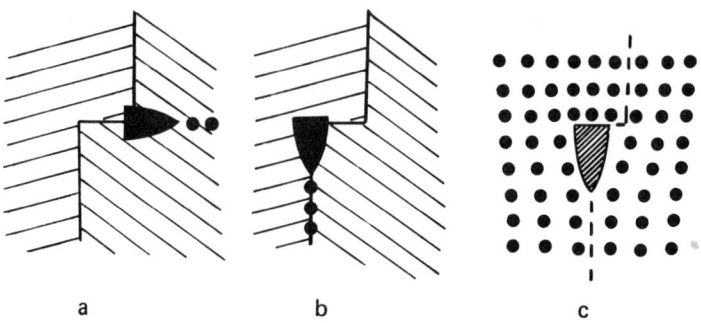

FIGURE 5.36 Microcrack initiation at grain boundary ledges. (a) Transgranular microcrack. (b) Intergranular microcrack. (c) Classical edge dislocation coalescence mechanism of Stroh[102] observed to resemble a grain boundary ledge. The wedge-shaped crack nucleus is shown shaded. Note also that the microcrack depicted in (a) and (b) can be considered as either tensile or shear depending on the dislocation character of the ledge—that is, edge or screw—or the character of dislocations passing through a grain boundary.

—will depend primarily on the cohesive strength of the grain boundary as compared with the lattice (grain interior). Thus, if $\sigma'_{max}/\sigma_{max}$ in Eq. (5.85) is small, intergranular microcracks would be preferential. It can also be noted in Fig. 5.36(c) that the classical wedge-type crack of Stroh[102] involving edge dislocation coalescence can be readily identified with the coalescence of GBD's to form a grain boundary ledge. In fact, it has been shown in the various models for dislocation-coalescence-stress initiation of a microcrack that the work done by the applied stress acting through the motion of m dislocations of Burgers vector b must be equal to the effective surface energy of the crack. Thus, the critical stress for tensile crack nucleation is expressed as

$$\sigma_F = \frac{\zeta(F_S - \gamma_{gb}/2)}{mb} \qquad (5.86)$$

where ζ is a constant with values ranging from 2 to 5.[102] Similarly, Das and Marcinkowski[107] have shown that, in the case of grain boundary ledges, the critical stress (fracture stress) is given by

$$\sigma_F(\text{ledges}) = \frac{2(F_S - \gamma_{gb}/2)}{B_L} \qquad (5.87)$$

which is, for all practical purposes, identical to Eq. (5.86), since $B_L = mb$ [Eq. (4.24b)]. Equation (5.87), however, applies to both tensile and shear microcracks, and the critical stress decreases inversely with an increase in the ledge height (or ledge strength), B_L.

It is possible, in the context of Eq. (5.87), to draw some important conclusions regarding the role of grain size and grain structure on grain boundary embrittlement in polycrystalline metals and alloys. It is obvious, for example, that fracture will be facilitated by ledge height. In addition, as the ledge density along a unit area of interface increases, the number of grain boundary microcracks could increase in some proportion. Furthermore, as the grain size decreases with a fixed ledge density (number of ledges per unit area), the frequency of occurrence of microcracks might increase. However, an increase in ledge density can, as was discussed in Section 5.8.1, cause an increase in the flow stress so that, in the absence of a segregation of solute atoms along the interface causing reduced cohesion, the microcracks may not propagate. In addition, at increased flow stresses and in the absence of reduced interfacial cohesion, ledge strain is normally relieved by the emission of dislocations from the ledge into the grain interior; that is, ledges function as effective sources of dislocations. Thus, grain boundary structure, particularly ledge structure, may well be a controlling factor in embrittlement, and the alteration of ledge density and/or grain size by heat treatment or thermomechanical processing may have important consequences in the embrittlement and fracture of metals and alloys. The implications of these statements are illustrated in Fig. 5.37, which shows for comparison examples of beryllium grain boundary structures that differ in grain size and interfacial structure as a result of fabrication and heat treatment. The larger-grained material with a statistically greater ledge density [Fig. 5.27(b)]

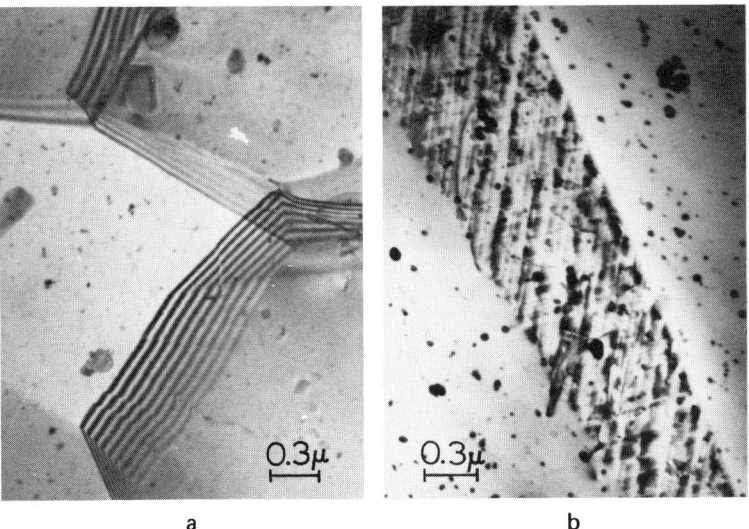

FIGURE 5.37 Comparison of grain boundary ledge structure in polycrystalline beryllium differing in fabrication heat treatment. (a) Pressed, roll-reduced beryllium; average grain size of 6 microns. (b) As-pressed beryllium; average grain size of 30 microns. Note that smaller-grain boundary structure lacks recognizable ledge structure.

has been observed experimentally to exhibit a greater degree of internal spallation as evidenced by a greater density of spall cracks when the material is deformed by explosive shock loading at a peak pressure (stress level) identical to that to which the material in Fig. 5.37(a) was subjected.[108]

Although detailed studies are not presently available, it would seem logical that the propagation of intergranular cracks would favor interfaces of high ledge density. As a consequence, grain boundaries that exhibit decohesion as shown in Fig. 5.35 may possess a range of misorientations within which the ledge density, for example, is a maximum, as illustrated experimentally in Fig. 4.33. Therefore, the embrittlement of many materials may depend not so much on grain boundary impurity or solute segregation as on the intrinsic structure of the grain boundaries which are characterized by unique ranges of misorientation or large ledge strengths (or both). A specific case is the embrittlement of iridium at room temperature (Fig. 5.35), which has been explained as being due to oxygen segregation to the grain boundaries.[109] However, more recent work by Auger electron spectrometry has revealed the presence of phosphorus in some samples,* and detailed examination of iridium grain boundaries in the transmission electron microscope as well as the field-ion microscope indicates that, unlike the somewhat normal distribution of grain boundary misorientations in ductile fcc metals and alloys [Fig. 4.42(a) and (c)], the misorientation distributions for iridium grain boundaries seem to lack values of Θ below about 65°, prompting the conclusion that this feature may contribute to the uncharacteristic embrittlement of relatively pure iridium.[110]

Liquid Metal Embrittlement of Grain Boundaries. Smith[111] has shown that if γ_{SL} is less than half the grain boundary free energy, γ_{gb}, in a polycrystalline metal, then grain boundary penetration embrittlement occurs—that is, $2\gamma_{SL} < \gamma_{gb}$. The embrittlement, as was pointed out above for solute segregation, is caused by a drastic reduction in the interfacial free energy, as manifested by the loss of interfacial cohesion.[112] However, although interfacial energy reduction is a necessary condition for embrittlement, it is not in general a sufficient one.

Prominent embrittlement characteristics have been observed for polycrystalline metals and alloys wetted by mercury,[100,113] the penetration of copper by bismuth and lead,[113] zinc by gallium,[99] and iron in molten cadmium.[114] These characteristics also typically illustrate the effect of temperature on the embrittlement process as shown in the data of Dityatkousky et al.[114] reproduced in Fig. 5.38. Temperature can operate in two ways to alter the embrittlement or the range of embrittlement. First, it can, through the adsorption equation [Eq. (1.37)], enhance segregation, etc., up to the point of or outside the range of precipitation. Second, temperature can influence the ledge density (number per unit length or area of grain boundary) as well as the ledge strength. In addition, temperature has a controlling effect on diffusion which is a significant aspect of transport along the grain boundary.

*L. E. Murr and H. Marcus, unpublished data.

PROPERTIES OF INTERFACES

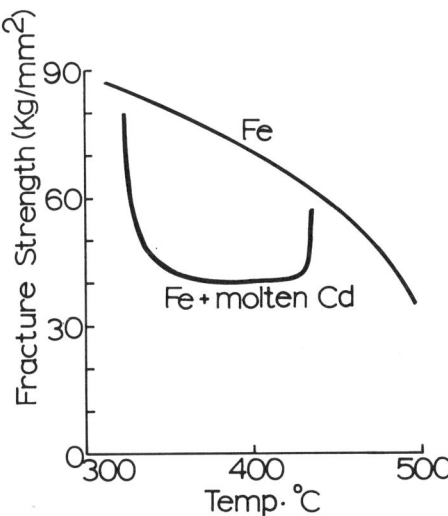

FIGURE 5.38 Embrittlement of polycrystalline iron by molten cadmium metal. (From data of Dityatkousky et al.[114])

Grain Boundary Corrosion and Stress Corrosion Cracking. Corrosion of grain boundaries is similar to liquid-metal embrittlement in that fracture is initiated by microcrack formation and propagated by decreased cohesion as a result of structural changes in grain boundary brought about by erosion or reaction at the interface, primarily at the crack tip. Corrosion at a grain boundary differs from liquid-metal embrittlement and also from hydrogen embrittlement[115] in that, in corrosion, chemical or electrochemical dissolution at the interface is involved.[116-118] These processes are enhanced by the systematic depletion of certain elements at a grain boundary as in the case of chromium depletion in steels as a result of precipitation.

In stress corrosion cracking, a crack tip may act as a local anode, with metal ions going into solution at the anode (crack tip) and electrons flowing from the anode region to some other portion of the matrix which acts as a cathode. With the application of a tensile stress on the crack, propagation is aided by the stress intensity that develops at the crack tip. Thus, decohesion at the crack tip occurs by dissolution under the action of an applied tensile stress which physically breaks metal bonds at the crack tip. The crack so initiated continues to propagate until the stress intensity decreases to some critical level or until the dissolution or reaction (corrosion) ceases. Under certain conditions stress corrosion cracking can also occur in a transgranular mode.

In the context of corrosion, hydrogen embrittlement is also considered in some instances to induce a chemical reaction with the formation of interfacial hydrides which possess a lower interfacial cohesion than the grain boundary. Preferential hydrogen adsorption, segregation, or precipitation to grain boundaries therefore serves to decrease the interfacial cohesion—in the same way that other impurities or solutes affect embrittlement phenomena as discussed above. In addition, there are other possible effects of hydrogen leading to hydrogen embrittlement—for example, its effect on dislocation mobility, inclusion, and decohesion.

5.6 SOLID–LIQUID TRANSITIONS: SOLIDIFICATION AND MELTING

Solidification of metals and alloys from their melts can be approximated by the assumption of equilibrium at the solid–liquid interface during growth; that is, there is only a negligible barrier to transport of atoms across the interface, and there is complete diffusion in both the solid and liquid states at the interface. The structure of the interface will have an important effect on the residual crystal growth because the solid–liquid transition can be influenced by the interfacial structure, which in turn influences nucleation. Solidification morphologies are generally characterized by faceted or nonfaceted morphologies, and melting is observed essentially as a reversible process. There is, in fact, a symmetry in cases of melting and solidification which has been referred to as a principle of microscopic reversibility.

Woodruff[119,120] has pointed out that, when facets form, the phase continues to grow with a faceted interface in the case of solidification. If a faceting crystal is partially melted in such a way that it displays an atomically rough and curved interface, growth will occur rapidly on the curved, high-index portions on resolidification until no further growth can take place without the formation of additional facets. Thus, the ability of a metal to facet is displayed in both solidification and melting, and the atom movements during solidification are essentially reversed during melting.*

5.6.1 Nucleation from the Melt

Solidification occurs at the solid–liquid (melt) interface as a result of nucleation in a manner similar to that discussed in Section 5.2.1. Homogeneous nucleation from the melt can be characterized by the formation of an ordered spherical nucleus of radius r, having a total excess free energy of

$$\Delta G = -\frac{4}{3}\pi r^3 \frac{\Delta H_f}{T_m} \Delta T + 4\pi r^2 \gamma_{SL} \tag{5.88}$$

---

*Grain boundary formation from melt has been simulated by a molecular dynamics method making it possible to observe both structure and migration of a grain boundary. See Cotterill et al.[121]

where ΔH_f is the latent heat of fusion, T_m is the equilibrium melting temperature, and ΔT is the temperature below the equilibrium melting point (an undercooling). The critical radius, r^* (as depicted ideally in Fig. 5.14), corresponds to the condition $\partial \Delta G / \partial r = 0$, which from Eq. (5.88) gives

$$r^* = 2\gamma_{SL} T_m / (\Delta H_f \Delta T) \qquad (5.89)$$

which is identical to Eq. (3.65), the Gibbs-Thomson relation applicable to the solid-liquid interface in solidification and melting. The critical radius in Eq. (5.89) is the radius of a spherical "particle" in equilibrium with the liquid melt supercooled by an amount ΔT. It should also be noted that, with reference to Fig. 5.14, the nucleus formed can be considered a free particle floating in the melt, since its formation does not require wetting (contact) of the solid surface.

Referring to Fig. 5.14, it can be argued that the necessary condition for heterogeneous nucleation is that the particle (the critical nucleus cap) should wet the nucleant in the presence of the liquid phase of the nucleating material. Energetically, this will occur when

$$\gamma_I < \gamma_{LV} + \gamma_{SL} \qquad (5.90)$$

where γ_I is the interfacial energy of the nucleus on the solid surface, γ_{LV} is the liquid surface energy, and γ_{SL} is the solid-liquid interfacial energy. In principle, $\gamma_I = \gamma_{gb}$ for the solid. It can be observed from Fig. 5.14 that the significance of heterogeneous nucleation is that the number of atoms required for the creation of a nucleus of radius r^* [Eq. (5.89)] will be considerably less than the number required for homogeneous nucleation. As a consequence, the amount of undercooling, ΔT, required for heterogeneous nucleation will be considerably less than that required for homogeneous nucleation.

Metals and alloys rarely undercool by more than a few degrees[122] before solidification (crystallization) occurs. Crystallization usually begins on impurity particles or mold walls, and the nucleation, while essentially identical to homogeneous nucleation, will be reduced by an amount depending on the contact angle as expressed in the contact angle function of Eq. (5.35). By purposely and systematically adding inoculating agents to metal or alloy melts, it is possible to increase the incidence of heterogeneous nucleation and thereby to increase the grain refinement (decrease the grain size). Such grain refiners should produce a small contact angle, Ω_C (Fig. 5.14), which implies a high interfacial energy between the particle and the melt and a low interfacial energy between the particle and the solid. The surface structure and crystallography of such nucleation catalysts have also been demonstrated experimentally,[123] and a number of workers have observed that the undercooling required to initiate heterogeneous nucleation increases with increasing lattice mismatch between the nucleant and the solidifying particle.[123-125]

5.6.2 Growth Mechanisms and Kinetics

Although nucleation can be treated phenomenologically as either the homogeneous or heterogeneous formation of an ordered (spherical) collection of atoms composing a critical nucleus, the ease of attachment of nuclei to a surface as well as the attachment of atoms to growing nuclei and consequently to the growing solid surface during solidification will depend ultimately on the structure of the interface. As was pointed out in Chapter 4, the solid–liquid interface is diffuse, with the interfacial phase characterizing the transition from the unordered liquid phase (melt) to the fully ordered crystalline phase (solid). As is well known,[122] both diffusion and convection also play an important role during solidification.

Solidification phenomenologically occurs by the propagation of the solid surface into the melt. Such surface propagation can occur by lateral growth of surface steps through the addition of atoms to kinks or ledges [Fig. 4.01(d)], by growth through the propagation of screw dislocations (spiral growth) [Fig. 4.14(a)], or by the continuous formation of two-dimensional nuclei on a flat surface which propagate laterally, forming a new (flat) layer upon which the process repeats itself.

The rate at which any one of these processes can occur or the rate of change of any single process upon the surface undergoing solidification will account for the residual growth morphology. Facets develop on low-index faces when solidification occurs by lateral spreading as a result of the more rapid growth of higher-index faces which disappear, leaving the low-index faces behind. When, during lateral (or layer) spreading, the edges grow in preference to the corners, so-called hopper crystals form which are characterized by crystallographic voids in the low-index faces. This feature is illustrated typically in Fig. 4.10(c). Faceted growth is not observed for most metals and alloys because they possess a low entropy of melting.[120] The normal solidification morphology for metals and alloys is a cell or columnar growth, a tree-like or dendritic growth, or a eutectic growth which is characterized by a two-phase solidification instead of the normal single-phase solidification process.

Dendritic Growth. In general, interfacial equilibrium during solidification is the controlling feature of solidification morphology, and the interfacial equilibrium will depend strongly on the form of thermal diffusion fields near the interface, which, as is implied above, will depend on diffusion (both thermal and material) and the degree of convection. During solidification of a metal or alloy melt, the temperature must first be reduced by some value, ΔT, in order to form a solid nucleus. In the case of a spherical nucleus, the temperature fields will be radial and the growing interface will follow this symmetrical isotherm. Since the heat source in the system is the latent heat from the interface, and the sink is outside the casting, a negative temperature gradient will exist around the nucleus. If a perturbation disturbs the spherical shape of the nucleus, those parts of the interface furthest from the center of the nucleus will be in a region of greater ΔT than those closer to the center. As a result, those regions experiencing a greater ΔT will grow faster, and the perturba-

PROPERTIES OF INTERFACES   317

tion, as reflected in nonsymmetric growth, will increase with time as depicted
schematically in Fig. 5.39. The condition giving rise to this effect is known as
constitutional supercooling.[126,127] As shown in Fig. 5.39, a new metastable interface morphology evolves which can be characterized as a cellular morphology as a
result of its appearance in a transverse section of the solidified mass. The breakdown of a planar interface into a cellular morphology in a binary alloy due to
constitutional supercooling was first explained by Rutter and Chalmers.[128]

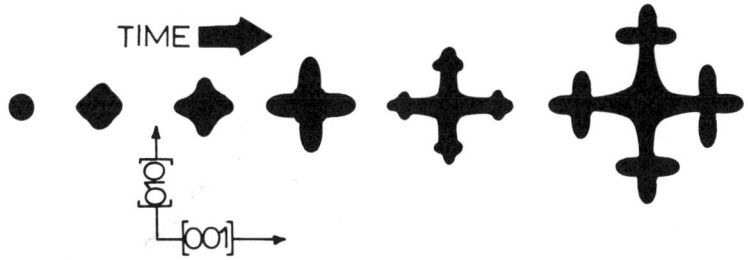

FIGURE 5.39   Development of systematic dendrite structure in the solidification of cubic
(fcc or bcc) metals and alloys undergoing directional (crystallographic) supercooling. Dendritic
growth is shown in two dimensions. In a true melt, dendrites also form in the [100] and [$\bar{1}$00]
directions in addition to the [010], [0$\bar{1}$0], [001], and [00$\bar{1}$] directions shown.

It should be apparent from Fig. 5.39 that, if the solidification is uniaxial as a
result of uniaxial supercooling, then the nuclei will evolve as elongated cells
(columnar dendrites). Similarly, under the same conditions, a planar interface will
break up into a cellular one. These cells can further break up into dendrites as
depicted in Fig. 5.39 because instabilities can also occur on the sides of the original
spikes (cells) as they grow into the melt. When the undercooling is symmetrical or
multiaxial, an equiaxed cell or grain structure results. The characteristic growth
directions for dendrites (as depicted in Fig. 5.39), while phenomenologically
ascribed to the anisotropies in growth rate and interfacial energetics in crystals
(minimization of energy), is one of the still unresolved problems of solidification
phenomena.[120,127]

We observe from the foregoing brief descriptions that growth from the melt,
unlike growth from the vapor or growth by electrodeposition [as depicted ideally in
Fig. 4.15(b)], occurs in a manner unique to the properties of the solid-liquid interface which, because of its transitional nature, is dominated by small deviations from
the temperature defining the solid-liquid transition—that is, the melting point.
Although faceted growth does not normally occur in melt growth, it can be induced
by catalysis or perturbations of growth as shown typically in the sequence of Fig.
5.12. The final structure of a solidified metal or alloy is not determined by the

simple branching characteristics depicted schematically in Fig. 5.39, but depends to a large extent on subsequent growth conditions such as coarsening which occurs as a result of the driving force which results from surface free energy minimization.

Eutectic Growth. If a liquid-metal eutectic composition solidifies, both the α- and β-phases should be produced together in equilibrium separated by an $\alpha\beta$-phase boundary. Furthermore, no change in composition should occur, as it may in single-phase solidification as a result of interfacial segregation. When a eutectic composition is solidified unidirectionally, the two phases that separate out may each grow as columnar cells or lamellae which in principle represent a single-stage composite.

A formal solution to the mechanism of lamellar eutectic growth and the associated interfacial properties has been given by Jackson and Hunt.[129] In this treatment, a eutectic alloy composition grows as depicted schematically in Fig. 5.40. The B atoms are rejected into the liquid as the α-phase grows, and the β-phase rejects the A atoms in the same way. Thus, the B atoms tend to concentrate in front of the α lamellae while being depleted in front of the β lamellae as shown in Fig. 5.40(a). This variation in composition across the interface produces a variation in local melting across the interface which must be offset by the boundary curvature [see, for example, Eq. (3.65)]. The total undercooling at the interface is constant and is given by

$$\Delta T = \Delta T_D + \Delta T_r = T_E - T^* \qquad (5.91)$$

where ΔT_D is the undercooling due to compositional variations from the eutectic composition as a result of solute diffusion, ΔT_r is the undercooling that results from capillarity, T_E is the eutectic temperature, and T^* is the actual interface temperature. The undercooling that results from capillarity is due to the radius of curvature, and, in the context of Eq. (3.65), is written

$$\Delta T_r = \gamma_{SL} T_E / (\rho_S \Delta H_f r) \qquad (5.92)$$

where ρ_S is the solid density, ΔH_f is the latent heat of fusion, and r is the radius of curvature. The convention in solidification is to consider a radius of curvature positive where liquidus temperature is depressed. Consequently, the characteristic undercoolings indicated by Fig. 5.40(a) are depicted in Fig. 5.40(b); and the interface shape shown in Fig. 5.40(c) is that which produces the requisite values of ΔT_r along the interface.

The curve for ΔT_D over the interface as depicted schematically in Fig. 5.40(b) can be determined quantitatively by specifying the lamellar spacing, λ, shown in Fig. 5.40(c), and the growth rate. Similarly, ΔT_r can be determined by specifying the angles between tangents to the lamellar surfaces at the lamellar edges and the growth direction at equilibrium. Equilibrium at the lamellar boundary juncture with the liquid [with reference to Fig. 5.40(c)] is expressed by

PROPERTIES OF INTERFACES   319

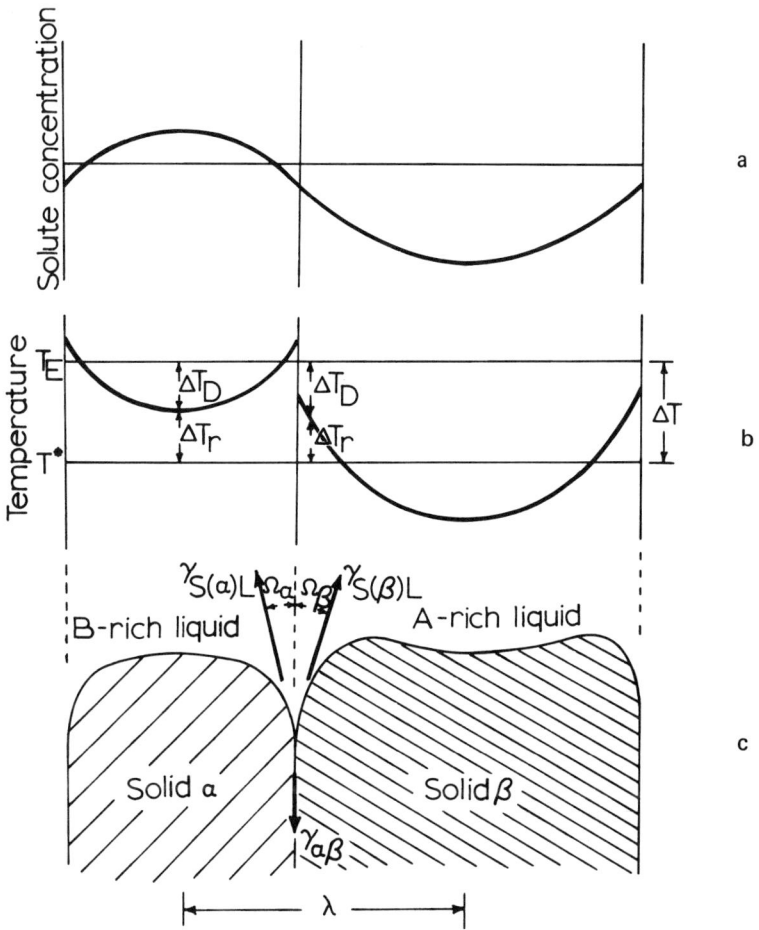

FIGURE 5.40   Lamellar curvature in eutectic growth. (a) Solute concentration in the liquid at the interface. (b) Interfacial undercoolings across the interface. (c) Predicted shape of the lamellar solid–liquid interface. (After Hunt and Jackson.[130])

$$\gamma_{S(\alpha)L} \cos \Omega_\alpha + \gamma_{S(\beta)L} \cos \Omega_\beta = \gamma_{\alpha\beta} \qquad (5.93)$$

where $\gamma_{S(\alpha)L}$ and $\gamma_{S(\beta)L}$ are the solid–liquid interfacial free energies, and $\gamma_{\alpha\beta}$ is the eutectic interfacial free energy (solid α–solid β interfacial free energy). Equation (5.93) is trigonometrically identical to Eq. (2.50), and, if the solid–liquid interfacial free energies are known, the eutectic interfacial free energy, $\gamma_{\alpha\beta}$, can be determined by measuring the dihedral angles at the lamellar boundary groove as was done in previous work by Hunt and Jackson.[130] Finally, the lamellar spacing, λ, the undercooling, ΔT, and the interface velocity (growth rate, R) have been shown to be related by[129]

$$\Delta T = aR\lambda + (b/\lambda) \tag{5.94}$$

where a and b are constants depending on the particular alloy system.

5.6.3 Melting of Metals and Alloys

We noted in Chapter 2 that the equilibrium equation for a solid–liquid interface can be written ideally as

$$\gamma_{LS} \equiv \gamma_{SL} = F_S - \gamma_{LV} \tag{2.52}$$

For most metals and alloys it is also possible to write

$$F_S > \gamma_{LV} + \gamma_{SL} \tag{5.95}$$

which means that, when the solid surface reaches the bulk melting point, T_m, then it is energetically favorable for a molten (liquid) interface phase to form on the surface. From Eq. (5.89) we also observe that for a planar surface or a surface facet characterized by a radius of curvature, r, very large or approaching infinity, ΔT becomes zero, and there is no nucleation barrier preventing surface melting from occurring spontaneously (at T_m). This feature accounts for the lack of observation of superheating in metals and alloys that melt at or very near T_m. However, it must be cautioned that this characteristic applies strictly to bulk solid metals and alloys and not to very tiny single-crystal particles as previously discussed in Chapters 1 and 3 [Eqs. (1.22), (3.64), and (3.65)]. In addition, Eq. (5.95) is strictly applicable to an ideal solid surface containing no defects and melting in a perfectly homogeneous manner. Obviously, this situation rarely occurs, and a solid containing defects either on the free surface or within the material therefore contains built-in heterogeneities which, in the context of a melting surface, are fixed in time and position. This could in principle allow for some superheating.

The interesting condition which describes melting is the fact that the solid will melt at the surface preferentially. As a consequence, this is an interfacial property. Woodruff[120] has pointed out that most theories of melting largely deny even the necessity for the existence of an interface, and one would have to conclude that such theories are therefore, in the main, of little value. The most relevant attempt to provide a reasonable theory of melting has been that of Kuhlmann-Wilsdorf[131] on the basis of lattice dislocations, which at the outset demonstrates the importance of defect structure in the phenomenological process of melting. Kuhlmann-Wilsdorf[131] has shown that at some well-defined temperature (which is very close to the experimentally determined melting point) dislocation cores (loops) should be generated spontaneously throughout the solid lattice, at which point the crystal will lose its shear strength; and this transition will be accompanied by the absorption of latent heat. Agreement with experiment is indeed quite good, and Kotze and Kuhlmann-Wilsdorf[132] have in fact shown on the basis of these concepts that the value of the solid-liquid interfacial free energy, γ_{SL}, can be calculated from the energy associated with these dislocation arrays, and that excellent agreement is found with experimental results from homogeneous nucleation studies.[120,133] It must be cautioned, however, that some reservation should be noted with regard to the original theory of Kuhlmann-Wilsdorf,[131] because there are a number of approximations in the theory and there is no direct experimental proof for it.

Grain Boundary Melting. If and when a grain boundary melts in preference to the lattice (or grain matrices), a region of liquid will in effect connect two solid-liquid interfaces. In this simple picture of a preferentially melting grain boundary, we could argue that the grain boundary energy must be composed not only of the two solid-liquid interfacial free energies, but also for a finite thickness of the melt region an interaction energy will be involved as a result of the overlap of the two solid-liquid interfaces. In addition, there must be a contribution due to the excess energy of the undercooled liquid. Consequently, we could write

$$\gamma_{gb} = 2\gamma_{SL} + t\,\Delta H_f^*(\Delta T/T_m)V^{-2/3} + \phi(t)V^{-2/3} \tag{5.96}$$

where t is the thickness of the interface phase, ΔH_f^* is the latent heat per atom, ΔT is the undercooling, T_m is the melting point, V is the atomic volume, and $\phi(t)$ represents an interaction potential for the two solid-liquid interfaces.[134]

An analysis of Eq. (5.96) by Woodruff[120] has shown that the molten layer thickness, t, would be only about 100 Å for $\Delta T \simeq 0.01°$. Glicksman and Vold,[135] who investigated the shape of the solid-liquid interfaces at grain boundaries in thin bismuth films in the transmission electron microscope, did in fact observe preferential melting at the grain boundaries very close to or at the melting point, with a melt region of roughly 50 Å from either side of the original grain boundary positions.

Gleiter and Chalmers[136] have shown, in their analysis of the theoretical aspects of grain boundary melting discussed by Shewmon[137] and Li,[106] that at the melting point of the boundary, T_{gb} (where $T_{gb} = T_m - \Delta T$), the solid and molten boundaries are in equilibrium, and hence

$$\frac{\Delta T}{T_m} = \frac{\gamma_{gb} - \gamma_{SL}}{M\,\Delta H_f} \tag{5.97}$$

where M is the amount of material per unit area of grain boundary region which melts at T_{gb}, and ΔH_f is the latent heat of fusion of the metal or alloy in question. Certainly, were it possible to accurately measure γ_{SL} and M from an experimental arrangement such as the melting of thin foils inside the electron microscope as in the work of Glicksman and Vold,[135] it would indeed be possible to measure ΔT and as a consequence determine whether $T_{gb} < T_m$—that is, whether or not a grain boundary will melt in preference to a free surface. It can be intuitively observed from Eq. (5.97) that grain boundary melting will not be possible if $\gamma_{gb} < \gamma_{LV}$ and $d\gamma_{gb}/dT \simeq d\gamma_{LV}/dt$, or if $\gamma_{gb} \simeq \gamma_{LV}$ and $d\gamma_{gb}/dT > d\gamma_{LV}/dT$. Since, for most pure metals, $\gamma_{LV} > \gamma_{gb}$, and for many specific cases (for example, Al, Cu, Au, Ag, δ-Fe, Ni), $d\gamma_{LV}/dT > d\gamma_{gb}/dT$ (see Tables 3.4 and 3.6), grain boundary melting is indeed to be expected. This further implies that ΔT is finite and that $T_{gb} < T_m$, indicating a distinction between the grain boundary and the lattice, a condition implicit in the structural view of a grain boundary presented in Chapter 4.

While, as was pointed out by Gleiter and Chalmers,[136] there seems to be no single, unambiguous experimental evidence or observation which proves the existence of grain boundary melting in pure metals, there apparently is a certain force tending in that direction. The problem in carrying out such an experiment rests primarily with the difficulty in completely eliminating grain boundary segregation, which in itself could alter the latent heat of fusion or the pure metal grain boundary free energy in such a way that $T_{gb} > T_m$ is observed. It must also be borne in mind that in the context of Eq. (5.89) grain boundary curvature could also influence the observations of preferential melting.

5.7 GRAIN BOUNDARY MIGRATION AND SLIDING

Some of the most important properties of metals and alloys involve recovery, recrystallization, and grain growth, all of which depend on or are characterized by the migration of grain boundaries—that is, the displacement of the interface perpendicular to its tangent plane by the transfer of atoms from one crystal to the other across the interface phase. In a general sense, recovery can be considered to be any modification of properties during annealing. These modifications usually involve the migration of point defects to dislocations and subgrain boundaries, the eventual rearrangement and annihilation of dislocations and the growth of subgrains by the migration of low-angle boundaries, and finally the migration of high-angle grain boundaries. Recrystallization, on the other hand, involves both the nucleation and growth of new strain-free grains, with growth characterized as synonymous with the migration of boundaries. The driving force for grain boundary migration is the strain energy stored in a crystal lattice during plastic deformation; and the rate of migration depends on various parameters including the misorientation of the grains separated by the migrating boundary, the orientation of the grains participating in the migration process, the temperature, the concentration and character of impurities in the lattice, and other more subtle features.

In addition to grain growth by the displacement of a grain boundary by migration, two grains can be translated with respect to one another by a shear movement parallel to the interface plane. Such grain boundary sliding also depends on grain orientation and misorientation of the boundary, and boundary structure, as well as the temperature, the concentration of impurities (including vacancies) or solute atoms, and the applied stress.

PROPERTIES OF INTERFACES   323

5.7.1 Boundary Migration during Recovery, Recrystallization, and Grain Growth

The major recovery of mechanical properties takes place during recrystallization, and the isothermal kinetics of recrystallization are typical of a nucleation and growth process. In the simplest consideration, nucleation of a new phase (a new crystal) within a plastically deformed phase occurs by the formation of an embryo having a stable spherical nucleus size of

$$r^* = -2\gamma_{gb}/\Delta G_v \qquad (5.98)$$

and a free energy of formation given by Eq. (5.28). Burke and Turnbull[138] have defined the critical radius of the embryo as

$$r^* = -2\gamma_{gb}/Z \qquad (5.99)$$

where Z is the difference in strain energy per unit volume between the plastically deformed state and the fully recrystallized condition.

When the radius r^* is attained, the embryo begins to grow, and the migrating boundary separates the recrystallized embryonic phase from the unrecrystallized material. If the growing grain is also considered to be spherical with a radius of curvature, r, the free energy per unit volume will be proportional to $\gamma_{gb} r^2/r^3$ or γ_{gb}/r as in Eq. (5.98). Thus, for a curved boundary, one side of the boundary can be considered to be convex, while the other matching side is considered to be concave with respect to any fixed reference point; and the radius of curvature on one side will equal the negative of the radius of curvature on the other side in order to achieve complete space filling. The free energy difference per atom between two sides of the boundary is then written as

$$\Delta G = KV\gamma_{gb}\left(\frac{1}{r_1} - \frac{1}{r_2}\right) \qquad (5.100)$$

where V is the atomic volume, and $K = K_0 \exp(-Q_G/RT)$, where K_0 is a constant and Q_G is the activation energy for boundary migration. If, as we stated above, all space is filled, then $r_1 = -r_2 \equiv r$, and Eq. (5.100) becomes

$$\Delta G = 2KV\gamma_{gb}/r \qquad (5.101)$$

The reader should note the similarity of Eqs. (5.100) and (3.1) as well as Eqs. (5.101) and (3.3), and recognize that Eq. (5.101) expresses the driving force for grain growth.

Burke and Turnbull[138] have derived the following general expression for the rate of grain growth, G_{gb}, based on absolute reaction-rate theory assuming the migration of any particular boundary to be driven by the isotropic grain boundary free energy:

$$G_{gb} = \left(\frac{2.72kT}{h}\right) \frac{\Delta t K V \gamma_{gb}}{rRT} \exp\left(\frac{\Delta S_A}{R}\right) \exp\left(\frac{-Q_G}{RT}\right) \quad (5.102)$$

where h is Planck's constant, k is Boltzmann's constant, T is the temperature, Δt is the thickness of the interface, R is the gas constant, ΔS_A is the difference in entropy between an atom in the activated state and in a site in the shrinking grain, Q_G is the activation energy of the migration process, and r, K, V, and γ_{gb} are as defined for Eq. (5.100). If it is assumed that $r \simeq D$, where D is the grain diameter and $dD/dt \simeq G_{gb}$, then Eq. (5.102) yields

$$D^2 - D_0^2 = K\gamma_{gb} Vt \quad (5.103)$$

where D_0 is the grain size at a time $t = 0$, and D is the grain size at some time t greater than zero. When $D_0 \ll D$, Eq. (5.103) can be simplified to

$$D = \sqrt{K\gamma_{gb} Vt} \quad (5.104)$$

Equation (5.104) is generally not fulfilled in grain growth in metals and alloys, and the observed isothermal data of the continuous grain growth in metals may be better represented by an equation of the form[139]

$$D = (Kt)^n \quad (5.105)$$

where n for metals is usually much less than 0.5 (about 0.1) and is sometimes dependent on temperature,[139,140] except near the melting point where $n \simeq 0.5$ for pure metals.

Byrne[141] has outlined another way of looking at grain growth where the growing grain contains a lower density of imperfections than its neighbors. If the imperfections are considered to be subgrain boundaries, or low-angle grain boundaries, then the recovery interface or recrystallization front can be considered as a high-angle boundary intersected repeatedly by low-angle boundaries which are consumed as the interface migrates into the subgrain boundary region. If the deformed matrix portion has n subgrain boundaries ending on the migrating grain boundary, then the migrating interface can be considered to be an n-sided polygon with inside dihedral angles equal to

$$\Omega = \pi\left(1 - \frac{2}{n}\right) \quad (5.106)$$

If we substitute this value of Ω into Eq. (4.20), we obtain

$$\frac{\gamma_{gb}}{\gamma_{gb}'} = \frac{1}{2\cos\left(\frac{\pi}{2} - \frac{\pi}{n}\right)} = \frac{1}{2\sin\left(\frac{\pi}{n}\right)} \simeq \frac{n}{2\pi} \quad (5.107)$$

PROPERTIES OF INTERFACES 325

provided n is large, where γ_{gb} is the high-angle grain boundary free energy (the interfacial energy for the migrating boundary), and γ_{gb}' is the average low-angle (subgrain) boundary free energy. If d is the subgrain boundary diameter, then

$$\pi D = nd \qquad (5.108)$$

where, as before, D is the average grain diameter for the growing grain. Substituting for n/π from Eq. (5.108) in Eq. (5.107) then results in

$$\gamma_{gb}/\gamma_{gb}' = D/2d \qquad (5.109)$$

Thus, when D/d becomes greater than $2\gamma_{gb}/\gamma_{gb}'$, the boundary will migrate into the subgrain matrix portion (the deformed matrix). Equation (5.109) also illustrates a rough approximation for measuring the ratio of average high angle to low angle grain boundary free energies by recrystallization and grain-growth experiments.

The processes of recovery and recrystallization often overlap, making a sharp distinction difficult if not impossible. This is particularly true for deformation processes at low strain, such as in explosive shock deformation,[142] where little if any distortion of the original grain boundaries occurs. In this case, as illustrated in Fig. 5.41, the migrating interface separating the deformed region of the grain from the relatively undeformed (recovered) grain portion is both the recovery and the recrystallization front. Figure 5.41(b) also illustrates phenomenologically the growth concept characterized by Eq. (5.109).

A grain boundary usually migrates toward its center of curvature during grain growth, whereas at the onset of recrystallization the growth of an embryo is away from its center of curvature. This is consistent in describing Fig. 5.41(b), since the recrystallizing grain is opposite the recovered grain. These and other features of recovery, recrystallization, and grain growth have been reviewed by a number of investigators[138,141,143,144] and will not be pursued in more detail.

The Effect of Orientation and Misorientation on Grain Boundary Migration. There have been a number of attempts to illustrate the effect of the orientation relationship between two grains on boundary mobility. Tiedema *et al.*[145] observed that the orientation of small crystals occluded in large recrystallized grains either were twin-related or possessed an orientation relationship nearly identical to that of the surrounding matrix, and the mobility for these low-angle or low-energy relationships appeared to be very small, leading to a high degree of stability for these small crystals. Figure 5.42 shows a typical example of such an occluded crystal. Selected-area electron diffraction patterns taken on the interface between the occluded crystal and the recrystallized matrix indicated little if any detectable orientation difference. Furthermore, the boundary is composed entirely of a systematic dislocation array characteristic of a low-angle grain boundary as discussed in Chapter 4. The straight boundary portion was also observed to lie near a coincidence lattice orientation. In this regard, numerous other results have been obtained which indicate the stability of coincidence-related boundaries as a result

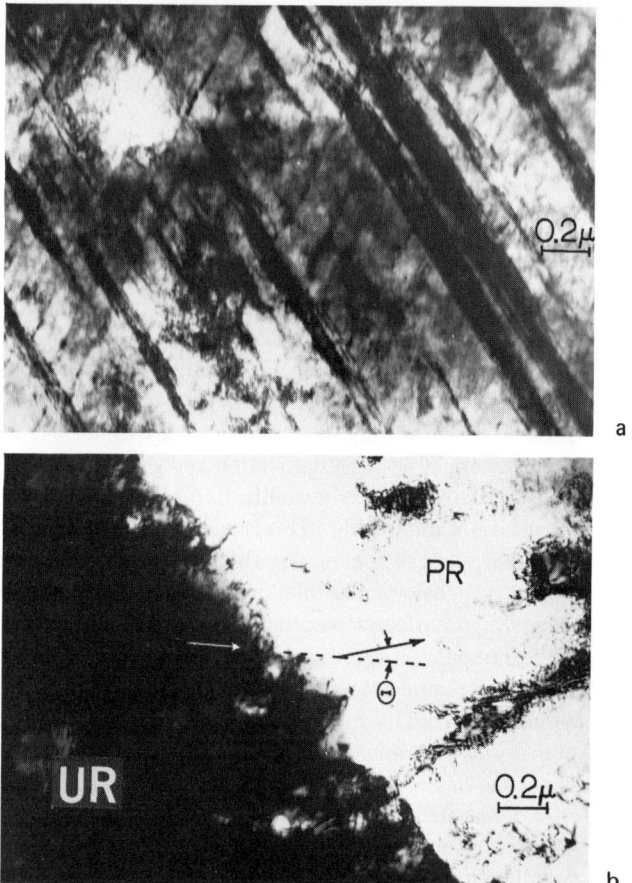

FIGURE 5.41 Recovery and recrystallization by boundary migration in plastically deformed 304 stainless steel. (a) Transmission electron micrograph showing deformation twins and dislocations in a (110) grain following explosive shock deformation at 425 kilobars. (b) Similar (110) orientation as in (a) undergoing recovery and recrystallization after 1 hour at 800°C. Note the misorientation of the migrating boundary and the subgrain structure in the deformed matrix portion. Partially recovered and unrecovered matrix portions are denoted PR and UR, respectively.

of their low mobility.[146-150] This variation in grain boundary mobility with misorientation is illustrated for aluminum grain boundary migration[151] in Fig. 5.43(a), and changes in the activation energy for grain boundary migration [Q_G in Eq. (5.102)] as a function of boundary misorientation in zone-refined lead[152] are shown in Fig. 5.43(b).

PROPERTIES OF INTERFACES 327

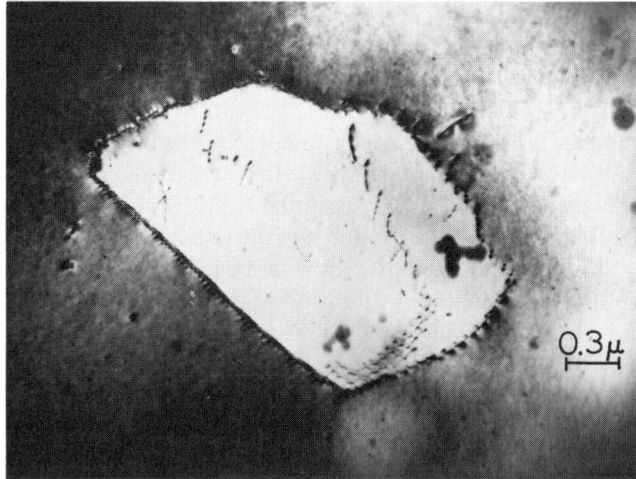

FIGURE 5.42 Occluded grain in a recrystallized stainless-steel matrix enclosed by a low-angle boundary having low mobility.

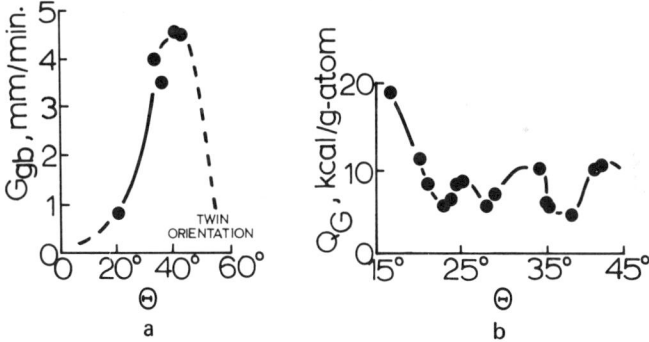

FIGURE 5.43 Boundary migration as a function of misorientation. (a) Migration rate, G_{gb}, versus misorientation in commercial aluminum (rotation about <111>). (From Liebmann and Lucke.[151]) (b) Activation energy for boundary migration, Q_G, versus misorientation in zone-refined lead (rotation about <001>). (From Rutter and Aust.[152]).

It can be observed in Fig. 5.43 that, not only is the boundary misorientation an important parameter in the migration of the interface, but also the orientation of the boundary itself—that is, whether it has a coincidence or a noncoincidence (general) orientation relationship. Further evidence for the influence of the orientation relationship on grain boundary mobility is available from measurements of boundary migration rates and activation energies for grain boundary migration in solid solutions from observations on idealized bicrystals[153,154] and polycrystals.[155,156] The influence of small amounts of impurities on the grain boundary mobility has also been demonstrated.[157] These observations indicate that the grain boundaries of highest mobility are those associated with high-density coincidence boundaries, because, as illustrated typically in Fig. 5.44, these boundary orientations exhibit a minimum of interaction with solute atoms or impurities. These features are consistent with the interfacial structural phenomena discussed in Chapter 4, and they attest to the importance of grain boundary structure in processes involving grain boundary migration.

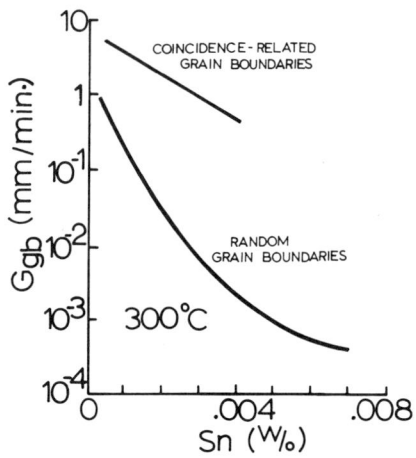

FIGURE 5.44 Grain boundary migration rate versus tin concentration in a Pb-Sn alloy. (Based on data of Aust and Rutter.[153,154])

Effect of Impurities and Other Inclusions on Grain Boundary Mobility. A comparison of Figs. 5.43 and 5.44 shows that, while intrinsic differences in grain boundary mobility occur as a result of interfacial structure (as evidenced by variations of G_{gb} with the misorientation angle, Θ), the interaction of solute atoms or impurities caused by boundary segregation can have an important effect on the mobility. This happens primarily because of the structural modifications imposed on the interface which normally cause a reduction in the interfacial free energy and a subsequent reduction in the driving force [Eq. (5.102)], as well as the retardation of migration by the drag imposed on the moving interface. The imposed drag results by individual atoms or other impurity inclusions, including precipitates which can form in the interface, or inclusions or precipitates in the matrix which

PROPERTIES OF INTERFACES   329

can block the migration of the boundary by pinning it as illustrated typically in Fig. 5.45 for a migrating twin boundary in a thoria (ThO$_2$)-dispersed metal (pure nickel).

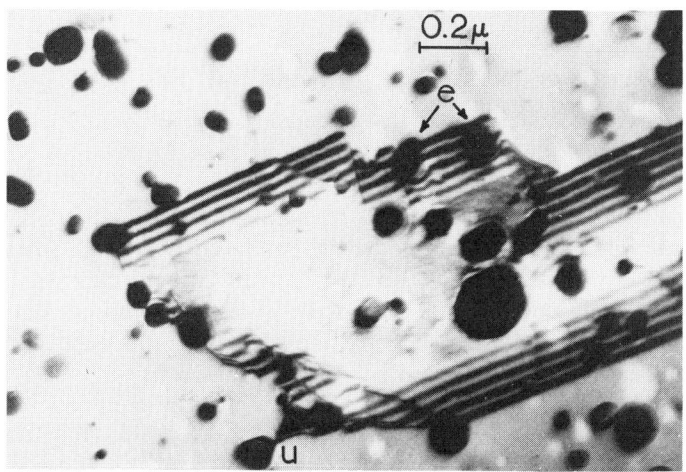

FIGURE 5.45 High-angle (noncoherent) boundary portions (mobile boundary portions) of occluded annealing twins in a thoria-dispersed nickel matrix pinned by noncoherent thoria particles. The mobile high-angle boundaries are migrating in an effort to annihilate the twin. Note boundary portion about to become unpinned at *u* and the envelopment of inclusions by the boundary at *e* (arrows).

As illustrated in Fig. 5.45 (arrows), if the driving force for grain boundary migration [Eq. (5.101)] exceeds the pinning force exerted by a particle or matrix inclusion, the boundary will pull free and migrate through the lattice without altering the particle or inclusion distribution. Ashby *et al.*[158,*] have shown that, when the grain boundary free energy inside a coherent inclusion is lower than that outside the inclusion, the boundary will in effect enter the inclusion. On the other hand, when the inclusion is noncoherent, or when the interfacial free energy for a boundary inside the inclusion exceeds the energy of the boundary outside the inclusion, the boundary tends to envelop the inclusion, changing the particle-matrix interfacial free energy. The envelopment of particles by the boundary in this manner is clearly visible in Fig. 5.45 [see also Fig. 1.12(*a*)].

Unlike the drag, pinning, and related retardation effects of solute atoms, inclusions, and precipitates, etc., on grain boundary migration, vacancies tend to enhance boundary mobility,[159-161] and high vacancy concentrations have been

*This treatment has been reviewed in Gleiter and Chalmers,[136] p. 170.

associated with the rapid recovery in deformed materials by significantly increasing the boundary mobility.[142,162,163] As we shall discuss in the next section, the principal reason for vacancy-enhanced grain boundary mobility is that the interface phase is made more mobile as a result of vacancy absorption within it, thereby permitting a more flexible structural arrangement within which the transfer of atoms necessary for migration can take place more easily.

5.7.2 Mechanisms of Grain Boundary Migration*

The migration of a grain boundary phenomenologically involves the transfer of atoms of one grain or crystal lattice to another grain or crystal lattice. This can be accomplished ideally by the transfer of groups of atoms by a simultaneous movement of atoms from the boundary region to the ordered lattice,[164] or the movement of dislocations by glide or climb into the lattice.[106] Boundary migration can also occur by the transfer of single atoms across the interface from one grain to the other either by a mass transfer process where each atom moves from a site in the boundary associated with one lattice into a new site in the interface associated with the other lattice[165] or by a vacancy-diffusion mechanism.[166]

Lin and Murr[167] have pointed out that it is as difficult to rationalize one single mechanism of grain boundary migration as it is to rationalize one single model for grain boundary structure. Indeed, it can be argued that any particular mechanism of grain boundary migration might be dependent on the grain boundary structure, and could be expected to vary as the structure of the boundary is altered. The atomistic step mechanism of Gleiter[165] could apply in the case of a low-angle tilt boundary or a coincidence-related interface which lacks step structure, since this would, consistent with experimental observations (Figs. 5.43 and 5.44), ensure a low or negligible migration rate. However, in the case of complex, high-angle grain boundaries which possess a high density of grain boundary ledge structure and related step features, this mechanism certainly appears tractable. One of the attractive features of the step mechanism[165,168] is the fact that the transfer of atoms across the interface during grain boundary migration can be represented by a process that is phenomenologically the same as the nucleation and growth of atoms at a free surface, as was discussed previously. It is this phenomenological consistency in both structure and process which, along with several experimental observations,[168,169,*] make this mechanism as illustrated in Fig. 5.46 a functionally accurate one for many cases of high-angle grain boundary migration. In the step mechanism illustrated in Fig. 5.46, migration of the boundary occurs by the emission of atoms from the steps associated with the surface of the grain being

---

*Migration of a grain boundary and subsequent grain growth involve not only the parallel displacement of the interface itself but also a translation of the triple line representing the line along which three grains meet. The movements of the triple line necessitates the simultaneous movement of three connected interface planes during grain growth. Similarly, grain boundary sliding must also involve the movement of triple lines which can occur through both migration and sliding. In a general sense, triple point translation may ultimately impose an overall rate control on both boundary migration and sliding.

PROPERTIES OF INTERFACES   *331*

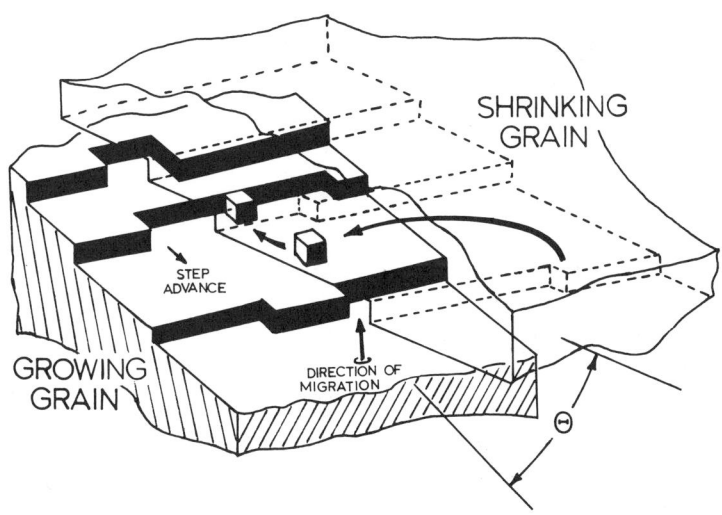

FIGURE 5.46 Schematic representation of grain boundary migration by the systematic movement of interfacial steps as a result of atom transfer from the shrinking grain to the growing grain. Arrows indicate the direction of transfer and step advance. The misorientation of the two grains is denoted by Θ. Note the similarity of the advancing surface component (the growing grain) to that shown in Fig. 4.10(*d*).

consumed and the absorptive transfer ("nucleation") of the same atoms at the steps associated with the growing grain. This process, like the growth of a free surface, could also occur by the propagation of spiral ramps associated with screw dislocations terminating on the interface. These spiral structures in the grain boundary as well as their rotation during migration have in fact been experimentally observed.[165] The systematic movement of steps coincident with specific lattice planes—for example, the $\{111\}$ planes in fcc metals and alloys by the mechanism illustrated in Fig. 5.46—is also consistent with the structural model in Fig. 4.37. It would indeed be desirable to test the general validity of this mechanism—for example, by experimentally measuring the rates of grain boundary migration or grain growth as a function of grain boundary ledge density.

5.7.3 Mechanisms of Grain Boundary Sliding

In the continuum treatment of grain boundary sliding,[170] the boundary is thought of as a surface that can slide upon itself in a continuous manner, totally ignoring atomistic structure, and avoiding the complexity that appears when dislocations, ledges, protrusions, facets, and other structural features are introduced. While, as experimentally demonstrated in Fig. 5.47, grain boundary sliding may appear macroscopically as the translation of two grains over a common planar interface as

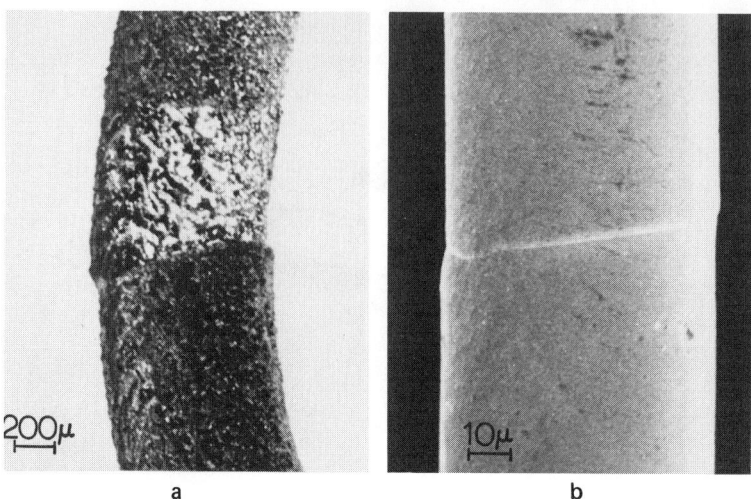

FIGURE 5.47 Relative translation of two grains by a shear movement parallel to the interface separating the two cylindrical crystals along a wire stressed at high temperature. (a) Grain boundary sliding in a Ti-Mo alloy wire (optical micrograph). (b) Grain boundary sliding in a nickel wire (scanning electron micrograph). Note the step in both examples where the grain boundary meets the free surface.

a result of resolved shear stresses acting parallel to the interface, it can occur microscopically only by systematic displacements on an atomic level involving the structural features of the grain boundary described in Chapter 4. Phenomenologically, it is difficult to rationalize the sliding motion of a grain boundary on the atomic level without invoking some degree of boundary migration involving a mechanism of migration that will allow for a relaxation of boundary stresses and maintain some structural integrity by atomistic transfer across the interface or diffusion within the interface plane. As a consequence, grain boundary sliding is most prominent at elevated temperatures, although sliding can take place at very high resolved shear stresses and low temperatures, and some evidence has been reported for grain boundary sliding at very low temperatures.[171,172] These phenomenological aspects of macroscopic grain boundary sliding have been reviewed extensively[173-175] and will not be pursued in any further detail.

It is interesting to compare the translation of solid matter by grain boundary sliding in Fig. 5.47 with any similar example of stress-induced slip of a perfect crystalline matrix (for example, Fig. 9-10 of Guy[85] showing macroscopic slip of a single-crystal metal rod). Indeed, this comparison suggests an immediate compulsion to draw a strong parallel between the two similar phenomena, since, as we have already discussed (Section 4.5.1), the grain boundary also possesses, in

PROPERTIES OF INTERFACES   333

conjunction with its structural features, unique interfacial dislocations (Table 4.7) which, like crystal lattice dislocations, can account for the systematic sliding of the boundary plane. Many of these features have been recognized and discussed by Ashby,[176] Gleiter and Chalmers,[136] and Gates;[177] and it must be concluded that grain boundary sliding can be consistently described by the movement of grain boundary dislocations (GBD's) in the interface.

It can be observed with reference to Table 4.7 that the GBD's associated with $\Sigma = 3$ coincidence site boundaries—that is, fcc coherent twin boundaries—contain dislocations which are identical to fcc crystal lattice partial dislocations, and would therefore not slide in preference to conventional slip. Sliding of a twin boundary is thus extremely difficult, as all atoms would have to move at the same time.[178] In addition, as we noted in Section 4.5.1, grain boundaries corresponding exactly to coincidence-site lattice arrays should contain few if any grain boundary dislocations or dislocations of any character accommodating to the coincidence array. As a result, grain boundary sliding should be sensitive to grain misorientation in a manner similar to the sensitivity of grain boundary free energy to misorientation as shown in Figs. 4.22 and 4.65. This feature is convincingly demonstrated in the experimental data of Biscondi and Goux[179] reproduced in Fig. 5.48.

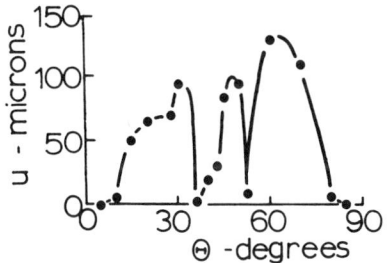

FIGURE 5.48 Grain boundary sliding displacement, u, versus misorientation, Θ, for <001> aluminum tilt boundaries subjected to a shear stress of 40 gf/mm² at 560°C. (After Biscondi and Goux.[179])

As in conventional lattice slip, both the rate of sliding and the sliding displacement will depend on the number of grain boundary dislocations. Ashby[176] has derived the following simple expression for the rate of grain boundary sliding:

$$\dot{U} = \frac{8 b_b D_{gb} \Delta t \, \tau}{kT} (\rho_b \lambda) \qquad (5.110)$$

where b_b is the Burgers vector of the grain boundary dislocations, D_{gb} is the grain boundary diffusion coefficient, Δt is the thickness of the interface phase, τ is the resolved shear stress, k is Boltzmann's constant, T is the temperature, ρ_b is the density of mobile boundary dislocations per unit length or per unit area of boundary, and λ is the translational periodicity in the boundary, and consequently the unit displacement of the grains. As we noted in Chapter 4 (Fig. 4.31), dislocations can exist in a grain boundary as a result of grain growth, or they can be formed on penetration by lattice dislocations. In addition, since GBD's form grain boundary ledges, it is only logical that under certain conditions ledges would act as sources of GBD's. Ledges can also act as barriers to the movement of GBD's, resulting in pile-ups and stress concentrations which can cause yielding of the ledge resulting in an avalanche of sliding,[180] or the formation of lattice dislocations or microcracks as illustrated in Fig. 5.36. Gifkins and Snowdon[181] have also proposed that grain boundary sliding at low stresses is controlled by grain boundary diffusion involving the transport of matter from ledges in the boundary under tensile stress to ledges under compressive stress, which in essence means that sliding phenomenologically occurs by the movement of double ledges along the interface.

As we indicated initially in this section, triple lines are effective in rate control of both migration and sliding. Ultimately, the sliding boundary can be envisioned as the movement of GBD's along the boundary plane which pile up at the triple line in much the same way that lattice dislocations pile up against some obstacle, thereby reducing the value of τ in Eq. (5.110), and decreasing the sliding rate. If the applied stress is increased sufficiently, lattice dislocations may be generated at the boundary, reducing the number of GBD's in the interfacial pile-up, and allowing further sliding to occur. GBD's piled up at a triple line can also set up a diffusive flux of matter from the compressed parts of the boundary to those parts of the boundary under tension, as outlined above for the diffusive glide of ledges.[181] This process may be phenomenologically viewed as the climb of GBD's from one interface to another around the triple line. As a result of an applied stress, we therefore observe that dislocations gliding into a grain boundary can generate GBD's, that, at high stresses, grain boundaries can act as a source of lattice dislocations, and that diffusive climb of GBD's can also aid in the sliding process. In addition, GBD's could conceivably be generated in the interface by a Frank-Read or other suitable dislocation source, as well as the dissociation of ledges. Furthermore, ledge movement itself can contribute to sliding by a diffusive glide process.

There emerges, therefore, a somewhat complex picture of the atomistic mechanisms of grain boundary sliding which may vary in a specific sense depending on the structure of the boundary, the temperature, and the applied stress, with composition being an implicit feature of the interfacial structure. Within the context of this complex picture, the effects of vacancy absorption and emission, solute segregation, and related phenomena can be dealt with quite satisfactorily, particularly when the coupling of sliding and migration are recognized;[176,182] however, it must be stressed that no single mechanism or model feature is capable of explaining all the experimental results in a completely satisfactory manner.

5.7.4 Grain Boundary Sliding and Creep in Polycrystals

Grain boundary sliding, as discussed above, is an integral part of diffusional creep in polycrystalline metals and alloys. Creep represents the deformation of polycrystals involving either the flow of ions through the grains, a process known as Nabarro-Herring creep,[183,184] or the flow of ions within the interface phase (boundary) separating the grains, a process known as Coble creep.[185] As Ashby pointed out,[177] the total creep process may be regarded either as deformation by grain boundary sliding with diffusion maintaining the integrity of the interface, or as deformation by mass transport with grain boundary sliding accommodating the incompatibilities which would otherwise appear in the interface. The essence of these features is illustrated in Fig. 5.49, from which it is observed that the transport process is essentially one involving grain boundary migration. It is relatively easy to envision

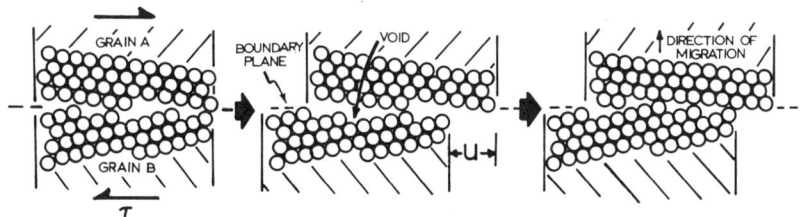

FIGURE 5.49 Idealized schematic of creep process involving a rigid translation (sliding) along the grain boundary and migration (or transport of matter) into the voids created at the interface. The sliding displacement is denoted by u. Solid-shaded atoms represent those atoms transported from crystal A to crystal B, causing an apparent migration of the boundary.

the formation of grain boundary voids during creep processes in which there is insufficient transport to accommodate the maintenance of the interfacial phase and/or by vacancy diffusion (Coble creep), which can be envisioned as the mechanism of void growth. This would be particularly true at high stresses and low temperature.[186,187] Voids can also form in the boundary as a result of ledge-induced microcrack formation (Fig. 5.36), which would also favor diffusion-induced growth at high temperatures. Figure 5.50 illustrates for comparison both the classical picture and the equivalent dislocation picture of diffusional creep along with the appropriate diffusional creep equations for each. The value of D represents the average grain diameter, while D_{gb} and D_v represent the grain boundary and lattice (volume) diffusion coefficients, respectively; σ, the applied stress, represents the equal and opposite tensile and compressive stresses; Ω in the creep-rate (strain-rate) equations represents the normal atomic volume; M is the dislocation mobility; and Δt is the grain boundary thickness. It is to be recognized

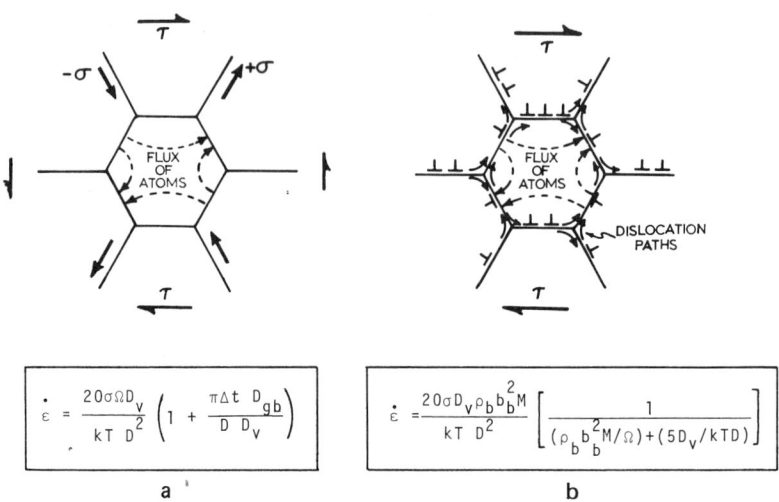

FIGURE 5.50 Schematic picture of classical diffusional creep (a) and the boundary dislocation view of the process (b). The associated creep-rate equations associated with each are shown. (After Ashby.[176])

that the term M in the equation associated with Fig. 5.50(b) accounts for the reduced sliding as a result of drag or pinning of dislocations in the boundary by solute atoms, precipitates, or other inclusions, as well as the effect of vacancy absorption. It should also be pointed out that the classical diffusional creep equation depicting the situation illustrated in Fig. 5.50(a) will take the form of the Herring[184] or Coble[185] strain-rate equations depending on whether $(\pi \Delta t D_{gb}/DD_v)$ is much less than 1 or much greater than 1, respectively.

The theory of diffusional creep and the phenomenology of creep processes are unique among the mechanical properties of metals, and certainly rely most intimately upon interfacial properties. While an attempt has been made in the brief account presented here to illustrate this intimacy, it has not been possible to give even a reasonable accounting of the considerable amount of work that has been done with regard to creep in polycrystalline metals and alloys. The serious reader is encouraged to consult the more detailed works of Garofalo[188] and Kennedy.[189]

5.7.5 Grain Boundary Sliding and Superplasticity

The general relation for creep rate can be written simply as

$$\dot{\epsilon} = K\sigma^n D^m \exp(-Q/RT) \tag{5.111}$$

where k is a material constant, σ is the applied stress, D is the grain size, Q is the activation energy for diffusion, R is the gas constant, T is the temperature, and n

PROPERTIES OF INTERFACES   337

and m are 1 and -2 for Nabarro-Herring creep and 1 and -3 for Coble creep (Fig. 5.50). In the general case, superplasticity or superplastic deformation involves both the stress dependence of the strain rate as expressed in Eq. (5.111) and the strain-hardening exponent, with Eq. (5.111) dominating for elevated temperature superplastic behavior.[190] Normally, superplastic behavior occurs for $n \leqslant 3$ and $m \leqslant 3$.[191,192] Phenomenologically, superplastic deformation involves plastic flow accompanied by extensive grain growth during deformation, and grain boundary sliding is also extensive. Thus, although superplastic deformation is similar to creep deformation, it differs mainly as a result of extensive grain boundary migration which must occur as a necessary condition for grain growth. Chaudhari[192] has developed a model for superplasticity wherein grain boundary sliding is blocked by triple lines, and the rate-controlling mechanism is not grain boundary sliding but rather the rate at which accommodation occurs at the triple lines. This process can be simulated by a dislocation pile up in the boundaries as discussed in Section 5.7.4, with climb along the grain boundaries. In effect, diffusional creep (primarily matter transport) is replaced by dislocation climb. A more phenomenological set of theories with the same general feature of coupled boundary and grain deformation has been presented by Hayden et al.[193]

Superplastic deformation, as shown schematically in Fig. 5.51, becomes prominent in ultrafine single-phase and two-phase alloys as a result of the increased density of triple lines and the structural properties of the interfaces. The grain size dependence of the strain rate for a fine-grained alloy is, as depicted in Eq. (5.111), a matter of compensation where deformation occurs at lower stresses. Consequently, the strain (elongation) attainable at some fixed stress can be increased by a decrease in the initial small grain size. It may also be speculated that, as the triple lines are reduced in density and oriented in the direction of the applied stress (Fig. 5.51), boundary sliding predominates, and long-range displacements occur, adding to the overall elongation which characterizes superplasticity. In addition, superplasticity must be recognized as a high-temperature phenomenon, since it is found at elevated temperatures where enhanced grain growth and grain

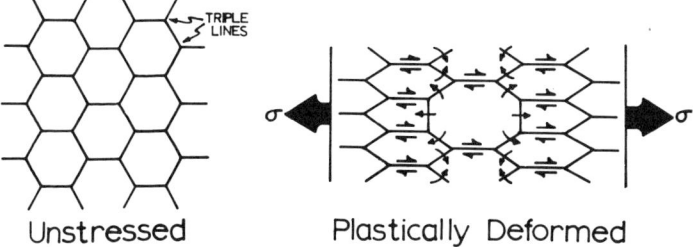

FIGURE 5.51 Grain boundary sliding and grain alteration (growth) by grain boundary migration during superplastic deformation. Single-headed arrows indicate the sense of shear (sliding); double-headed arrows represent the transport or migration paths.

orientation can occur to make accommodation of sliding more favorable during deformation.

5.8 INFLUENCE OF GRAIN AND INTERPHASE BOUNDARIES ON MECHANICAL PROPERTIES

Although we have already dealt with mechanical properties of metals and alloys—metallic friction and adhesion in Section 5.3, fracture of metals and alloys in Section 5.5, and recovery, recrystallization, and grain growth as well as creep and superplastic deformation in Section 5.7—we have not yet discussed the general features of elastic and plastic compatibility at solid-solid interfaces and the strengthening of solids as a direct result of the presence of internal surfaces, or the attendant properties of yielding and plastic flow. By compatibility we mean, as Hirth pointed out in an excellent review lecture,[194] the accommodation of incompatibility stresses and strains introduced at grain or phase boundaries as a result of the elastic and plastic anisotropies, which are maximum at the interface and decrease continuously with distance from the interface. As a consequence, a slip band impinging on a grain boundary requires either accommodating slip or accommodating elastic strains for local compatibility.

Individual dislocations interact with the compatibility stress fields of an interface in the same way as a slip band, where, in addition, repulsive forces acting on the dislocation exist. As we noted in Chapter 4, the dislocation can penetrate a grain or phase boundary with a resulting change in its Burgers vector and the production of a step (or ledge) in the interface plane (Fig. 4.31). We have also discussed some of the properties of metals and alloys associated with the conservative glide of interfacial (grain boundary) dislocations (in creep and superplasticity), and we have observed that grain boundary ledges formed either by heterogeneous shear of the boundary or by the coalescence of grain boundary dislocations can also act as dislocation sources (Fig. 4.34) or as nuclei for cracks directed intergranularly or transgranularly (Fig. 5.36). In this section, we shall reexamine the interaction of dislocations with grain and interphase boundaries in slightly more detail, the role of boundaries as both sources and sinks for dislocations, and the resulting macroscopic effects occurring primarily at low temperature, which include yielding and flow as well as the strength of metals and alloys and strengthening mechanisms involving interfacial impedance to dislocation motion by pinning or glide obstruction.

5.8.1 Grain and Interphase Boundaries as Sources for Dislocations

It is apparent from many studies of the early stages of deformation or plastic flow that dislocations arise mainly from the grain boundaries. There have been very few direct observations of the operation of classical Frank-Read type dislocation sources in metals and alloys,[195,196] and few observations of initial matrix sources of dislocations or dislocation multiplication. The majority of the initial dislocations that appear during the onset of plastic flow arise at grain boundaries or other interphase interfaces.

It was noted in Section 4.5.1 that dislocation ledge structures formed in a grain boundary by several possible mechanisms[106,197,198] could act as sources of dislocations, and many direct observations have confirmed this property of grain boundaries.[197-209] In addition to glide dislocations observed emanating from grain boundary ledges as shown in Fig. 4.34, climb loops as well as prismatic loops generated from incoherent interfaces have also been observed.[210-217] However, most of the observations reported for the generation of dislocation loops indicate that the phenomenon is due to a specific property of an interphase interface. While dislocation nucleation at a particle-matrix interphase interface can occur by stress relief envisioned from Fig. 4.61(a) as the loss of coherency with the movement of a dislocation from the interface, prismatic punching can be compared with interfacial ledge mechanisms which produce dislocation loops in a manner similar to singular ledges, as illustrated for comparison in Fig. 5.52. In effect, the formation of dislocation loops at an interface as shown in Fig. 5.52(b) can be considered as simply a double ledge by comparison with Fig. 5.52(a), and could in principle operate at either a grain or interphase boundary. The reader should also note the similarity of the initial ledge in Fig. 5.52(a) to that in Fig. 5.36(c). The ledge mechanism illustrated in Fig. 5.52(c) is that originally proposed by Li,[106] while the prismatic punching mechanism in Fig. 5.52(b) has been discussed by Brown and Stobbs[218] and Ashby[219] and demonstrated with considerable experimental clarity by Humphreys and Stewart.[220] Figure 5.53 also convincingly illustrates many of the features of interfacial dislocation sources alluded to in Fig. 5.52. In addition, Fig. 4.47(b) illustrates the emission of stacking faults from grain boundaries by the propagation of a Shockley partial dislocation from a ledge; similar observations have been made by Warlimont.[221] The formation of deformation twins at grain boundaries has also been reported.[222]

Whereas a limited number of dislocations can be emitted from an interface by the mechanisms illustrated in Fig. 5.52, the mechanisms proposed by Orlov,[223] Baro et al.,[197] and Gleiter et al.[224] can account for the formation of a practically unlimited number of dislocations. These involve the operation of monatomic steps as a dislocation source in two-slip systems, the generation of dislocations in a one-slip system emanating from a boundary at the line of intersection with an impinging array of piled-up dislocations in another slip system, and the generation of dislocations at triple lines and grain boundary kinks as a result of grain boundary sliding; these have been described in the review by Gleiter.[225] Price and Hirth[226] have also described a continuous screw dislocation source operating from ledge-type dislocations in a grain boundary, and the role of grain boundary dislocations and ledges in dislocation emission has been reviewed recently by Hirth and Balluffi[227] and Hirth.[194] While the mechanisms of Orlov,[223] Baro et al.,[197] and Gleiter et al. et al.[224] would be more favorable after plastic flow has occurred, and dislocations begin to pile up against grain boundaries, the mechanism of Price and Hirth[226] would, along with those illustrated in Fig. 5.52, constitute the principal features of dislocation emission from interfaces during the onset of plastic flow.

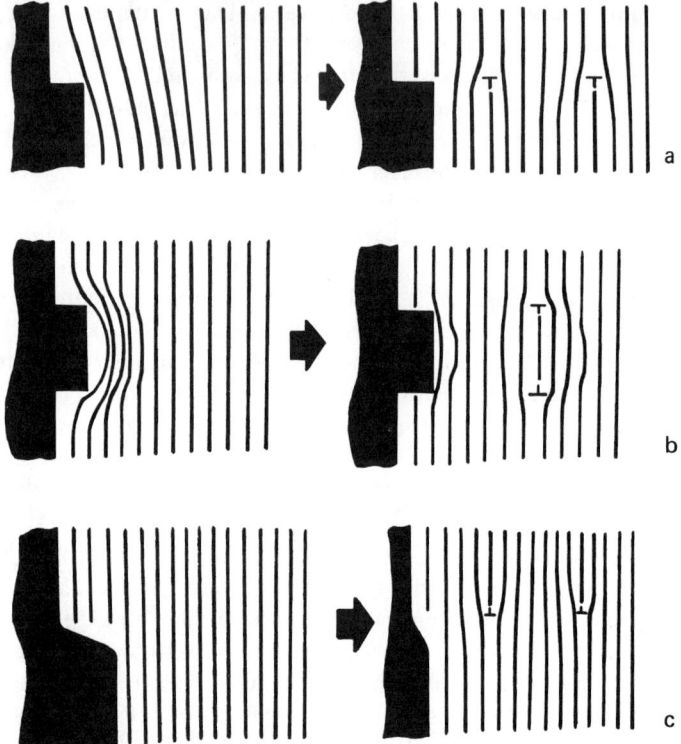

FIGURE 5.52 Interfacial ledges as sources of dislocations. (a) Single, coherent ledge emitting dislocations into the matrix. (b) Double ledge emitting dislocation loop (by prismatic punching). (c) Noncoherent ledge emitting dislocations. The interface step can emit dislocations of opposite sign into both neighbor grains [cf. Fig. 2(d) of Murr.[209]] Note that the interface step can be composed of microsteps which can each emit dislocations on different (parallel) slip planes.

5.8.2 Yield, Plastic Flow, and the Hall-Petch Relation

A relation between the yield or flow stress, σ, and the grain size for polycrystalline metals and alloys, D, was first proposed by Hall[228] in the form

$$\sigma = \sigma_0 + K_p D^{-1/2} \qquad (5.112)$$

where σ_0 is the lattice frictional stress, and K_p is the Petch[229] slope. The microscopic mechanism originally envisioned by Hall[228] and later by Petch and co-workers[229-231] involves a pile-up of dislocations at the grain boundaries such that

PROPERTIES OF INTERFACES	341

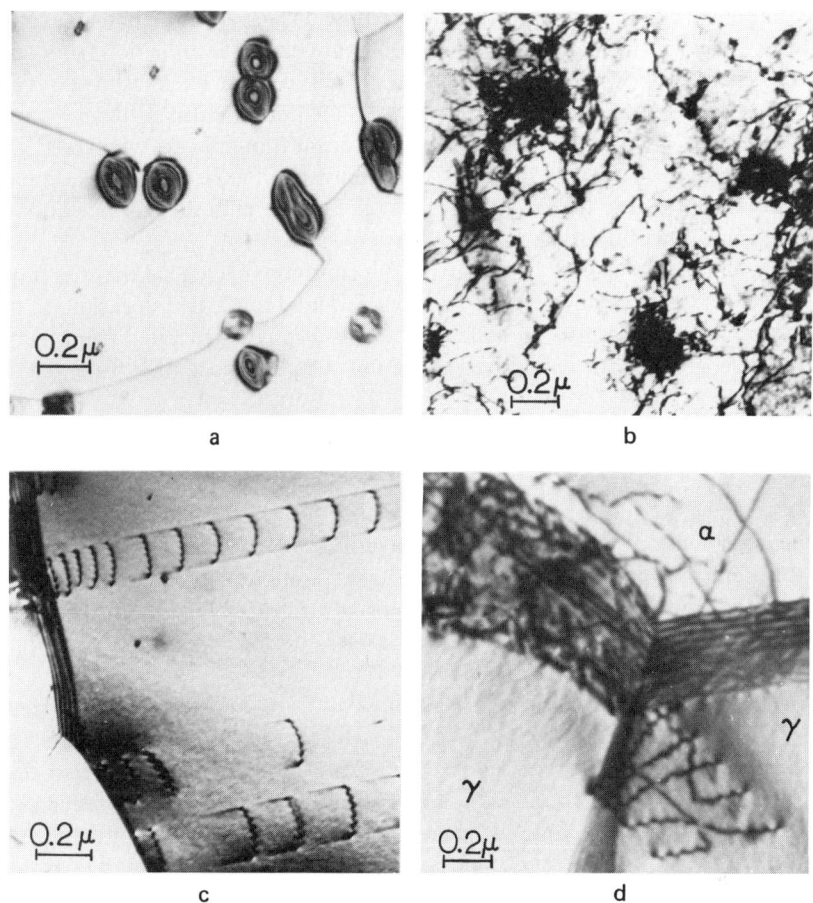

FIGURE 5.53 Dislocation generation at interfaces. (a) Coherent precipitates in Inconel.
(b) Same alloy as (a) after shock deformation at 150 kilobars. Precipitates have lost coherence, and large numbers of dislocations have been generated at the interface. (c) Glide dislocations emanating from grain boundary ledges in stainless steel. (d) Ledge structure and dislocation emission from grain and interphase boundaries in a two-phase stainless steel.

$$K_p = \left[\frac{\sigma_i GbD}{\pi(1-\nu)l}\right]^{1/2} \tag{5.113}$$

where σ_i is referred to as the average strength of the grain boundary, G is the shear modulus, D is the average grain size, ν is Poisson's ratio, and l is the average length of the pile-up of dislocations having a Burgers vector b. Equation (5.112) presumes that, on substituting Eq. (5.113), yielding or flow occurs when the pile-up exerts sufficient stress at the grain boundary to propagate plastic deformation from one grain to another. However, flow can be considered to begin when dislocations are emitted from the grain boundaries, and pile-ups are not observed with any certainty at the onset of yielding. In addition, dislocation pile-ups as such disappear with an increase in stacking-fault energy, and the Hall-Petch relation is generally inappropriate for describing yielding phenomena in alloys, since Eq. (5.113) approaches infinity as l approaches zero.

Li[232] and Li and Chou,[233] recognizing this shortcoming, argued that the flow stress at the time of yielding will be the stress required to move dislocations in the forest (or dislocation arrays) formed by all the dislocations generated from grain boundary ledges, and thereby arrived at the following expression for the Petch slope:

$$K_p = \alpha Gb(8m/\pi)^{1/2} \qquad (5.114)$$

where α is a constant which depends on the dislocation arrangement (or the stacking-fault free energy of the metal or alloy),[234] and usually has a value of the order of 0.4, and m is the grain boundary ledge density (see Fig. 4.33). The strength of the grain boundary therefore becomes

$$\sigma_i = 8\alpha^2 Gb(1 - \nu)ml/D \qquad (5.115)$$

which is roughly the stress required to push a dislocation through the stress field of the ledges if the ledges are assumed to form a wall of dislocations. Consequently, the stress required to penetrate the grain boundary by a pile-up is essentially identical to the stress necessary to initiate plastic deformation by generating dislocations from grain boundary ledges to form a forest array near the interface.

If we substitute Eq. (5.114) into Eq. (5.112), we obtain an expression of the form

$$\sigma = \sigma_0 + \alpha Gb(8m/\pi)^{-1/2}D^{-1/2} \qquad (5.116)$$

which is consistent with the work-hardening phenomenon where the flow stress and average dislocation density, ρ, are related by

$$\sigma = \sigma^* + \alpha Gb\rho^{1/2} \qquad (5.117)$$

Provided σ_0 and σ^* in Eqs. (5.116) and (5.117) are equal, it is observed that

$$\rho = 8m/\pi D \qquad (5.118)$$

Equation (5.118) is certainly a physically sound representation, since we would expect the average dislocation density to increase with the number of ledges

PROPERTIES OF INTERFACES   343

(sources) and that, since m represents the number per unit area or length, the boundary length or area would increase as D decreases. In addition, it should be noted that Eqs. (5.116) and (5.118) are based on the presumption that each ledge is essentially a unit ledge. If the average ledge height is in fact B_L, then Eq. (5.118) becomes

$$\rho = 8mB_L/\pi D \qquad (5.118a)$$

where B_L is given by Eq. (4.24a), which is certainly a reasonable description of the dislocation density at the onset of plastic flow (initial stage I work hardening) characterizing the microyield region.

It can of course be recognized that, following the initial generation of dislocations at the grain or interphase boundaries to form forests or tangled arrays, subsequent dislocation multiplication can occur within each grain by a double cross-slip mechanism, for example.[235,236] As a result, the effective value of either m or B_L in Eq. (5.118a) will increase, and the dislocation density, ρ, will in turn also increase.

Grain Boundary Hardening. Because the initial production of dislocations at the boundaries and their subsequent densification can induce image stresses which repel dislocations from the interfaces or promote cross-slip,[194] an effective hardening can be observed at the grain boundaries. The boundary region (region adjacent to the interface plane) becomes hardened because the internal elastic stresses resisting flow are greater there, and because the local flow stress to penetrate the dislocation forest and tangles adjacent to the interface can be relatively larger there than in the matrix. Hook and Hirth[237] have reviewed the experimental observations which show grain boundary hardening, and Braunovic and Haworth[238] have convincingly demonstrated that grain boundary hardening is definitely related to the interface as shown in their data for iron reproduced in Fig. 5.54.

Since, as we have indicated in Eq (5.43), hardness is simply related to the yield stress, σ_y ($\sigma = \sigma_y$), the data of Fig. 5.54 demonstrate rather convincingly that the flow stress near the interface plane is larger than that in the grain interior away from the interface. Similar tangles and dislocation forest arrays are clearly evident at the interphase interfaces associated with the precipitates in Fig. 5.53(b). It should also be pointed out that a direct correlation of boundary hardness with dislocation forests has not been established experimentally and has in fact been generally attributed to the migration of some species (vacancies or vacancy–solute complexes) to the boundary during annealing or cool-down (quenching).[239,240] In addition, grain boundary softening has been observed,[241,242] and this has been attributed to vacancy annihilation. However, softening could occur as a result of a very high ledge (source) density in the grain boundaries which gives rise to an initial easy glide which acts as a relief of the stress concentration at the boundary. In many cases of interfacial denuding and recognizable structural differences [Fig. 4.64(b)] which occur in the vicinity of the boundary plane, the observed hardening or softening cannot in general be attributed to dislocation forests. It is hoped that

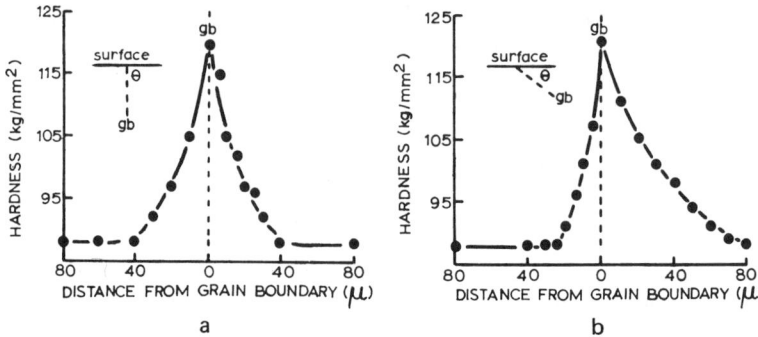

FIGURE 5.54 Boundary profiles for microhardness in iron for different values of tilt angle, θ (inclination of the interface plane with the surface). (a) θ = 90°; interface plane perpendicular to the surface. (b) θ = 34°. Note skewing of hardness along the plane of the interface as compared with (a). (From Braunovic and Haworth.[238])

future investigations will afford a more critical assessment to be made with regard to the role that dislocation forests and generation of dislocations at the interface play in grain boundary and interphase interface hardening. Furthermore, as was pointed out by Murr,[208,209] thermomechanical treatment of metals and alloys to attain specific grain boundary ledge densities and concomitant grain sizes may also contribute significantly to the enhancement of strength and flow properties through more or less conventional metal working techniques.

5.8.3 Interphase Interface Strengthening

The presence of an interface can, as noted previously, serve as both a source and a sink for dislocations, and the flow or work hardening of a metal or alloy can be related to the efficiency of dislocation generation or obstruction at the interface. The strengthening of a metal or alloy by the growth, agglomeration, precipitation, or dispersion of a second phase in the matrix is well known.[243,244] As a shear stress is applied in such systems, plastic flow begins when $\tau > \sigma_y$ (where τ is the shear stress), and dislocation lines are forced against the interphase interfaces of the included particles, causing a bowing of the dislocation lines which depends on the mean interparticle spacing, P; the yield stress of the strengthened matrix is given by

$$\tau = \tau_0 + (Gb/P) \tag{5.119}$$

where τ_0 is the yield stress (in shear) for the original matrix prior to the inclusion of the second phase. In effect, the incorporation of finer and finer dispersions increases τ in a manner similar to the increase of yield stress with decreasing grain size. However, here the structural refinement occurs in the matrix, and the role of

PROPERTIES OF INTERFACES   345

the interphase interface is simply one of providing a stationary object to block dislocation glide motion. However, Fig. 5.53(b) illustrates that in many cases the interface also provides additional dislocations which can amplify the strengthening by providing a mechanism for dislocation multiplication.

When precipitation or dispersion (solute segregation) occurs in the grain boundaries, the boundaries can be strengthened in several ways. First, the interphase interface created may contain ledges or other structural features which are conducive to dislocation generation. Second, for deformation at high temperature where grain boundary sliding (creep) would contribute to plastic flow, such interfaces when encountered by grain boundary dislocations would provide an essentially rigid wall behind which GBD's would pile up. Third, grain boundary phases can also act as pinning points for the grain boundary plane and prevent or retard boundary migration by a drag mechanism as outlined previously. To a large extent many of these features will depend upon the structure of the particle interface—that is, whether it is coherent or noncoherent. This would also bear upon the ability of the included phase to generate dislocations, primarily as illustrated in Fig. 5.52.

The role of interfaces in metal matrix composites as well as eutectic systems which possess aligned boundaries of considerable area is similar to that played by other more extensive interphase interfaces and grain boundaries. Again, it must be emphasized that, not only does the interface play a role in blocking dislocation motion, but it may also, as a result of its structure, play an important role in generating dislocations. Such continuous interfaces may also promote extensive directional sliding at the interface as a consequence of high densities of interfacial dislocations or ledge structures such as those illustrated in Fig. 5.55. This would become increasingly prominent at elevated temperature deformation. The chemical

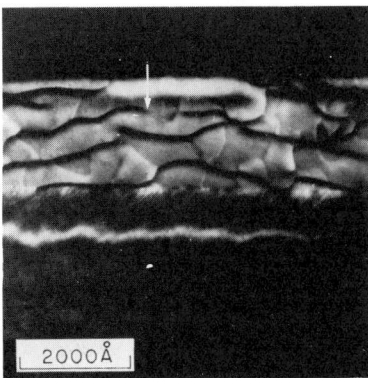

FIGURE 5.55   Ledge structure of interlamellar interface of directionally solidified Al-CuAl$_2$ eutectic composite. Dark-field electron micrograph, g_{CuAl_2} = [20$\bar{2}$]. (Courtesy Garmong and Rhodes.[247])

compatibility of such interfaces is another important consideration, particularly at high temperatures.[245,246]

Stacking-Fault Strengthening. When a dislocation shears an interface or a particle in a matrix—for example, a coherent precipitate—the yield stress of the metal or alloy is determined by the work done by the applied stress on the slip plane of the dislocation. We observed in Chapter 3 that the addition of solute to most metals (particularly fcc metals) alters the stacking-fault free energy. Hirsch and Kelly[248] have pointed out that, if the distribution of solute is not uniform, then a dislocation will be attracted to those regions where the stacking-fault energy is lower, provided the elastic moduli remain the same. For example, if a precipitate or included phase has a lower stacking-fault energy than the surrounding matrix, then a dislocation would tend to pass through as many particles as possible [see Fig. 5.53(a)], and work must be done to pull the dislocation from the particles, increasing the yield strength. If, on the other hand, the stacking-fault energy of the particles is higher than the surrounding matrix, then the dislocation would be repelled and would avoid passing through them. In most cases of coherent precipitates, the former phenomenon predominates. As a consequence, the strengthening due to such inclusions may be regarded as a stacking-fault strengthening as a result of energy reduction, as opposed to a simple pinning phenomenon as generally specified in precipitation hardening mechanisms.

5.8.4 Effect of Stacking-Fault Energy on Mechanical Properties.

Read,[249] more than two decades ago, pointed out that stacking faults extended completely through a crystal will act as barriers to slip on intersecting planes. In addition, dislocations that split up into partials separated by a stacking-fault interface are restricted to a single glide plane and can cross-slip only if the partials recombine. Since the mean width of separation is given by Eq. (3.61) for close-packed crystals, cross-slip becomes increasingly difficult as the stacking-fault free energy, γ_{SF}, decreases. Thus, extended dislocations can harden intersecting slip systems by providing a kind of grain refinement which occurs after deformation has initiated dislocations and stacking faults, which tend to form pile-ups extending across the grains, intersected by other pile-ups and stacking faults. In the case of mobile dislocations which are not measurably extended, as in high stacking-fault energy metals, the tendency to form linear arrays (pile-ups) diminishes in favor of an arrangement whereby the overall energy is reduced. Such arrangements involve the formation of forest dislocation arrays which, as a result of large deformations, are transformed into subboundaries or dislocation cells. These cells are rigidly rotated (by small misorientations, Θ) with respect to the surrounding matrix, so that the cell is always surrounded by a complete net of low-angle (low-energy) boundary.[250,251] Such dislocation cells can constitute a subgrain hardening mechanism similar to that of ordinary grain structures, except that the low misorientation and associated interfacial energies render the effect weaker and of shorter range than the high-angle, high-energy interfaces. An equation similar to Eq. (5.112) of the form

$$\sigma = \sigma_0 + K'd^{-n} \tag{5.120}$$

where d is the average subgrain size and

$$K' \simeq 3.34Gb \tag{5.121}$$

has been found to be applicable to subgrain boundaries in highly strained iron alloys[252,253] for $n = 1$, where solute distribution plays an important role in the residual hardening. Because of the associated low interfacial energy of stacking faults, effective subgrain refinement of a matrix by extended stacking faults would also be expected to be described by an equation similar to Eq. (5.120). As a consequence, the residual hardening and/or strengthening following deformation which causes either dislocation cell formation or planar dislocation arrays (or extended stacking faults) can be expected to increase as the number of intersecting faults and planar arrays in low stacking-fault energy alloys increases, or dislocation cell size in high stacking-fault energy metals and alloys decreases. Figure 5.56 illustrates the somewhat systematic change in residual microstructure and matrix refinement caused by the creation of low-energy interfaces with variations in stacking-fault energy for metals and alloys subjected to a uniaxial stress under essentially identical compressive-shock loading conditions, where essentially only the stacking fault energy varies with changes in composition as outlined in Section 3.4.5.

It is generally observed that strengthening is more efficient in low stacking-fault energy alloys, and that in fact the strengthening decreases with an increase in the stacking-fault energy. This means that stacking faults and planar arrays of dislocations are more effective in increasing the yield strength, σ, than are dislocation cells. Marcinkowski *et al.*[254,255] have recently demonstrated that, below some critical interplanar spacing, the mutual interaction between extended dislocations is sufficient to bring about coalescence, followed by spontaneous cross-slip and subsequent mutual annihilation of dislocations. Just above this critical interplanar spacing, the extended dislocations may pass one another with a maximum interaction. At this maximum interaction, the stress is maximum, and consequently the residual strengthening is maximum. This maximum stress has in turn been shown to decrease with increasing stacking-fault free energy; it therefore provides an alternative theory to the pile-up model for work hardening behavior in fcc metals and alloys, with particular reference to stage III.[256]

5.8.5 Deformation-Induced Interfaces and Residual Strengthening

It should by now be apparent that solid interfaces of essentially any structure and energy can act as effective barriers to dislocation motion, can generate dislocations, and can otherwise alter the mechanical properties of metals and alloys. In most of our treatments of such systems in this section we have alluded to or implied interfaces resulting primarily from nucleation and growth, except in the brief description of deformation-induced stacking faults in Section 5.8.4. In Section 4.5.3 it was pointed out that, in fcc alloys in particular, overlapping stacking faults

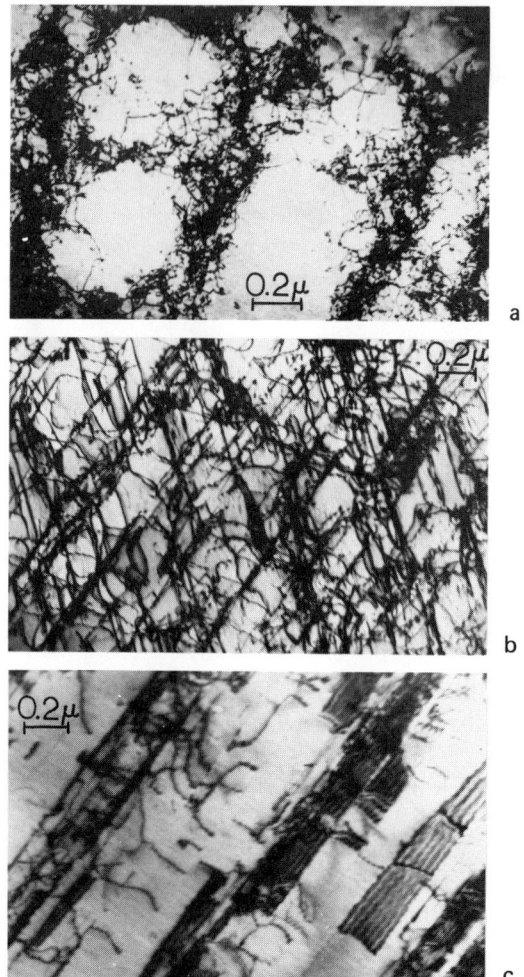

FIGURE 5.56 Residual microstructure following uniaxial, constant-temperature deformation by explosive shock loading of metals and alloys of varying stacking-fault energies. (a) Ni, γ_{SF} = 128 ergs/cm^2, σ = 80 kilobars; (b) 80 Ni–20 Cr, γ_{SF} = 40 ergs/cm^2, σ = 80 kilobars; (c) 8 Ni–16 Cr–71 Fe (stainless steel), γ_{SF} = 21 ergs/cm^2, σ = 150 kilobars. Pulse durations were constant at 2 μsec.

could produce twinned or phase-transformed regions (ϵ-phase). Such lattice shear processes can also produce other phase transformations, as was discussed in Chapter 4, the most prominent being the martensitic transformation in steels. It is well known that twinning and martensitic transformations in many fcc alloys result from thermal and mechanical treatments. It is also well known that, in martensitic transformations, the resulting alloy structure can be strengthened considerably by the interphase hardening which occurs, as well as the structural refinement which leaves new interphase interfaces to block dislocation glide motion.

Although deformation-induced twinning is not a common feature of alloys—in particular, of metals at ambient temperature—it is a common characteristic of shock-deformed alloys.[257-261] Twinning is prominent at other high rates of deformation of fcc alloys in particular.[262,263] In a recent study of twinning and martensitic transformations in stainless steel by shock deformation and reduction by cold rolling, Murr and Grace[261] demonstrated that the formation of twins and ϵ-phase at high rates of deformation ($\gamma \to \epsilon \to$ twins) followed a similar sequence of partial dislocation movements in the formation of α'-martensite at low rates of deformation ($\gamma \to \epsilon \to \alpha'$). More recently, Murr and Staudhammer[264] have demonstrated the formation of martensite in stainless steel, shock-loaded at higher pulse durations where α'-martensite originates at the intersection of deformation twins or within twin packets as previously proposed.[261] These features are of interest because they demonstrate a phenomenon that might aptly be called interfacial transformations wherein, for example, twin boundaries associated with intersecting deformation twins can give rise to the formation of a new crystallographic phase and an associated interphase interface. This may occur by systematic movements of of partial or fractional-partial (half-partial) dislocations, and phenomenologically amount to systematic overlapping of stacking faults in fcc alloys as illustrated previously in Figs. 4.45 and 4.47. Figure 5.57 illustrates somewhat unambiguously the formation of deformation twins as a result of overlapping intrinsic stacking faults in shock-loaded 304 stainless steel.

The formation of intersecting deformation twins in the manner illustrated in Fig. 5.57 can, as was indicated previously for intersecting stacking faults, contribute to residual strengthening of metals and alloys by promoting an effective matrix grain refinement. The refined twinned-grain structure will tend to decrease as the twin density increases; this has been shown for a number of alloys to depend on the magnitude of the applied stress (or pressure in the specific case of shock-loaded alloys).[260,261] This feature is somewhat dramatically illustrated in Fig. 5.58, which shows the annealed grain structure before shock deformation, and the appearance of a grain following shock deformation at 370 kilobars, which contains a regular intersecting array of deformation twins which act as a more-or-less regular subgrain structure. In line with the predictions of Eq. (5.120), an effective increase in the flow stress, σ, would be expected from Fig. 5.58(a) to Fig. 5.58(b). This has in fact been observed on comparing the residual microhardness of 470(VHN)* for the material of Fig. 5.58(b) with the value of 230(VHN) for the annealed material of Fig. 5.58(a).[260] Obviously, not all of this hardness difference is due to the presence of deformation twins. A greater portion may simply be a result of the high dislocation density which occurs during shock loading.

*Vickers Hardness Number; 1 VHN $\simeq$ 1 kg/mm^2.

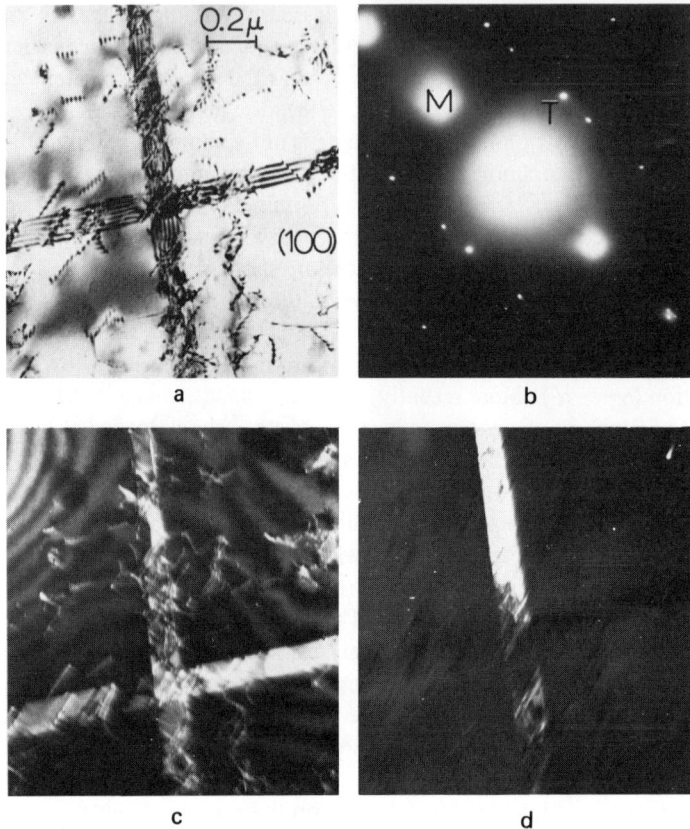

FIGURE 5.57 Stacking fault→deformation twin formation in shock-loaded 304 stainless steel. (a) Bright-field transmission electron micrograph of intersecting stacking faults and twin bundles in sample shock-loaded at 150 kilobars (0.5-μsec pulse). (b) Selected-area electron diffraction pattern of (a). (c) Dark-field image of (a) using the (002) matrix reflection (M) indicated in (b) which gives rise to stacking-fault contrast. (d) Dark-field image of (a) using twin reflection (T) indicated in (b) which illustrated twinned region coincident with overlapping stacking faults shown in (c).

5.9 SUMMARY

Following the chronology of interfacial phenomena established in Chapters 3 and 4, an attempt has been made in this final chapter to deal with the properties of interfaces that control or influence the behavior of metals and alloys under various conditions. An effort has also been made to illustrate some of the more pertinent applications of the interfacial phenomena that influence the properties and uses of metals and alloys, and of associated systems such as metal oxides and composite

PROPERTIES OF INTERFACES   351

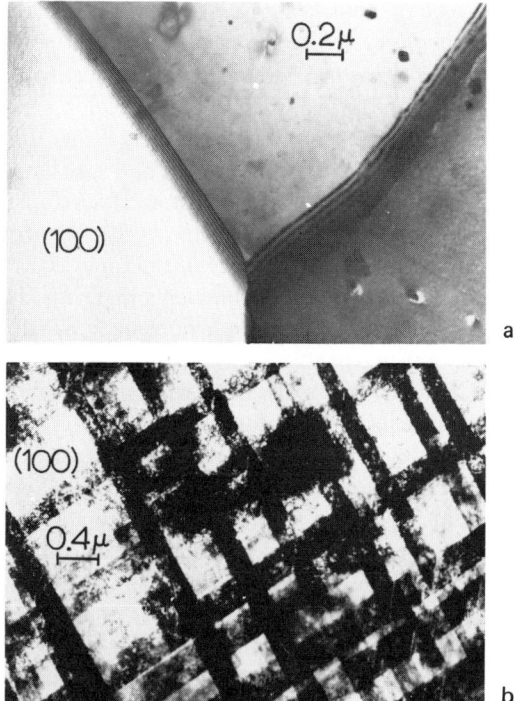

FIGURE 5.58   Deformation twinning as a subgrain refinement. (a) Annealed Inconel 600 showing grain boundary triple junction (15 microns average grain size). (b) Inconel 600 following explosive shock loading at 370 kilobars showing intersecting deformation twins having a mean thickness of 150 Å. [(b) is from Murr and Foltz.[260]]

systems. In his 1972 Campbell Memorial Lecture to the American Society for Metals, Hirth[194] stated: "The topic of interfaces in materials is a very broad one: indeed, if one excludes single crystal research one could claim that the balance of the field of metallurgy relates to interfacial properties." Indeed, including free surfaces and single-crystal research as we have done here leaves little if anything in the science of metals and alloys or more generally materials science that is independent of interfacial phenomena.

In this chapter I have tried to interconnect the first chapter through Chapter 4, in defining interfacial properties and in presenting examples of interfacial applications. Certainly many of the properties discussed in this chapter are arbitrary with regard to their importance or with respect to the level of development of the concept rendered. Nonetheless, a consistent generalization of the importance of interfacial phenomena must surely be an impression left with the reader. Within this framework of generalizations, the reader must also realize that this chapter has not been intended to be rigorous in any facet of the interfacial properties it has explored, and the level of development, including derivations of pertinent equations, was done more on a personal basis than on a subjective one. In this regard, there has been little attempt to treat the historical development of any particular area, and references cited have been chosen as timely rather than chronological or historical. A sufficient number of review articles or books have been included, however, so that the reader can pursue such historical developments, or investigate the necessary introductory background in some specific property or application area to render a judgment of the significance of the topic as discussed in the text.

There are, finally, several important comments that should be made in summarizing the linking of the interfacial phenomena in metals and alloys which has been the overriding theme of this book. First, the reader should now recognize a kind of canonical harmony that exists in interfacial structure. That is, the structure of a free surface really is not fundamentally different from that of a solid-solid interface, and it is characterized for the most part by facets, ledges, terraces, and related dislocation structures. Energetically, of course, there is a difference caused primarily by the difference in the atomic binding coordination at the interface. Second, the reader should recognize that, within the framework of this structural harmony, ledges or steps have been shown to play a dominant role not only in the growth and expansion of a surface, but also in the way an interface becomes involved in the determination and/or control of the properties of metals and alloys. It is this fact, emphasized in Chapter 5, that leads to the obvious conclusion that a considerable effort must be made in future research to investigate the structural features of interfacial dislocations and ledge structures in much greater detail. A third and final comment is that the structural and energetic similarities of many interfacial phenomena described in this book should be recognized in the fabrication of new alloys and metallic systems of the future, and their implications, incorporated in the development of certain specifically desirable properties, insofar as such properties may in fact involve interfacial phenomena. This certainly is my charge to both the student and the researcher.

REFERENCES

1. G. Ehrlich, in "Metal Surfaces," ASM, Metals Park, Ohio, p. 222 (1963).
2. A. W. Adamson, "Physical Chemistry of Surfaces," 2nd ed., Interscience Publishers, New York, p. 670 (1967).
3. F. H. Buttner, E. R. Funk, and H. Udin, *J. Phys. Chem.*, 56, 657 (1952).
4. U. M. Ahmad and L. E. Murr, *Scripta Met.*, 8, 631 (1974).
5. I. Langmuir, *J. Amer. Chem. Soc.*, 40, 1361 (1918).
6. D. O. Hayward and B. M. W. Trapnell, "Chemisorption," Butterworths, London (1964).
7. E. Drauglis, R. D. Gretz, and R. I. Jaffee (eds.), "Molecular Processes on Solid Surfaces," McGraw-Hill Book Co., New York (1969).
8. J. H. de Boer (ed.), "Reactivity of Solids," Elsevier Publishing Co., Amsterdam (1960).
9. E. K. Rideal and B. M. W. Trapnell, *Proc. Roy. Soc., Ser. A*, 205, 409 (1951).

10. B. M. W. Trapnell, *Proc. Roy. Soc., Ser. A*, **206**, 39 (1951).
11. O. Beeck, W. A. Cole, and A. Wheeler, *Discussions Faraday Soc.*, 8, 314 (1950).
12. O. Beeck, *Discussions Faraday Soc.*, 8, 118 (1950).
13. J. R. Schrieffer, *J. Vac. Sci. Technol.*, 9, 561 (1972).
14. N. D. Lang, *Solid State Phys.*, **28**, 225 (1973).
15. B. J. Thorpe, *Surface Sci.*, **33**, 306 (1973).
16. J. W. Gadzuk, *Surface Sci.*, **43**, 44 (1974).
17. H. D. Hagstrum, *Science*, **178**, 75 (1972).
18. J. W. Gadzuk and E. W. Plummer, *Rev. Mod. Phys.*, **45**, 487 (1973).
19. W. M. H. Sachtler and P. Van der Plank, *Surface Sci.*, **18**, 62 (1969).
20. J. Bernard, J. Oudar, and F. Cabane-Brouty, *Surface Sci.*, **3**, 359 (1965).
21. F. Delamare and G. E. Rhead, *Surface Sci.*, **35**, 172, 185 (1973).
22. B. J. Hopkins and K. R. Pender, *Surface Sci.*, **5**, 155 (1966).
23. J. M. Blakely, "Introduction to the Properties of Crystal Surfaces," Pergamon Press, New York (1973).
24. O. Kubaschewski and B. Z. Hopkins, "Oxidation of Metals and Alloys," Academic Press, New York (1962).
25. K. R. Lawless, in "Energetics in Metallurgical Phenomena," Vol. 1, W. M. Muller (ed.), Gordon and Breach, New York, p. 345 (1965).
26. F. Gronlund, *J. Chim. Phys.*, **53**, 660 (1956).
27. F. W. Young, J. V. Cathcart, and A. T. Gwathmey, *Acta Met.*, **4**, 145 (1956).
28. E. A. Gulbransen and K. F. Andrews, *J. Electrochem. Soc.*, **106**, 511 (1959).
29. W. W. Harris, F. L. Ball, and A. T. Gwathmey, *Acta Met.*, **5**, 574 (1957).
30. M. Otter, *Z. Naturforsch.*, **14a**, 355 (1959).
31. E. Menzel and W. Stossel, *Naturwissenschaften*, **41**, 302 (1954).
32. J. Bardolle and J. Benard, *Rev. Met.*, **49**, 613 (1952).
33. F. W. Young, *Acta Met.*, **8**, 117 (1960).
34. V. A. Phillips, *J. Appl. Phys.*, **33**, 712 (1962).
35. L. E. Murr, *Phys. Status Solidi*, **24**, 135 (1967).
36. A. Clark, "The Theory of Adsorption and Catalysis," Academic Press, New York (1970).
37. H. H. Storch, N. Golumbic, and R. B. Anderson, "The Fischer Tropsch and Related Synthesis," John Wiley & Sons, New York (1951).
38. E. K. Rideal, *Proc. Cambridge Phil. Soc.*, **35**, 130 (1938).
39. P. H. Emmett (ed.), "Catalysis," Vols. I and II, Reinhold Publishing Co., New York (1955); Vols. III and IV (1956).
40. M. Prager, *J. Cryst. Growth*, **22**, 6 (1974).
41. J. Horiuti, G. Okamoto, and K. Hirota, *Sci. Pap. Inst. Phys. Chem. Res.* (Tokyo), **29**, 223 (1936).
42. A. T. Gwathmey and R. E. Cunningham, *Advan. Catal.*, **10**, 57 (1958).
43. H. M. C. Sosnovsky, *J. Phys. Chem. Solids*, **10**, 304 (1959).
44. J. W. Hall and H. F. Rase, *Nature*, **199**, 585 (1963).
45. J. Bagg, H. Jaeger, and J. V. Sanders, *J. Catal.*, **2**, 449 (1963).
46. C. F. Powell, J. H. Oxley, and J. M. Blucher, Jr. (eds.), "Vapor Deposition," John Wiley & Sons, New York (1966).
47. M. Volmer and H. Flood, *Z. Phys. Chem.*, **A170**, 373 (1934).
48. R. Becker and W. Doring, *Ann. Physik*, **24**, 719 (1939).
49. G. M. Pound, M. T. Simnad, and L. Yang, *J. Chem. Phys.*, **22**, 1215 (1954).
50. J. P. Hirth and G. M. Pound, *Progr. Mater. Sci.*, **11**, (1965).
51. J. H. Holloman and D. Turnbull, *Progr. Met. Phys.*, **4** (1953).
52. J. P. Hirth, K. L. Moazed, and V. Ruth, in "Epitaxie-Endotaxie," H. G. Schneider (ed.), VEB Deutscher Verlag für Grundstoffindustrie, Leipzig, p. 25 (1969).

53. J. Lothe and G. M. Pound, *J. Chem. Phys.,* **36,** 2080 (1962).
54. G. M. Pound, in "Energetics in Metallurgical Phenomena," Vol. 2, M. M. Mueller (ed.), Gordon & Breach, New York (1965).
55. J. P. Hirth and K. L. Moazed, in "Physics of Thin Films," Vol. 4, G. Hass and R. E. Thun (eds.), Academic Press, New York, p. 97 (1967).
56. K. L. Moazed and G. M. Pound, *Trans. AIME,* **230,** 234 (1964).
57. J. L. Kenty, in "Advances in Epitaxy and Endotaxy: Physical Problems of Epitaxy," H. G. Schneider and V. Ruth (eds.), VEB Deutscher Verlag für Grundstoffindustrie, Leipzig, p. 83 (1971).
58. H. P. Singh and L. E. Murr, *Scripta Met.,* **6,** 333 (1972).
59. J. A. Venables, *Phil. Mag.,* **27,** 697 (1973).
60. H. P. Singh and L. E. Murr, *Phil. Mag.,* **26,** 649 (1972).
61. D. W. Pashley, *Advan. Phys.,* **14,** 327 (1965).
62. K. L. Chopra, "Thin Film Phenomena," McGraw-Hill Book Co., New York (1969).
63. L. E. Murr, *Thin Solid Films,* **7,** 101 (1971).
64. L. E. Murr and M. C. Inman, *Phil. Mag.,* **14,** 135 (1966).
65. L. E. Murr, O. T. Inal, and H. P. Singh, *Thin Solid Films,* **9,** 241 (1972).
66. E. Rabinowicz, "Friction and Wear of Materials," John Wiley & Sons, New York (1965).
67. J. Ferrante and J. R. Smith, *Surface Sci.,* **38,** 77 (1973).
68. R. M. Pilliar and J. Nutting, *Phil. Mag.,* **16,** 181 (1967).
69. L. E. Murr, *Mater. Sci. Eng.,* **12,** 277 (1973).
70. L. E. Murr, *J. Mater. Sci.,* **9,** 1309 (1974).
71. L. E. Murr, *Appl. Mater. Res.,* **4,** 31 (1965).
72. M. Kerridge, *Proc. Phys. Soc.* (London), **66B,** 400 (1955).
73. M. M. Kruschov and M. A. Babichev, "Investigations into the Wear of Metals" (in Russian), Academy of Sciences, USSR, Moscow (1960).
74. F. P. Bowden, J. N. Gregory, and D. Tabor, *Nature,* **156,** 97 (1945).
75. F. P. Bowden and D. Tabor, "The Friction and Lubrication of Surfaces, II," Clarendon Press, Oxford (1964).
76. F. Llewellyn Jones, *Brit. J. Appl. Phys.,* **12,** 318 (1961).
77. E. W. Müller, S. B. McLane, and O. Nishikawa, "Proceedings of the International Congress for Electron Microscopy, Kyoto," Vol. 1, Academic Press, New York, p. 235 (1966).
78. E. W. Müller and O. Nishikawa, in "Adhesion or Cold Welding of Materials in Space Environment," STP No. 431, ASTM, Philadelphia, p. 67 (1967).
79. E. W. Müller and T. T. Tsong, "Field-Ion Microscopy Principles and Applications," American Elsevier Publishing Co., New York (1969).
80. W. W. Mullins, *J. Appl. Phys.,* **28,** 333 (1957).
81. W. W. Mullins, *Trans. TMS AIME,* **218,** 354 (1960).
82. B. C Allen, *Trans. TMS AIME,* **236,** 903 (1966).
83. P. G. Shewmon, "Diffusion in Solids," McGraw-Hill Book Co., New York (1963).
84. F. R. Tolmon and J. G. Wood, *J. Sci. Instrum.,* **33,** 236 (1956).
85. A. G. Guy, "Introduction to Materials Science," McGraw-Hill Book Co., New York (1972).
86. S. Yukawa and M. J. Sinnott, *Trans. AIME,* **203,** 996 (1955).
87. W. R. Upthegrove and M. J. Sinnott, *Trans. ASM,* **50,** 1031 (1958).
88. M. Jurisch, Ph.D. Dissertation, Freiberg (Sachsen) (1969).
89. K. E. Easterling and A. R. Thölen, *Acta Met.,* **20,** 1001 (1972).
90. M. F. Ashby, *Acta Met.,* **22,** 275 (1974).
91. T. L. Wilson and P. G. Shewmon, *Trans. AIME,* **236,** 48 (1966).
92. E. R. Kimmel and W. Scheithauer, Jr., *Met. Trans.,* **5,** 1495 (1974).
93. A. A. Griffith, *Phil. Trans. Roy. Soc.,* **A221,** 163 (1920).

94. E. Orowan, "Fatigue and Fracture of Metals," John Wiley & Sons, New York, p. 139 (1950).
95. G. R. Irwin, *Trans. ASM,* **40,** 147 (1948).
96. M. J. Marcinkowski and E. S. P. Das, *Phys. Status Solidi (a),* **8,** 249 (1971).
97. E. D. Hondros and D. McLean, *Phil. Mag.,* **29,** 771 (1974).
98. N. S. Stoloff and T. L. Johnston, *Acta Met.,* **11,** 251 (1963).
99. A. R. C. Westwood and M. H. Kamdar, *Phil. Mag.,* **8,** 787 (1963).
100. N. S. Stoloff, in "Surfaces and Interfaces II," J. J. Burke, N. L. Reed, and V. Weiss (eds.), Syracuse University Press, Syracuse, New York, p. 157 (1968).
101. R. J. Stokes and C. H. Li, in "Fracture of Solids," D. C. Drucker and J. J. Gilman (eds.), Interscience Publishers, New York, p. 289 (1963).
102. A. N. Stroh, *Advan. Phys.,* **6,** 418 (1957).
103. A. N. Stroh, *Phil. Mag.,* **3,** 597 (1958).
104. A. H. Cottrell, *Trans. AIME,* **212,** 192 (1958).
105. J. J. Gilman, *Trans. AIME,* **212,** 783 (1958).
106. J. C. M. Li, *J. Appl. Phys.,* **32,** 525 (1961).
107. E. S. P. Das and M. J. Marcinkowski, *J. Appl. Phys.,* **43,** 4425 (1972).
108. A. L. Stevens and L. E. Pope, in "Metallurgical Effects at High Strain Rates," R. W. Rohde, B. M. Butcher, J. R. Holland, and C. H. Karnes (eds.), Plenum Press, New York, p. 459 (1973).
109. M. A. Fortes and B. Ralph, *Acta Met.,* **15,** 707 (1967).
110. L. E. Murr and O. T. Inal, *Metals Eng. Quart.,* **12,** 29 (1972).
111. C. S. Smith, *Trans. AIME,* **175,** 15 (1948).
112. J. J. Gilman, *J. Appl. Phys.,* **31,** 2208 (1960).
113. W. Rostoker, J. M. McCaughey, and H. Markus, "Embrittlement by Liquid Metals," Reinhold Publishing Corp., New York (1960).
114. Y. M. Dityatkousky, I. V. Andreyev, and K. F. Gorshkov, *Phys. Metal. Metallogr. (USSR),* **15,** 94 (1963).
115. J. P. Fidelle and M. Rapin (eds.), "L'Hydrogene dans les Metaux," Colloque-Valduc, Centre D'Etudes de Bruyeres, Le-Chatel (1967).
116. H. Liebowitz (ed.), "Fracture," Vols. II and VI, Academic Press, New York (1966-1968).
117. H. L. Logan, "The Stress Corrosion of Metals," John Wiley & Sons, New York (1967).
118. M. G. Fontana and N. D. Greene, "Corrosion Engineering," McGraw-Hill Book Co., New York (1967).
119. D. P. Woodruff, *Phil. Mag.,* **18,** 123 (1968).
120. D. P. Woodruff, "The Solid–Liquid Interface," Cambridge University Press, London (1973).
121. R. M. J. Cotterill, T. Leffers, and H. Lilholt, *Phil. Mag.,* **30,** 265 (1974).
122. M. C. Flemings, "Solidification Processing," McGraw-Hill Book Co., New York (1973).
123. M. E. Glicksman and W. J. Childs, *Acta Met.,* **10,** 925 (1962).
124. P. B. Crosley, A. W. Douglas, and L. F. Mondolfo, in "Solidification of Metals," Iron and Steel Institute Publication No. 110, p. 10 (1968).
125. B. L. Bramfitt, *Met. Trans.,* **1,** 1987 (1970).
126. W. A. Tiller, K. A. Jackson, J. W. Rutter, and B. Chalmers, *Acta Met.,* **7,** 428 (1953).
127. B. Chalmers, "The Principles of Solidification," John Wiley & Sons, New York (1964).
128. J. W. Rutter and B. Chalmers, *Can. J. Phys.,* **31,** 15 (1953).
129. K. A. Jackson and J. D. Hunt, *Trans. AIME,* **236,** 1129 (1966).
130. J. D. Hunt and K. A. Jackson, *Trans. AIME,* **236,** 843 (1966).
131. D. Kuhlmann-Wilsdorf, *Phys. Rev.,* **140A,** 1599 (1965).
132. I. A. Kotze and D. Kuhlmann-Wilsdorf, *Appl. Phys. Lett.,* **9,** 96 (1966).
133. D. Turnbull, *J. Appl. Phys.,* **21,** 1022 (1950).

134. G. F. Bolling, *Acta Met.*, **16**, 1147 (1968).
135. M. E. Glicksman and C. L. Vold, *Acta Met.*, **15**, 1409 (1967).
136. H. Gleiter and B. Chalmers, *Progr. Mater. Sci.*, **16** (1972).
137. P. G. Shewmon, *Acta Met.*, **5**, 335 (1957).
138. J. E. Burke and D. Turnbull, *Progr. Met. Phys.*, **3**, 220 (1952).
139. P. A. Beck, J. C. Kremer, L. J. Demer, and M. L. Holzworth, *Trans. AIME*, **175**, 372 (1948).
140. P. A. Beck, J. Towers, and W. O. Manley, *Trans. AIME*, **175**, 1621 (1948).
141. J. G. Byrne, "Recovery, Recrystallization, and Grain Growth," The Macmillan Co., New York (1965).
142. L. E. Murr and J. A. Korbonski, *Met. Trans.*, **1**, 3333 (1970).
143. J. E. Burke, "The Fundamentals of Recrystallization and Grain Growth," ASM, Cleveland (1949).
144. P. Gordon, in "Energetics in Metallurgical Phenomena," W. M. Mueller (ed.), Vol. I, Gordon & Breach, New York, p. 207 (1965).
145. T. Tiedema, W. May, and W. G. Burgers, *Acta Crystallogr.*, **2**, 151 (1949).
146. P. A. Beck, P. R. Sperry, and H. Hu, *J. Appl. Phys.*, **21**, 420 (1950).
147. C. J. Simpson, K. T. Aust, and W. C. Winegard, *Scripta Met.*, **3**, 171 (1969).
148. S. Kohara, M. N. Parthasarathi, and P. A. Beck, *J. Appl. Phys.*, **29**, 1125 (1958).
149. M. L. Kronberg and F. H. Wilson, *Trans. AIME*, **185**, 501 (1949).
150. H. Yoshida, B. Liebmann, and K. Lucke, *Acta Met.*, **7**, 51 (1959).
151. B. Liebmann and K. Lucke, *J. Metals*, **8**, 1413 (1956).
152. J. W. Rutter and K. T. Aust, *Acta Met.*, **13**, 181 (1953).
153. K. T. Aust and J. W. Rutter, *Trans. AIME*, **215**, 119 (1959).
154. K. T. Aust and J. W. Rutter, *Trans. AIME*, **215**, 820 (1959).
155. S. R. Goodman and H. Hu, *Trans. AIME*, **236**, 710 (1966).
156. H. Hu, R. S. Cline, and B. Rath, *Trans. AIME*, **239**, 1103 (1967).
157. R. F. Mehl, "Recrystallization and Grain Growth," Metals Handbook, ASM, Cleveland, p. 248 (1948).
158. M. F. Ashby, J. Harper, and J. Lewis, Harvard University Division of Engineering and Applied Physics, Report No. 547 (1967).
159. W. W. Mullins, *Acta Met.*, **4**, 421 (1956).
160. J. Talbot, "Recovery and Recrystallization of Metals," Wiley-Interscience, New York, p. 269 (1963).
161. M. Feller-Kniepmeier and K. Schwartzkopf, *Acta Met.*, **17**, 497 (1969).
162. L. E. Murr and H. R. Vydyanath, *Scripta Met.*, **4**, 183 (1970).
163. L. E. Murr and H. R. Vydyanath, *Acta Met.*, **18**, 1047 (1970).
164. N. F. Mott, *Proc. Roy. Soc.*, **60**, 391 (1948).
165. H. Gleiter, *Acta Met.*, **17**, 565 (1969).
166. W. P. In Der Schmitten, P. Haasen, and F. Haessner, *Z. Metallk.*, **51**, 101 (1960).
167. W. N. Lin and L. E. Murr, *Scripta Met.*, **3**, 799 (1969).
168. H. Gleiter, *Acta Met.*, **17**, 853 (1969).
169. C. Crussard and R. Tamhankar, *Trans. AIME*, **212**, 718 (1958).
170. R. Raj and M. F. Ashby, *Trans. AIME*, **2**, 1113 (1971).
171. F. E. Hauser, P. R. Landon, and J. E. Dorn, *Trans. ASM*, **48**, 986 (1956).
172. G. Y. Chin, W. F. Hosford, and W. A. Backofen, *Trans. AIME*, **230**, 437 (1964).
173. F. Weinberg, *Progr. Met. Phys.*, **8**, 105 (1959).
174. R. C. Gifkins, in "Fracture," B. L. Averbach, D. K. Felbeck, D. T. Hahn, and D. A. Thomas (eds.), John Wiley & Sons, New York, p. 579 (1959).
175. R. N. Stevens, *Met. Rev.*, **11**, 129 (1966).
176. M. F. Ashby, *Surface Sci.*, **31**, 498 (1972).
177. R. S. Gates, *Acta Met.*, **21**, 855 (1973).

178. J. L. Walter and H. E. Cline, *Trans. AIME*, **242**, 1823 (1968).
179. M. Biscondi and C. Goux, *Mem. Sci. Rev. Met.*, **65**, 167 (1968).
180. D. McLean and M. H. Farmer, *J. Inst. Metals*, **85**, 41 (1956/57).
181. R. C. Gifkins and K. U. Snowdon, *Trans. AIME*, **239**, 910 (1967).
182. R. N. Stevens, *Surface Sci.*, **31**, 543 (1972).
183. F. R. N. Nabarro, in "Report of a Conference on the Strength of Solids," Physical Society, London, p. 75 (1948).
184. C. Herring, *J. Appl. Phys.*, **21**, 437 (1950).
185. R. L. Coble, *J. Appl. Phys.*, **34**, 1979 (1963).
186. R. P. Skelton, *Phil. Mag.*, **15**, 405 (1967).
187. R. C. Gifkins, *Acta Met.*, **4**, 655 (1956).
188. F. Garofalo, "Fundamentals of Creep and Creep Rupture in Metals," The Macmillan Co., New York (1965).
189. A. J. Kennedy, "Proceedings of Creep and Fatigue in Metals," John Wiley & Sons, New York (1963).
190. E. W. Hart, *Acta Met.*, **15**, 1545 (1967).
191. R. H. Johnson, *Met. Rev.*, **146** (1970).
192. P. Chaudhari, *IBM Res. Rept.*, No. 2496 (1969).
193. H. W. Hayden, S. Floreen, and P. D. Goodell, *Met. Trans.*, **3**, 833 (1972).
194. J. P. Hirth, *Met. Trans.*, **3**, 3047 (1972).
195. H. G. F. Wilsdorf, in "Structure and Properties of Thin Films," John Wiley & Sons, New York, p. 156 (1959).
196. A. Mascanzoni and G. Buzzichelli, *Phil. Mag.*, **22**, 857 (1970).
197. G. Bäro, H. Gleiter, and E. Hornbogen, *Mater. Sci. Eng.*, **3**, 92 (1968).
198. M. J. Marcinkowski, "Conference on Fundamental Aspects of Dislocation Theory, Gaithersburg, Md., April 21-25, 1969," NBS Special Publication No. 317 (1969).
199. A. Berghezan and A. Fourdeux, *J. Appl. Phys.*, **30**, 1913 (1959).
200. E. Hornbogen, *Trans. ASM*, **57**, 120 (1964).
201. W. Bonfield, *Trans. AIME*, **233**, 1719 (1965).
202. G. S. Ansell and J. Weertman, *Trans. AIME*, **215**, 838 (1959).
203. J. T. McGarth and W. J. Bratina, *Acta Met.*, **15**, 329 (1967).
204. M. E. Golden and W. D. Nix, *Phil. Mag.*, **18**, 217 (1968).
205. D. L. Douglas and T. W. Barbee, *J. Mater. Sci.*, **4**, 138 (1969).
206. H. Gleiter, E. Hornbogen, and G. Baro, *Acta Met.*, **16**, 1053 (1968).
207. L. E. Murr, R. J. Horylev, and W. N. Lin, *Phil. Mag.*, **22**, 515 (1970).
208. L. E. Murr, *Appl. Phys. Lett.*, **24**, 533 (1974).
209. L. E. Murr, *Met. Trans.*, **6A**, 505 (1975).
210. D. A. Jones and J. W. Mitchell, *Phil. Mag.*, **3**, (1958).
211. A. S. Paransnis and J. W. Mitchell, *Phil. Mag.*, **4**, 171 (1959).
212. A. Fourdeux and A. Berghezan, *C. R. Acad. Sci. (Paris)*, **252**, 1462 (1961).
213. G. C. Weatherly, *Metal Sci. J.*, **2**, 237 (1968).
214. R. A. Huggins, *Acta Met.*, **7**, 357 (1959).
215. A. Eikum and G. Thomas, *Acta Met.*, **12**, 537 (1964).
216. R. S. Barnes and D. J. Mazey, *Acta Met.*, **11**, 281 (1961).
217. K. H. Westmacott, R. S. Barnes, and R. E. Smallman, *Phil. Mag.*, **7**, 1585 (1962).
218. L. M. Brown and W. M. Stobbs, *Phil. Mag.*, **23**, 1201 (1971).
219. M. F. Ashby, *Phil. Mag.*, **14**, 1157 (1966).
220. F. J. Humphreys and A. T. Stewart, *Surface Sci.*, **31**, 389 (1972).
221. H. Warlimont, Iron and Steel Institute Special Report, No. 93, London, p. 73 (1965).
222. J. M. Oblak and D. S. Duvall, in "Proceedings of the Electron Microscopy Society of America," C. J. Arceneaux (ed.), Claitor's Publishing Division, Baton Rouge, Louisiana, p. 114 (1971).

223. L. G. Orlov, *Sov. Phys.-Solid State*, **9**, 1836 (1968).
224. H. Gleiter, E. Hornbogen, and G. Baro, *Acta Met.*, **16**, 1053 (1968).
225. H. Gleiter, *J. Less-Common Metals*, **28**, 297 (1972).
226. C. W. Price and J. P. Hirth, *Mater. Sci. Eng.*, **9**, 15 (1973).
227. J. P. Hirth and R. W. Balluffi, *Acta Met.*, **21**, 929 (1973).
228. E. O. Hall, *Proc. Phys. Soc.*, **1364**, 747 (1951).
229. N. J. Petch, *J. Iron Steel Inst.*, **174**, 25 (1953).
230. A. Cracknell and N. J. Petch, *Acta Met.*, **3**, 186 (1955).
231. J. Heslop and N. J. Petch, *Phil. Mag.*, **3**, 1128 (1958).
232. J. C. M. Li, *Trans. AIME*, **227**, 239 (1963).
233. J. C. M. Li and Y. T. Chou, *Met. Trans.*, **1**, 1145 (1970).
234. J. E. Bailey and P. B. Hirsch, *Phil. Mag.*, **5**, 485 (1960).
235. E. Orowan, "Dislocations in Metals," AIME, New York, p. 103 (1954).
236. W. G. Johnston and J. J. Gilman, *J. Appl. Phys.*, **31**, 632 (1960).
237. R. E. Hook and J. P. Hirth, *Trans. Japan Inst. Metals*, **9**, 778 (1968).
238. M. Braunovic and C. W. Haworth, *J. Mater. Sci.*, **9**, 809 (1974).
239. K. T. Aust, in "Surfaces and Interfaces," Vol. II, J. J. Burke, N. L. Reed, and V. Weiss (eds.), Syracuse University Press, Syracuse, New York, p. 213 (1968).
240. A. U. Seyboldt, J. H. Westbrook, and D. Turnbull, *Acta Met.*, **12**, 1456 (1964).
241. K. T. Aust, A. J. Peat, and J. W. Westbrook, *Acta Met.*, **14**, 1469 (1966).
242. K. T. Aust, R. E. Hanneman, P. Niessen, and J. H. Westbrook, *Acta Met.*, **16**, 291 (1968).
243. A. Kelly and R. B. Nicholson, *Progr. Mater. Sci.*, **10**, No. 3 (1963).
244. L. J. Broutman and R. H. Krock (eds.), "Modern Composite Materials," Addison-Wesley Publishing Co., Reading, Massachusetts (1967).
245. K. P. Staudhammer, L. E. Murr, C. Marinoff, and V. Reineking, *Microstructures*, **3**, 17 (1972).
246. A. G. Metcalfe (ed.), "Composite Materials," Vol. 1: Interfaces in Metal Matrix Composites, Academic Press, New York (1974).
247. G. Garmong and C. G. Rhodes, *Acta Met.*, **22**, 1373 (1974).
248. P. B. Hirsch and A. Kelly, *Phil. Mag.*, **12**, 881 (1965).
249. W. T. Read, Jr., "Dislocations in Crystals," McGraw-Hill Book Co., New York (1953).
250. J. P. Hirth and J. Lothe, "Theory of Dislocations," McGraw-Hill Book Co., New York (1968).
251. D. Kuhlmann-Wilsdorf, *Met. Trans.*, **1**, 3173 (1970).
252. H. J. Rack and M. Cohen, *Mater. Sci. Eng.*, **6**, 320 (1970).
253. H. J. Rack and M. Cohen, in "Frontiers in Materials Science: Distinguished Lectures," L. E. Murr and C. Stein (eds.), Marcel Dekker, New York (1975).
254. M. J. Marcinkowski, K. Sadananda, and R. J. Taunt, *Acta Met.*, in press.
255. M. J. Marcinkowski, R. T. Taunt, and K. Sadananda, *Acta Met.*, in press.
256. J. P. Hirth and J. Weertman (eds.), "Work Hardening," Gordon & Breach, New York (1968).
257. "Response of Metals to High Velocity Deformation," Metallurgical Conferences, Interscience Publishers, New York (1961).
258. E. G. Zukas, *Metals Eng. Quart. (ASM)*, **6**, 16 (1966).
259. L. E. Murr and F. I. Grace, *Exp. Mech.*, **9**, 145 (1969).
260. L. E. Murr and J. V. Foltz, *J. Appl. Phys.*, **40**, 3796 (1969).
261. L. E. Murr and F. I. Grace, *Trans. AIME*, **245**, 2225 (1969).
262. L. E. Murr, G. Wong, and J. V. Foltz, *Mater. Sci. Eng.*, **7**, 278 (1971).
263. L. E. Murr, J. V. Foltz, and F. D. Altman, *Phil. Mag.*, **23**, 1011 (1971).
264. L. E. Murr and K. P. Staudhammer, in "Proceedings of the Electron Microscopy Society of America," C. J. Arceneaux (ed.), Claitor's Publishing Division, Baton Rouge, Louisiana, p. 364 (1974).

PROBLEMS

(5.1) Referring to Fig. 5.3, calculate the pressure that induces 40% adsorption of hydrogen on tungsten at 0°C. What is the corresponding volume of hydrogen needed for full coverage of the surface?

(5.2) Assume that three nuclei have shapes approximated by Fig. 2.27(a), (b) and (c). If the nuclei represent the same metal under varying conditions, calculate the ratios of free energy of formation as follows: (a)/(b); (c)/(a); (b)/(c). Which nucleus [(a), (b), or (c)] would be expected to dominate if nucleation of all shapes can occur?

(5.3) The microhardness of a series of Cu and Cu-Al alloys cold-rolled 25% were determined as illustrated below. Assuming all proportionality constants relating to microhardness to be unchanged from sample to sample, make a plot of variations of surface free energy with concentration at room temperature. Discuss any inconsistencies that arise in terms of the assumptions you have made.

Sample	Microhardness (kg/mm^2)
Cu	148
Cu-1.5 a/o Al	156
Cu-5 a/o Al	216
Cu-16 a/o Al	306

(5.4) Referring to Fig. 5.27, calculate the shear modulus for ThO_2. How does this compare with the shear modulus for pure nickel? Calculate the shear modulus of pure nickel from Fig. 5.29(a) and discuss the calculated value with the accepted value.

(5.5) Calculate the critical fracture stress associated with grain boundary ledges in nickel at room temperature if the ledge height is 100 Å. Compare this value with the shear modulus. How many dislocations of the type $a/2$ <110> could such a grain boundary ledge emit?

(5.6) Referring to Fig. 5.56(a), calculate the average interfacial free energy associated with the subgrain boundaries (dislocation cell walls) if the grain size was observed to be 15 microns (average grain diameter). Estimate the number of dislocations to be associated with a unit area of cell wall to account for this energy. Draw an equilibrium configuration representing a dislocation cell intersecting a grain boundary, and indicate the magnitude of the associated dihedral angle. Would you expect to see this configuration in Fig. 5.56(a) if a grain boundary could be observed? Discuss your answer.

(5.7) Discuss, on the basis of flow-stress considerations, the observed variation in microhardness measured between Fig. 5.58(a) and (b). Is the increased hardness a consistent feature? If not, suggest microstructural features which could phenomenologically account for this inconsistency.

(5.8) Calculate the misorientation associated with the straight boundary portion of the occluded grain in Fig. 5.42 if the dislocations are of the type $a/2 <110>$. Assuming the boundary to be described by a curve identical to that in Fig. 5.43(a), estimate the boundary migration rate. Would you expect this boundary portion to migrate faster or slower than the curved portion closest to the magnification marker? Discuss your answer.

(5.9) Plot a curve of grain boundary cohesive stress versus temperature for copper and 80/20 NiCr between room temperature and 1000°C. What is the ratio of grain boundary cohesive stress to that associated with a single crystal of copper at room temperature?

(5.10) Discuss the application of nucleation theory to recrystallization. Discuss also the addition of solute in the recrystallization process. Show how activation energy for recovery and recrystallization are influenced by solute addition, and indicate which process is most influenced by solute addition.

AUTHOR INDEX

AUTHOR INDEX

Numbers in parentheses indicate numbers of references cited in the text. Numbers set in *italics* designate page numbers on which the complete literature citations are given.

Aaron, H. B., 18(36), 82(131), 130(133), 218(145), *28, 85, 161, 255*
Aaronson, H. I., 25(49), 66(95), 82(95, 131), 232(171), 234(182,183), 237(185), 239(190-193), *28, 84, 85, 255, 256*
Abraham, F., *165*
Adam, N. K., 90(8), *158*
Adams, J. C., 90(9), 91(9), 92(9), 95(9), 156(9), *158*
Adamson, A. W., 67(97), 260(2), *84, 352*
Addison, C. C., 101(44), *159*
Ahmad, U. M., 155(186), 156(186), 262(4), *106, 162, 352*
Aliev, I. M., 181(46), *252*
Allen, B. C., 43(34), 76(34), 79(34,123), 80(123), 95(22), 97(22), 100(50), 102(50,51), 103(50,51), 104(50,51), 105(50), 106(50), 112(50), 123(51), 124(107,108), 125(51,107,108), 126(51), 129(108), 130(108), 131(107,108), 132(50), 291(82), 292(82), *83, 85, 158-160, 354*
Altman, F. D., 349(263), *358*
Amelinckx, S., 45, *45*, 188(72), 227(166, 167), 229(169), *45, 252, 253, 255*
Anderson, J., 130(131), *161*
Anderson, J. S., 229(168), *255*
Anderson, R. B., 272(37), *353*
Andes, G. M., 67(100), *84*
Andreas, J. M., 95(19), *158*
Andrews, K. F., 269(28), *353*
Andreyer, I. V., 312(114), 313(114), *355*
Andreas, J. M., 95(19), *158*
Ansell, G. S., 339(202), *357*
Arakaleyan, V. S., 114(84), *160*
Ashby, M. F., 298(90), 299(90), 300(90), 301(90), 302(90), 329(158), 331(170), 333(176), 334(176), 336(176), 339(219), *354, 356, 357*
Ashcroft, N. W., 185(59), *253*
Astrom, H. V., 154, *160*, 161, 124(103), 132(103)
Aust, K. T., 20(42), 21(42), 66(90,91,92), 188(70,75), 225(156), 243(198), 326(147,152), 327(152), 328(153,154), 343(239,241,242), *28, 84, 253, 255, 256, 356, 358*
Averbach, B. L., 77(116), 179(28), *85, 252*
Avraamov, Y. S., 181(43), 182(43), 183(43), *252*

Babichev, M. A., 289(73), *354*
Backofen, W. A., 332(172), *356*
Baro, G., 200(102,103), 232(172), 233(172), 339(197,206,224), *254, 255, 357, 358*
Bagg, J., 275(45), *353*
Bailey, G., 76(112), 342(234), *85, 358*
Bakker, G., 10(25), *27*
Ball, F. L., 269(29), *353*
Balluffi, R. W., 201(110,113), 212(113), 213(110), 232(113), 339(227), *254, 358*
Barbee, T. W., 339(205), *357*
Barbour, J. P., 125(123), 132(123), *161*
Bardeen, J., 181(50), 184(50), *252*
Bardolle, J., 269(32), *353*
Barnes, R. S., 124(105), 339(216,217), *160, 357*
Baron, K., 179(34), *252*
Baroux, B., 245(209), 246(209,212,213), 247(209), *256*
Barrand, P., 79(125), 80(125), *85*
Barrett, C. R., 152(175), 153(175), *162*
Barrett, C. S., 36(17), 54(62), 226(162),

Barrett, C. S. (cont.)
227(162), 235(162), *83, 255*
Bartell, F. E., 102(58), *159*
Bashforth, F., 90(9), 91(9), 92(9), 95(9), 156(9), *158*
Basterfield, J., 9(18), 41(24), 55(75), *27, 83, 84*
Bayuzich, R. J., 208(124), *254*
Beck, P. A., 324(139), 326(146,148), *356*
Beckner, R., 176(20), 276(48), *252, 353*
Beeck, O., 263(11,12), 273(12), *353*
Behrndt, K. H., 7, *7*
Belforte, D. A., 102(53), *159*
Bellemans, A., 167(3), *251*
Beloguros, B. V., 183(57), *253*
Benard, J., 269(32), *353*
Bennendijk, N., 45, *45*
Berghezan, A., 339(199,212), *357*
Bering, B. P., 100(291), *159*
Bernard, G., 102(59), 104(59), 113(59), 114(59), *159*
Bernard, J., 266(20), *353*
Bernstein, I. M., 205(115), *254*
Bikerman, J. J., 67(98), 92(12), 93(12), 99(12), 108(12), 116(88), *84, 158, 160*
Bircumshaw, L. L., 101(42,43), 105(3), *159*
Biscondi, M., 194(90), 245(210), 246(210, 212,213), 247(210), 333(179), *253, 256, 357*
Bishop, G. H., 41(23), 80(130), 209(126), 210(126), 222(126), *42, 83, 85, 254*
Bitler, W. R., 18(41), 134(135), 135(135), *28, 161*
Blakely, J. M., 77(115), 179(34), 268(23), *85, 168, 252, 353*
Blucher, J. M., 276(46), 280(46), *353*
Boehme, W., 116(91), *160*
Bolling, G. F., 55(68), 62(68), 65(68), 78(68), 79(68), 218(145), 321(134), *84, 255*
Bollmann, W., 196(92,93), 210(128-130), 212(128), 233(128,129), *253, 254*
Bolton, M. J., 54(66), 55(66), *84*
Bondi, A., 123(124), *161*
Bonfield, W., 92(18), 93(18), 339(201), *158, 357*
Bonzel, H. P., 18(36), 130(133), *28, 161*
Boos, J. Y., 194(90), 245(210), 247(210), *253, 256*
Bouchard, M., 234(184), 235(184), *256*
Bowden, F. P., 290(74,75), *354*
Bowkett, K. M., 47(46), 148(171), 196(99), 197(99), *83, 162, 253*
Boyd, J. D., 154(179), 155(179), *162*
Bragg, W. L., 188(79), *253*
Brandon, D. G., 47(45), 196(96,97), 198(96), 199(96), 200(97), 225(97), *83, 253*
Bratina, W. J., 339(203), *357*

Bramfitt, B. L., 315(125), *355*
Braunovic, M., 343(238), *358*
Breger, A. Kh., 181(45), 183(45), *252*
Brillouin, M., 187(64), *253*
Broutman, L. J., 344(244), *358*
Brown, F. E., 97(23), *158*
Brown, L. M., 142(148), 143(149), 339(218), *161, 357*
Brown, M. H., 208(123), *254*
Brown, N., 62, *62*, 66, *66*
Bruggeman, G., 41(23), 80(130), *42, 83, 85*
Bryant, L. F., 124(109), 131(109), 138(109), *161*
Bufalini, P., 225(157), *255*
Buff, F., 167(2), 168(2), *251*
Burgers, J. M., 188(80), 325(145), *253, 356*
Burke, J. E., 62(80,84), 323(138), 325(138, 143), *84, 356*
Burmeister, J., 9(24), *27*
Burton, W. K., 176(18), *252*
Buttner, F. H., 77(114), 124(113), 131(113), 262(3), *85, 160, 352*
Buzzichelli, G., 207(118), 338(196), *254, 357*
Byrne, J. G., 324(141), 325(141), *356*

Cabane-Brouty, F., 266(20), *353*
Cabrera, N., 3(9), 176(18), *27, 252*
Cahill, J. A., 105(68), *159*
Cahn, J. W., 17(35), 18(35), 74(110), 82(110), *28, 84*
Cahn, R. W., 62(82), 78(121), *84, 85*
Calverley, A., 105(67), *159*
Cantor, M., 99(25), *158*
Carpenter, H. C. H., 62(83), *84*
Carter, G., 136(141), *161*
Cathcart, J. V., 269(27), *353*
Chadwick, G. A., 7(13), 74(109), *27, 84*
Chalmers, B., 43(29), 188(69,70,75,77), 193(88), 209(126), 210(126), 222(126), 248(217), 317(127,128), 321(136), 322(136), 329(136), 333(136), *83, 253-256, 355, 356*
Charbonnier, F. M., 125(123), 132(123), *161*
Chaudhari, P., 337(192), *357*
Childs, W. J., 315(123), *355*
Chin, G. Y., 332(172), *356*
Chopra, K. L., 232(174), 233(174), 281(62), *255, 354*
Chou, Y. T., 206(117), 342(233), *254, 358*
Christian, J. W., 25(50), 176(21), 235(21), *28, 252*
Churchmarev, S. K., 168(5), *251*
Cizeron, G., 66(92), *84*
Clark, A., 270(36), 272(36), 273(36), *353*
Cline, H. E., 234(180,181), 333(178), *255, 357*
Cline, R. S., 328(156), *356*

AUTHOR INDEX

Clough, W. R., 105(66), *159*
Coble, R. L., 335(185), 336(185), *357*
Cockayne, D. J. H., 144(158), *161*
Cohen, M., 77(116), 347(252,253), *85, 358*
Coldrey, J. M., 101(44), *159*
Cole, W. A., 263(11), *353*
Coll, J. A., 78(121), *85*
Copley, S. M., 25(47), *28*
Cracknell, A., 340(230), *358*
Crosley, P. B., 315(124), *355*
Cotterill, R. M. J., 246(214), 250(214), 314(121), *256, 355*
Cottrell, A. H., 191(85), 309(103), *253, 355*
Couchman, P. R., 2, *2*
Crois, C., 243(200), *256*
Crussard, C., 330(169), *356*
Cunningham, R. E., 275(42), *353*
Curie, P., 3(3), *27*
Cyrot-Lackmann, F., 250(219,220), *256*

Dabosi, F., 237(186), *256*
Daniels, F. W., 54(66), 55(66), *84*
Das, E. S. P., 214(137), 305(96), 309(107), 310(107), *254, 355*
Dash, J., 66, *66*
Dash, S., 62, *62*
Davies, H. A., 101(48), 103(48), 104(48), 105(48), *159*
Davies, I. G., 251(222), *256*
Davies, J. K., 102(58), *159*
Davies, V. de L., 101(38), *159*
daVinci, L., 88(5), *158*
deBoer, J. H., 261(81), *352*
deBruyn, P. L., 20(44), 25(44), *28*
Defay, R., 167(3), *251*
DeHoff, R. T., 36(11), *75*
DeJonghe, L. C., 227(165), *255*
Dekeyser, W., 45, *45,* 188(72), *45, 253*
Delamare, F., 266(21), *353*
Delavighette, P., 229(169), *255*
Demer, L. J., 324(139), *356*
dePlessis, J. C., 233(177), *255*
DeWitt, R., 214(142), *254*
Dillamore, I. L., 144(256), 147(156), *161*
Dimitrov, O., 243(200), *256*
Dityatkousky, Y. M., 312(114), 313(114), *355*
Dobson, P. S., 144(154), 145(154,159,160), 146(154), 148(154), *161, 162*
Dodd, R. A., 147(168), *162*
Dolan, W. W., 125(123), 132(123), *161*
Domain, H. A., 66(95), 82(95), *84*
Dooley, G. J., 130(130), *161*
Doring, W., 176(20), 276(48), *252, 353*
Dorn, J. E., 332(171), *356*
Dorsey, N. E., 92, *92*
Douglas, A. W., 315(124), *355*
Douglas, D. L., 339(205), *357*

Doyama, M., 246(214), 250(214), *256*
Drath, G., 100(34), 101(34), *159*
Drauglis, E., 261(7), 283(7), *283, 352*
Drechsler, M., 45(52), *83*
Ducastelle, F., 250(219,220), *256*
Dunn, C. G., 54(64,65,66), 55(64,66), 119(100), *83, 84, 160*
Dunning, W. J., 9(17), 177(24), 179(124), *27, 252*
Duvall, D. S., 339(222), *357*
Dwarakadasa, E. S., 214(136,140), *254*
Dyke, W. P., 125(123), 132(123), *161*
Dyson, B. R., 102(56), 103(56), 105(56), 113(56), *159*

Easterling, K. E., 296(89), *354*
Edmonds, T., 179(33), *252*
Eguchi, T., 16(33), *28*
Ehrlich, G., 124(112), 129(112), 131(112), 260(1), *160, 352*
Eichen, E., 66(95), 82(95), 234(183), *84, 255*
Eikum, E., 339(215), *357*
Elliott, R., 155(187), *162*
Emmett, P. H., 272(39), *353*
Eötvös, R. Von, 106(72), *159*
Eremenko, V. N., 101(39,40,45), 102(49, 54), 104(63), 114(83), *159, 160*
Ericsson, T., 145(162), *162*
Estermann, I., 225(160), *255*
Estrup, P. J., 130(128,131), 179(29), *161, 252*
Everett, D. H., 167(3), *251*
Ewen, D., 187(67), *253*
Ewing, R. W., 248(216,217), *256*
Ewing, W. W., 104(62), *159*

Farmer, M. H., 334(180), *357*
Farnsworth, H. E., 179(32), 180(37), *252*
Faulkner, R. G., 155(188), 239(194), *162, 256*
Feller-Kniepmeier, M., 329(161), *356*
Ferran, G., 66(92), *84*
Ferrante, J., 285(67), *354*
Fidelle, J. P., 313(115), *355*
Fisher, J. C., 253, 54(63), 62(86), 63(86), 64(86), 65(86), 119(100), 225(158), *83, 84, 160, 255*
Fleetwood, M. J., 135(140), *161*
Flemings, M. C., 170(9), 172(9), 176(9), 315(122), 316(122), *251, 355*
Floreen, S., 337(193), *357*
Flood, E. A., 179(30), 276(47), *252, 353*
Foltz, J. V., 349(260,262,263), 351(260), *358*
Fontana, M. G., 313(118), *355*
Fordham, S., 95(20), 96(20), *158*
Fortes, M. A., 18(38), 242(197), 312(109), *28, 203, 256, 355*

Fourdeux, A., 339(199,212), *357*
Fraikor, F. C., 16(34), *28*
Frank, F. C., 7(11), 176(15,16,18), 196(95), 212(95), *27, 252, 253*
Frenkel, J., 176(23), 181(49), 183(49), 185(49), *252*
Fresenko, V., 101(40), 114(83), *159, 160*
Friedel, G., 47(43), *83*
Fuller, H. W., 227(164), *255*
Fullman, R. L., 8(14), 35(10), 36(10), 55(67), 61(67), 62(86), 63(86), 64(86), 65(86), 76(10,111), 225(158), *27, 75, 84, 85*
Funk, E. R., 80(128), 125(121), 132(121), 262(3), *85, 161, 352*

Gadzuk, J. W., 264(16,18), *353*
Gale, B., 53(56), *83*
Gallagher, P. C. J., 144(153), 145(153,165), 146(153), 147(153,165), 148(153,165), 152(153), 153(153), *161, 162*
Garmong, G., 345(247), *358*
Garofalo, F., 336(188), *357*
Gates, R. S., 333(177), 335(177), *356*
Gauss, K. F., 88(3), *158*
Geld, P. V., 168(5), *251*
Germer, L. H., 180(35,36), *252*
Gibbs, J. W., 3(2), 4(2), 5(2), 6(2), 7(2), 9(2), 10(2), 17(2), 17(2), 25(2), 31(1), *27, 75*
Gibson, H., 92(13), *158*
Gifkins, R. C., 188(74), 199(74), 332(174), 334(181), 335(187), *253, 356*
Gilman, J. J., 189(83), 306(105), 312(112), 343(236), *253, 355, 358*
Gilmore, C. M., 55(71), 79(71), *84*
Givens, W. G., 168(6), *251*
Gjostein, N. A., 9(16,22), 31(4), 43(32), 50(52,53), 51(53), 52(52,53), 53(53,54, 55), 55(74), 66(95), 76(55), 82(95), 193(87), *27, 83, 253*
Gleiter, H., 149(173), 150(173), 188(76, 77), 192(76), 194(76), 200(102,103), 209(76), 210(76), 219(147), 225(159), 232(172), 233(172), 243(201), 244(208), 321(136), 322(136), 329(136), 330(165, 168), 331(165), 339(197,206,224,225), *62, 162, 253, 254, 255, 356, 358*
Glicksman, M. E., 74(107), 82(107), 155(181), 315(123), 321(135), *84, 162, 355, 356*
Golden, M. E., 339(204), *357*
Golumbic, N., 272(37), *353*
Good, R. J., 168(6), *251*
Goodell, P. D., 337(193), *357*
Goodhew, P. J., 145(159), *161*
Goodman, S. R., 328(155), *356*
Goodrich, R. S., 208(124), *254*

Gordon, P., 325(144), *356*
Gorshkov, K. F., 312(114), 313(114), *355*
Goux, C., 38(20), 40(20), 194(90), 210(107), 245(209,210), 246(212,213), 247(209,210,215), 333(179), *83, 253, 254, 256, 357*
Grace, F. I., 349(259,261), *358*
Grant, J. T., 130(130), *161*
Gratsiansky, N. N. 105(70), *159*
Greene, N. O., 313(118), *355*
Greenhill, E. B., 125(122), 132(122), *161*
Greenough, A. P., 45(40), 78(119), 79(124), 116(87), 117(96,97), 124(114), 125(118), 132(118), *83, 85, 160, 161*
Gregory, J. N., 290(74), *354*
Gretz, R. D., 261(7), 283(7), *283, 352*
Griffith, A. A., 304(93), 305(93), *354*
Grimmer, H., *203*
Gronlund, F., 269(26), *353*
Grosse, A. V., 108(73), *159*
Grube, W. C., 62(85), 65(85), 66(85), *84*
Gschneider, K. A., 250(218), *256*
Guggenheim, E. A., 10(27), 25(27), 108(75), 113(81), *27, 160*
Guillot, J. B., 245(209), 246(209), 247(209), *256*
Guinier, A., 237(187), *256*
Gulbransen, E. A., 269(28), *353*
Guy, A. G., 293(85), 294(85), 332(85), *354*
Gvozdev, A. G., 181(43), 182(43), 183(43), *252*
Gwathmey, A. T., 269(27,29), 275(42), *353*

Haas, T. W., 130(130), *161*
Haasen, P., 218(146), 330(166), *255, 356*
Haessner, F., 330(166), *356*
Hagstrum, H. D., 264(17), *353*
Hale, M. E., 227(164), *255*
Hall, E. O., 62(81), 340(228), *84, 358*
Hall, J. W., 275(44), *353*
Hall, M. G., 232(171), *255*
Hanneman, R. E., 20(42), 21(42), 343(242), *28, 358*
Hargreaves, F., 188(78), *253*
Harker, D., 32(7), 72(7), *75*
Harkins, W. D., 97(23), 104(62), *158, 159*
Harper, J., 329(158), *356*
Harris, L. A., 130(132), *161*
Harris, W. W., 269(29), *353*
Harrison, W. A., 182(56), 184(56), 245(56), 250(56), *253*
Hart, E. W., 134(136), 337(190), *161, 357*
Hartmann, R., 111, *111*
Hartt, W. H., 41(23), 80(130), *42, 83, 85*
Hasegawa, T., 201(109), *254*
Hasson, G., 194(90), 245(209,210),

247(209,210,215), *253, 256*
Hauser, E. A., 95(19), *158*
Hauser, F. E., 332(171), *356*
Hawksbee, 88(6), *158*
Hayden, H. W., 337(193), *357*
Hayward, E. R., 79(124), 125(118), 132(118), *85, 161*
Haywood, D. O., 261(6), *352*
Haworth, C. W., 343(238), 344(238), *358*
Hellawell, A., 251(222), *256*
Herbeural, I., 194(90), 246(210), 247(210), *253, 256*
Herring, C., 3(7), 7(7), 31(2), 33(2), 47(47), 117(98), 118(98), 121(98), 122(98), 335(184), 336(184), *6, 27, 83, 160, 357*
Heslop, J., 340(231), *358*
Hess, J. B., 42(26), 76(26), *83*
Hilliard, J. E., 17(35), 18(35), 74(110), 77(116), 82(110), *28, 84, 85*
Hills, R. J., 188(78), *253*
Hirota, K., 275(41), *353*
Hirsch, P. B., 36(14), 38(14), 145(164), 342(234), 346(248), *75, 162, 358*
Hirth, J. P., 2, 14(30,31), 16(34), 124(109), 131(109), 134(138), 138(109), 142(138), 144(155), 145(155), 220(148), 224(148), 226(148), 240(148), 249(148), 276(50), 277(52), 279(55), 338(194), 339(194, 226), 339(226,227), 343(194,237), 346(250), 347(256), 351(194), *2, 28, 160, 161, 255, 353, 354, 357, 358*
Hoar, T. P., 113(82), *160*
Hodges, C. H., 250(221), *256*
Hohenberg, P., 181(51), 185(51), *252*
Holl, H. A., 78(119), 124(114), *85, 160*
Holloman, J. H., 276(51), *353*
Holzworth, M. L., 324(139), *356*
Hondros, E. D., 116(89,95), 118(95), 119(89,95), 123(89), 124(95), 125(89, 95,115,120), 128(95), 129(95), 131(89, 95,115), 132(120), 134(95), 155(120), 156(120), 306(97), *90, 160, 161, 355*
Hook, R. E., 208(123), 343(237), *254, 358*
Hooker, M. P., 130(130), *161*
Hopkins, B. Z., 268(24), 269(24), *353*
Hopkins, B. J., 268(22), *353*
Hopkins, R. H., 154(178), 155(178), *162*
Horiuti, J., 275(41), *353*
Hornbogen, E., 200(102), 339(197,200, 206,224), *254, 357, 358*
Horylev, R. J , 40(22), 42(28), 45(38), 58(28,77), 59(22,28,38,77), 60(28), 61(28), 70(103), 74(108), 77(22,77), 78(22,28,120), 79(38), 122(102), 125(102,116), 131(116), 132(102), 133(116), 138(102,116), 139(116), 141(146), 142(146), 146(116), 147(102), 148(116), 191(84), 194(84), 195(84), 199(84), 207(84), 210(84), 211(84), 244(206), 339(207), *83-85, 160, 253, 256, 357*
Hosford, W. F., 332(172), *356*
Howie, A., 36(14), 38(14), 146(167), *75, 162*
Hren, J. J., 170(10,12), *251*
Hu, H., 326(146), 328(155,156), *356*
Huggins, R. A., 339(214), *357*
Humenik, M., 92(17), 94(17), 102(57), 103(60), 104(57,60), 113(17), 114(17), 115(85), 116(85), *158-160*
Humphrey, J. C. W., 187(66), *253*
Humphreys, F. J., 339(220), *357*
Hunt, J. D., 318(129), 319(130), 320(129, 130), *355*
Hyde, B. G., 229(168), *255*

Inal, O. T., 76(117), 125(117), 132(117), 138(117), 174(14), 179(14), 208(125), 233(14), 283(65), 312(110), *85, 160, 252, 254, 354, 355*
InDerSchmittem, W. P., 330(166), *356*
Inman M. C., 18(37,40), 45(39,41), 55(39, 41,69), 61(39), 76(39), 77(39,41), 78(39), 116(86), 124(86,111), 125(86), 128(111), 129(111), 131(86,111), 132(86), 134(111), 283(64), *28, 83, 84, 160*
Ioieleva, K. A., 100(29), *159*
Irwin, G. R., 305(95), *355*
Ishida, Y., 198(100), 199(100), 201(109, 114), 202(114), 203(114), 214(134), *203, 254*
Ives, L. K., 147(169,170), 152(170), *162*
Iwaschenko, N., 101(40), *159*

Jackson, A. G., 130(130), *161*
Jackson, K. A., 317(126), 318(129), 319(130), 320(129,130), *355*
Jaeger, H., 275(45), *353*
Jaffee, R. I., 261(7), 283(7), *283, 352*
Jesser, W. A., 2, *2*
Johnson, C. A., 6(10), *2, 27*
Johnson, R. H., 337(191), *357*
Johnson, W. C., 243, *243*
Johnson, W. G., 189(83), *253*
Johnston, T. L., 308(98), *355*
Johnston, W. G., 343(236), *358*
Jonas, J. J., 148(172), *162*
Jones, D. A., 339(210), *357*
Jones, E. R., 31(3), *75*
Jones, F. Llewellyn, 290(76), *354*
Jones, H., 121(101), 129(127), 130(127), *160, 161*
Joshi, A., 18(39), 135(139), *28, 161*
Jossang, T., 144(155), 145(155), *161*
Joyner, P. A., 92(16), *158*

Jurisch, M., 296(88), *354*

Kamdar, M. H., 308(99), 312(99), *355*
Kamenska, E., 78(118), *85*
Kasen, M. B., 21(45), *28*
Katayama, M., 108(74), *160*
Kaufaman, S. M., 102(52), 103(52), 104(52), 114(52), 115(85), 116(85), *159, 160*
Ke, T. S., 199(101), *254*
Kear, B. H., 136, 224(149), *136, 255*
Kelly, A., 25(48), 344(243), 346(248), *28, 358*
Kemball, C., 92(15), *158*
Kennedy, A. J., 336(189), *357*
Kenty, J. L., 279(57), 280(57), *354*
Kerridge, M , 287(72), *354*
Khan, A. R., 55(69), *84*
Kimmel, E. R., 301(92), *354*
King, R., 43(29), 188(69), *83, 253*
Kingery, W. D., 66(93), 92(17), 94(17), 103(60,61), 104(60,61), 113(17), 114(17), 123(51), 125(51), 126(51), *84, 158, 159*
Kinoshita, C., 16(33), *28*
Kinsman, K. R., 232(171), 234(182,183), 239(192), *255, 256*
Kirkwood, J. G., 167(2), 168(2), *251*
Kirshenbaum, A. D., 105(68), *159*
Klein, H. P., 149(173), 150(173), *162*
Klein Wassink, R. J., 105(65), *159*
Koch, E. F., 234(180,181), *255*
Kohara, S., 326(148), *356*
Kohn, W., 181(51,52,54), 182(54), 185(51, 52,54), 186(54), 187(54), *252*
Komen, Y., 201(113), 212(113), 232(113), *254*
Kondo, S., 22(46), *28*
Korbonski, J. A., 325(142), 330(142), *356*
Korolkow, M., 101(35), *159*
Koshevnik, A. Yu., 92(11), *158*
Kossel, W., 225(161), *255*
Kossowsky, R., 154(178), 155(178), *162*
Kotze, I. A , 321(132), *355*
Kramer, J. J., 154(177), 155(177), *162*
Krause, W., 102(55), 104(55), *159*
Kremer, J. C., 324(139), *355*
Kritzinger, S., 145(160), *162*
Krock, R. H., 344(244), *358*
Kronberg, M. L., 47(44), 62(44), 196(94), 198(94), 200(94), *83, 253*
Kruschov, M. M., 289(73), *354*
Kubaschewski, O., 268(24), 269(24), *353*
Kubitschek, Y., 101(37), *159*
Kuhlmann-Wilsdorf, D., 2, 320(131), 321(132), 346(251), *2, 355, 358*
Kusakov, M. M., 92(11), *158*

Lacombe, P., 225(153), *255*
Laforce, H. R., 18(39), 135(139), *28, 161*
Laird, C., 234(182), 237(185), 239(190, 192,193), 243, *243, 255, 256*
Lander, J. J., 130(129), *161*
Lando, J. L., 97(24), 98(24), *158*
Landon, P. R., 332(171), *356*
Lang, N. D., 181(54), 182(54), 185(54,58), 186(54), 187(54), 264(14), *252, 253, 353*
Langmuir, I., 261(5), *352*
Lange, F. F., 40(21), *83*
Laplace, P. S., 87(2), *158*
Lawless, K. R., 268(25), 269(25), *353*
Lazarev, V. B., 114(84), *160*
Leak, G. M., 78(122), 129(127), 130(127), *85, 161*
Leffers, T., 314(121), *355*
Lepie, M. P., 102(53), *159*
Levy, J., 201(107,108), 212(108), 213(107, 108), *254*
Lewis, J., 329(158), *356*
Li, C. H., 309(101), *355*
Li, J. C. M., 194(91), 195(91), 205(91), 206(117), 207(91), 215(143,144), 216(143) 243(199), 309(106), 321(106), 339(106), 342(232,233), *254, 255, 355, 358*
Liebmann, B., 326(150,151), 327(151), *356*
Liebowitz, H., 313(116), *355*
Lilholt, H., 314(121), *355*
Lin, T. L., 208(121), *254*
Lin, W. N., 40(22), 58(77), 59(22,77), 77(22,77), 78(22), 141(146), 142(146), 191(84), 194(84), 195(84), 199(84), 207(84), 210(84), 211(84), 330(167), 339(207), *83, 84, 161, 253, 356, 357*
Lionetti, F., 54(65), *84*
Liu, Y. C., 145(165), 147(165), 148(165), *162*
Livak, R. J., 234(184), 235(184), *256*
Loberg, B., 201(111,112), 213(111), 218(112), *254*
Logan, H. L., 313(117), *355*
Lorimer, G. W., 13(28), 16(28), 244(204), 245(204), *27, 256*
Lothe, J., 14(30), 134(138), 142(138), 220(148), 224(148), 226(148), 240(148), 249(148), 277(53), 346(250), *28, 161, 255, 354, 358*
Lubman, N. M., 92(11), *158*
Lucke, K., 326(150,151), 327(151), *356*
Lupis, C. H. P., 102(59), 104(59), 113(59), 114(59), *159*
Luton, M. J., 148(172), *162*
Lyon, L. B., 179(26), *252*

McCaughey, J. M., 312(113), *355*

McDonald, S. R., 125(122), 132(122), *161*
McGrath, J. T., 339(203), *357*
Mack, G. L., 102(58), *159*
McKinney, J. T., 31(3), *75*
McLane, S. B., 137(142), 290(77), *161, 354*
McLean, D., 16(32), 18(40), 20(32), 21(32), 45(41), 55(41), 77(41), 124(111), 128(111), 129(111), 131(111), 133(134), 134(111), 188(71), 241(71), 306(97), 334(180), *28, 160, 161, 253, 254, 355, 357*
McLean, M., 53(56,57), 54(56,57,60), 78(57), 80(127), 125(120), 132(120), 155(120), 156(120), 201(114), 203(114), *83, 85, 161, 203, 254*
Macleod, D. B., 108(76), *160*
McNutt, J. E., 67(100), *84*
MacRae, A. V., 31(3), *75*
Manley, W. O., 324(140), *356*
Marcinkowski, M. J., 200(104), 201(105), 204(105), 209(127), 214(136-140), 215(105), 305(96), 309(107), 310(107), 339(198), 347(254,255), *254, 355, 358*
Marcus, H., *312*
Marinoff, C., 346(245), *358*
Markus, H., 312(113), *355*
Martin, E. E., 125(123), 132(123), *161*
Mascanzoni, A., 207(118), 338(196), *254, 357*
Massalski, T. B., 36(17), 54(62), 226(162), 227(162), 235(162), *83, 255*
Matthews, J. W., 234(179), *255*
May, J. W., 180(35,36), 325(145), *252, 356*
Mazanec, K., 78(118), *85*
Mazey, D. J., 339(216), *357*
Mehl, R. F., 154(177), 155(177), 328(157), *162, 356*
Melford, D. A., 113(82), *160*
Menzel, E., 269(31), *353*
Metcalfe, A. G., 346(246), *358*
Metzger, G., 100(33), 104(64), *159*
Michalke, M., 102(55), 104(55), *159*
Miller, W. A., 7(13), 9(18), 41(24), 42(27), 55(75), 74(109), 76(27), *27, 83, 84*
Mills, B., 78(122), *85*
Missol, W., 181(48), 182(48), 183(48), 184(48), 185(48), *252*
Mitchell, J. W. 339(210,211), *357*
Mitchell, T E., 145(164), *162*
Moazed, K. L., 277(52), 279(55,56), *353, 354*
Mondolfo, L. F., 315(124), *355*
Moore, A., 155(187), *162*
Moore, A. J. W., 9(21), *27*
Mott, N. F., 187(68), 199(68), 330(164), *253, 356*
Mourey, M., 237(186), *256*

Müller, E. W., 137(142), 168(7), 170(7,11), 179(7), 290(77-79), *161, 251, 354*
Mullins, W. W., 3(8), 7(12), 43(30,31), 44(31), 181(38), 291(80,81), 329(159), *27, 83, 252, 354*
Murr, L. E. *28, 74-76, 83-86, 123-125, 131, 132, 138, 139, 146, 147, 160-162, 193, 251-256, 282, 312, 352-358*
Mykura, H., 45(35,37), 48(48,49), 49(49), 53(49,57), 54(47,49,60,61), 78(57), 79(49), 80(127,129), *83, 85*

Nabarro, F. R. N., 118(99), 121(99), 122(99), 214(141), 335(183), *160, 254, 357*
Nagata, F., 201(109), *254*
Naiditsch, Y. V., 101(39), 102(49), 102(54), *159*
Newman, R. W., 170(12), *251*
Nicholas, J. F., 45(42), 53(58), 187(60), *83, 253*
Nicholas, M. E., 92(16), *158*
Nicholson, R. B., 13(28), 16(28), 25(48), 36(14), 38(14), 154(179), 155(179), 244(203,204), 245(204), 344(243), *27, 28, 75, 162, 256, 358*
Niehenko, B. I., 101(40,45), *159*
Nielsen, J. P , 65(88), 66(88,89), *84*
Niessen, P., 20(42), 21(42), 343(242), *28, 358*
Nishida, M., 116(90), *160*
Nishikawa, O., 290(77,78), *354*
Nix, W. D., 339(204), *357*
Nolfi, F. V., 2, *2*
Norden, H., 201(111,112), 213(111), 218(112), *254*
Nordstrom, T. V., 152(175), 153(175), *162*
Nutting, J., 68(101), 155(185), 286(68), *84, 162, 354*

Oakley, H. T., 97(24), 98(24), *158*
Oblak, J. M., 224(149), 339(222), *255, 357*
Olson, M. D., 92(16), *158*
Okamoto, G., 275(41), *353*
Ono, S., 22(46), *28*
Oriari, R. A., 112(79), 126(79), *160*
Orlov, L. G., 339(223), *358*
Orowan, E., 305(94), 343(235), *355, 358*
Osika, L. M., 234(181), *255*
Otooni, M. A., 31(5), *75*
Otter, M., 269(30), *353*
Oudar, J., 266(20), *353*
Ouyang, S., 31(5), *75*
Oxley, J. H., 276(46), 280(461), *353*

Padday, J. F., 90(10), 97(10), 108(10), *158*
Page, T. F., 146(157), *161*
Palmberg, P. W , 179(27), 283, *252*

Panitz, J. A., 137(142), 241(196), *161, 256*
Paransnis, A. S., 339(211), *357*
Park, R. L., 179(32), 180(37), *252*
Parker, E. R., 32(7), 72(7), *75*
Parthasarathi, M. N., 326(148), *356*
Pashley, D. W., 36(14), 38(14), 232(173), 281(61), *75, 255, 354*
Peat, A. J., 343(241), *358*
Peck, R. L., 145(161), *162*
Pelzel, E., 101(41), 103(41), 104(41), 105(41), 106(41), *159*
Pender, K. R., 268(22), *353*
Pershikov, A. V., 114(84), *160*
Petch, N. J., 340(229-231), *358*
Pettersson, B., 146(166), *162*
Phillips, V. A., 224(150), 225(150), 269(34), *255, 353*
Pick, H. J., 176(17), *252*
Pilliar, R. M., 68(101), 155(185), 286(68), *84, 162, 354*
Pitkethley, R. C., 179(33), *252*
Plummer, E. W., 264(18), *353*
Poisson, S. D., 88(4), 89(4), *158*
Pokrovskii, N. L., 100(32), *159*
Pope, L. E., 312(108), *355*
Pound, G. M., 116(94), 124(94), 131(94), 154(177), 155(177), 276(49,50), 277(53, 54), *160, 162, 165, 353, 354*
Powell, C. F., 276(46), 280(46), *353*
Prager, M., 273(40), 274(40), *353*
Pranatis, A. L., 116(94), 124(94), 131(94), *160*
Preston, G. D., 237(188), *256*
Price, A. T., 78(119), 124(114), *85, 160*
Price, C. W., 339(226), *358*
Prigogine, I., 167(3), *251*
Pulham, L., 101(44), *159*
Pumphrey, P. H., 47(46), 196(99), 197(99), 214(131,132), 243(202), 244(202), *83, 253, 254, 256*

Quader, M. A., 147(168), *162*
Queeney, R. A., 55(73), 76(73), 77(73), 140(145), *84, 161*
Quincke, G., 187(65), *253*

Rabinowicz, E., 284(66), 285(66), 287(66), *354*
Rack, H. J., 347(252,253), *358*
Radcliffe, S. V., 80(126), 132(119), *85, 161*
Raj, R., 331(170), *356*
Ralph, B., 18(38), 146(157), 155(188), 196(96), 198(96), 199(96), 239(194), 242(197), 312(109), *28, 161, 162, 253, 256, 355*
Ranganathan S., 196(96,98), 198(96), 199(96), *253*
Rapin, M., 313(115), *355*

Rase, H. F., 275(44), *353*
Rath, B. B., 205(115), 328(156), *254, 356*
Raue, G., 104(64), *159*
Ray, I. L. F., 144(158), *161*
Rayabar, A. K., 105(70), *159*
Rayleigh (Lord), 89(7), *158*
Read, W. T., 189(81), 191(81), 200(81), 245(81), 246(81), 346(249), *253, 358*
Redding, G. B., 124(105), *160*
Reineking, V., 346(245), *358*
Reiss, H., 234(178), *255*
Rhead, G. E., 266(21), *353*
Rhee, S. K., 155(182), 156(182), *162*
Rhines, F. N., 36(11), 53(55), 76(55), 193(87), *75, 83, 253*
Rhodes, G. C., 345(247), *358*
Rideal, E. K., 263(9), 272(38), 273(38), *352, 353*
Robertson, W. M., 9(23), 43(33), 50(51), 54(59), 76(59), *27, 83*
Rodin, T. N., 179(27), *252, 283*
Rosenhain, W., 187(66,67), *253*
Rostoker, W., 312(113), *355*
Rouhani, M. D., 181(55), 186(55), *253*
Rouze, S. R., 62(85), 65(85), 66(85), *84*
Ruff, A. W., 144(152), 147(169,170), 152(170), *161, 162*
Ruth, V.I., 277(52), *353*
Rutter, J. W., 66(90), 243(198), 317(127,128), 326(152), 327(152), 328(153,154), *84, 256, 355, 356*
Ryan, H. F., 201(106), 208(120), 254

Saada, G., 144(151), *161*
Sacedon, J. L., 283, *283*
Sachtler, W. M. H., 264(19), *353*
Sadananda, K., 209(127), 210(127), 214(138,139), 347(254,255), *254, 358*
Saidov, M., 100(32), *159*
Sanders, J. V., 275(45), *353*
Sankaran, R., 243, *243*
Sargent, C. M., 225(154), *255*
Sastry, D. H., 148(172), *162*
Sauerwald, F. (H. C. F.), 100(34), 102(55), 104(55,64), *159*
Sawai, I., 166(90), *160*
Schattler, R., 181(55), 186(55), *253*
Schneider, R., 111, *111*
Schober, T., 210(110,113), 212(113), 213(110), 232(113), *254*
Schrieffer, J. R., 264(13), *353*
Schrodinger, R., 99(27), *159*
Schwarzkopf K., 329(161), *356*
Seeger, A., 218(146), *255*
Seidman, D. N., 179(28), *252*
Semenchenko, V. K., 3(5,6), 4(5,6), 7(5,6), 9(6), 20(6), 99(28), 101(28), 102(28), 103(28), 104(28), 105(28), 106(28),

AUTHOR INDEX 369

113(28), 130(28), 181(42), 27, 159, 252
Seyboldt, A. U., 343(240), 358
Shaler, A. J., 116(92), 124(92), 131(92), 160
Sham, L. J., 181(52), 185(52), 252
Shebzukhora, I. G. 181(46), 252
Sheithauer, W., 301(92), 354
Shewmon, P. G., 9(23), 47(50), 50(51), 54(59), 76(59), 291(83), 298(91), 321(137), 27, 83, 354, 356
Shockley, W., 189(81), 191(81), 200(81), 245(81), 246(81), 253
Shriver, B. C., 20(43), 30(43), 28
Shuttleworth, R., 1(1), 2(1), 43(29), 27, 83
Simnad, M. T., 276(49), 353
Simpson, C. J., 326(147), 356
Singh, H. P., 174(14), 177(14), 233(14), 279(58), 280(58,60), 281(58), 283(65), 252, 282, 354
Sinke, G. C., 112(80), 160
Sinnott, M. J., 295(86,87), 354
Skapski, A. S., 112(78), 125(126), 181(40), 182(40), 186(40), 187(40), 188(40), 160, 161, 252
Skelton, R. P., 335(186), 357
Smallman, R. E., 144(154), 145(154,159, 160), 146(154), 148(154), 339(217), 161, 162, 357
Smith, C. S., 31(16), 32(8), 54(8), 66(8), 71(8), 82(8), 312(111), 75, 355
Smith, D. A., 146(157), 148(171), 201(112), 214(134), 218(112), 161, 162, 208, 254
Smith, J. R., 181(53), 182(53), 185(53), 285(67), 252, 354
Smith, P. J., 55(71), 79(71), 84
Snowdon, K. V., 334(181), 357
Somorjai, G. A., 179(26,31,34), 135, 252
Son, U. T., 170(10), 251
Sosin, A., 227(163), 255
Sosnovsky, H. M. C., 275(43), 353
Southworth, H. M., 170(13), 252
Speiser, R., 124(109), 131(109), 138(109), 160
Sperry, P. R., 326(146), 356
Staudhammer, K. P., 167(4), 346(245), 349(264), 254, 358
Stauffer, C. E., 96(21), 158
Stein, D. F., 18(39), 135(139), 28, 161
Stevens, A. L., 312(108), 355
Stevens, R. N., 332(175), 334(182), 356
Stewart, A. T., 339(220), 357
Stobbs, W. M., 339(218), 357
Stokes, R. J., 309(101), 355
Stoloff, N. S., 308(98,100), 312(100), 355
Storch, H. H., 272(37), 353
Storey, G. G., 176(17), 252
Stossel, W., 269(31), 353

Stowell, M. J., 157(191), 162
Stratton, R., 181(39), 252
Stroh, A. N., 309(102,103), 310(102), 355
Stuart, L. E. H., 125(115), 131(115), 160
Stull, D. R., 112(80), 160
Sugden, S., 108(77), 160
Suiter, J., 201(106), 208(120), 254
Sundquist, B. E., 9(19), 68(102), 27, 84
Suzuki, H., 14(29), 150(174), 28, 162
Suzuki, T., 178(22), 252
Swann, P.R., 146(167), 162
Szostak, R. J., 180(35), 252

Tabor, D., 290(74,75), 354
Talbor, J., 329(160), 356
Tamhankar, R., 330(169), 356
Tamman, G., 116(91), 160
Tamura, S., 62(83), 84
Taunt, R. J., 347(254,255), 358
Taylor, J. W., 100(30,31), 103(31), 104(31), 188(63), 159, 253
Tay-Schon-Wej, L., 101(45), 159
Temrokov, A. I., 181(46), 252
Tessem, B. M., 92(16), 158
Thölen, A. R., 214(133), 296(89), 161, 254, 354
Thomas, G., 36(13), 230(170), 234(184), 235(184), 339(215), 75, 255, 256, 357
Thomson, R., 134(137), 161
Thornton, P. R., 145(164), 162
Thorpe, B. J., 264(15), 353
Tiedema, T., 325(145), 356
Tien, J. K., 25(47), 28
Tiller, W. A., 317(126), 355
Tipler, H. R., 18(37,40), 45(39,41), 55(39,41), 61(39), 76(39), 77(39,41), 78(39), 116(86), 124(86), 125(86), 124(86,111), 125(86), 128(111), 129(111), 131(86,111), 132(86), 134(111), 28, 83, 160
Tisone, T. C., 145(163), 162
Todd, C. J., 283
Tolansky, S., 36(18), 83
Tolman, R. C., 167(1), 251
Tolman, F. R., 45(36), 292(84), 83, 354
Towers, J., 324(140), 355
Trapnell, B. M. W., 261(6), 262(9,10), 352, 353
Trolan, J. K., 125(123), 132(123), 161
Tseng, W. F., 214(136,138), 254
Tsong, T. T., 168(7), 170(7,11), 179(7), 290(79), 251, 354
Tucek, S. C., 168(6), 251
Tucker, C. W., 233(176), 255
Tucker, W. B., 95(19), 158
Turnbull, D., 62(80), 155(180), 157(180, 190), 187(61,62), 188(61,62), 276(51), 321(133), 323(138), 325(138), 343(240),

Turnbull, D. (cont.)
 84, 162, 253, 353, 355, 356

Udin, H., 77(114), 80(128), 116(92,93), 124(92,113), 125(121), 131(92,113), 132(121), 262(3), *85, 160, 161, 352*
Underwood, E. E., 36(12), *75*
Unwin, P. N. T., 13(28), 16(28), 244(203, 204), 245(204), *27, 256*
Upthegrove, W. R., 295(87), *354*

Van der Merwe, J. H., 225(151), 232(175), 233(177), *255*
Van der Plank, P., 264(19), *353*
Van der Waals, J. D., 10(25), 106(71), *27, 159*
VanVlack, L. H., 66(94), 82(94), 155(183), *84, 162*
Vasiliu, M. I., 104(63), 114(83), *159, 160*
Vaughan, T. B., 176(17), *252*
Vaughn, D., 244(205,207), *256*
Venables, J. D., 153(176), 280(59), *162, 354*
Verschaffelt, I. E., 10(26), 99(26), *27, 158*
Vogel, F. L., 189(82), *253*
Vold, C. L., 74(197), 82(107), 155(181), 321(135), *84, 162, 356*
Volmer, M., 176(19), 225(160), 276(47), *252, 255, 353*
Vook, R. W., 31(5), *75*
Vonnegut, B., 157(189), *162*
Vydyanath, H. R., 330(162,163), *356*

Wagner, R. J., 193(88), *253*
Wald, M. S., 196(96), 198(96), 199(96), *253*
Walsh, J. M., 136, *136*
Walter, J. L., 234(180,181), 33(178), *255, 357*
Warlimont, H., 339(221), *357*
Warrington, D. H., *203, 255*
Watkins, H., 76(112), *85*
Weatherly, G.C., 9(18), 41(24), 239(189), 240(189), 339(213), *27, 83, 256, 357*
Webb, M. B., 31(3), *75*
Weber, R. E., *135*
Weertman, J., 191(86), 339(202), 347(256), *253, 357, 358*
Weertman, J. R., 191(86), *253*
Weinberg, F., 41(25), 188(73), 332(173), *83, 253, 356*
Weins, M. J., 245(211), 246(211), *256*
West, J. M., 101(38), *159*
Westbrook, J. H., 20(42), 21(42), 343(240,241), *28, 358*

Westmacott, K. H., 145(161), 339(317), *162, 357*
Westwook, A. R. C., 308(99), 312(99), *355*
Whalen, T. J., 102(52,57), 103(52), 104(52,57), 114(52), 115(52), 116(85), *159, 160*
Wheeler, A., 263(11), *353*
Whelan, M. J., 36(14), 38(14), 142(147), 144(158), *75, 161*
White, D. W. G., 101(47), 103(47), 105(69), 110(69), *159*
Whitwham, D., 225(153), *255*
Williams, T. M., 79(125), 80(125), *85*
Williams, W. M., 42(27), 76(27), *83*
Wilsdorf, H. G. F., 338(195), *357*
Wilson, F. H., 47(44), 62(44), 196(94), 198(94), 200(94), *83, 253*
Wilson, T. L., 298(91), *354*
Winegard, W. C., 55(68), 62(68), 65(68), 78(68), 79(68), 326(147), *84, 356*
Winslow, F. R., 49(50), *83*
Winterbottom, W. L., 9(15,16,20), 50(53), 51(53), 52(53), 53(53,54), 54(54), 66(96), 67(96), 68(96), 71(96), *27, 83, 84*
Wong, G. I., 45(38), 59(38), 79(3), 122(102), 125(102,116,117), 131(116), 132(102,117), 133(116), 138(102, 116,117), 139(116), 146(116), 147(102), 148(116), 208(125), 244(206), 349(262), *83, 85, 160, 252, 254, 256, 358*
Wood, J. G., 45(36), 292(84), *83, 354*
Woodruff, D. P., 74(106), 179(25), 314(119,120), 316(120), 317(120), 320(120), 321(120), *84, 252, 355*
Worthington, A. M., 92(14), *158*
Wulff, G., 3(4), 4(4), *27*
Wulff, J., 77(114), 80(128), 116(92), 124(92,113), 125(121), 131(92,113), 132(121), *85, 160, 161*
Wuttig, M., 20(43), 30(43), *28*
Wynblatt, P., 31(4), *75*

Yang, L., 276(49), *353*
Yoshida, H., 326(150), *356*
Young, T., 87(1), *185*
Young, F. W., 269(27,33), *353*
Young, Th., 67(99), *84*
Yukawa, S., 295(86), *354*

Zadumkin, S. N., 181(41,44,46,47), 183(44), *252*
Zhiviv, V. G., 101(36), *159*
Zhukhoritskii, A. A., 181(45), 183(45), *252*
Zukas, E. G., 349(258), *358*

SUBJECT INDEX

SUBJECT INDEX

Abrasive fragmentation, 288
Absorption, at solid–solid interfaces, 241, 242
Adhesion, 284-290
  contact, 297
  energy of, 285, 303, (table) 286
Adsorption, density (relative), 22
  interfacial, 241, 242
  isotherms, 261-263
  negative, 12
  positive, 12
  Suzuki, 14, 153
  to solid surfaces, 129
Amonton's law, 286
Antiphase (domain) boundary structure, 227-230
Asymmetry, 40
  angle of, 40
  grain boundary, 40, 189, 190
Atom-probe field-ion microscope, 137
Auger (electron) spectroscopy, 130, 135, 136, 312
Axis-angle pairs, 196
  for cubic coincidence lattice boundaries (table), 197

Bamboo structure, of wires, 120
Bicrystal section, (general) 40
  degrees of freedom, 40
Bubble at grain boundary, 71
Bubble in liquid, 90
  measurement of liquid metal surface tension, 98, 99

Cahn-Hilliard criterion, 134
Capillary tube, 88, 89
  liquid rise in, 89
  surface tension measurement, 89
Catalysis, 270-276
  effect of surface defects on, 275
Chemical potential, 4, 5, 50
Chemisorption, 259, 260, 263-266

Coincidence, lattice, 47
  site, *see* grain boundary structure
Contact angle, 67, 89
Corner twin, 59, 60, 62-64
Corrosion, grain boundary, 313
  stress, 313
  surface, 267
Crack, dislocations, 305
  elastic, 305
  initiation at grain boundary ledge, 309, 310
  length, 304, 305
  micro, 310
  nucleation at grain boundaries, 309
  plastic, 305
  propagation model (Griffith), 305
Creep, Coble, 335, 337
  diffusional, 335, 336
  Nabarro-Herring, 118, 335, 337
  rate, 376
  zero-creep method, 116
Critical temperature, 106-108
Crystalline particles, equilibrium shape of, 3
  on substrate, 67
  surface free energy of, 3
Curved interfaces, 25
Cusps, 194
  see also γ-plot

Deformation twinning, 348-351
Dendrite growth, 316
Dendrite structure, 317
Diffusion, activation energy for, 295
  coefficients, 121, 293
  grain boundary, 292
  interfacial structure, effect of, 293
  interdiffusion, 293
  lattice, 292, 299
  phenomena, 291
  self-diffusion coefficient, 121
  surface, 43-45, 291
  vacancy, 330-332

371

Diffusion (cont.)
  volume, 43-45, 291
Diffusional creep, see creep
Dihedral angle, 36
Disclination, model of grain boundary structure, 214-216
  wedge dipole, 215
Dislocation, Burgers vector, 142
  character angle, 142-143
  core cut-off radius, 191
  core structure, 145, (grain boundary), 195
  cross-slip, 39
  edge, 178
  emission from ledges, 207
  etch pits, 189, 190, 275
  Frank partial, 224
  misfit, 233
  models of grain boundary structure, 188-196
  networks (in coincidence boundaries), 200
  modes in stacking-fault energy measurements, 142-148
  partial, 142-144
  pile-up, 341, 342
  screw type, see Spirals
  Shockley partial, 223, 224
  spirals, 175-177, 178
  tilt boundary, 189, 190
  traces (as thickness guide), 39
  twist boundary, 190
  wall, 215, see also Disclination
Dispersed phase, 24, 70
  interface structure, 238
Dispersoid, see Dispersed phase
Domain boundary structure, antiphase, 227-230
  electric (ferroelectric, ferrielectric), 227, 228
  magnetic (ferromagnetic, ferrimagnetic), 227, 228
Drop weight, correction factors (table), 98
  determination of, 97
  method, 96
  see also Surface tension
Dupre equation, 34, 66, 67, 72

Electron/atom ratio, 148
Electron transmission microscopy, node measurements, 142-145
  of grain boundary ledges, 206
  of grain boundary structure, 190, 191, 200
  of twin boundary-grain boundary intersections, 54-66
Embrittlement, grain boundary, 308, 312
  liquid metal, 307, 308, 312
  surface, 306
Energy of adhesion, 285, 297, 298
  effective, 290
  table, 286
Energy of adsorption, 260
Energy ratios, 49, 54, 55, 59, 64, 70, 74, 325
  tables of, 76, 82
Entropy, coherent twin boundary (table), 138
  general interface, 11, 13
  grain boundary (table), 131, (calculation of), 248
  liquid metal, 100, (table), 101
  solid surface, 2, (table), 124
  stacking fault (table), 145
  see also Temperature coefficients
Eötvös constant, 107
  equation, 106, 107
Epitaxial, growth, 281-284
  interface structure, 232-234
Epsilon (ϵ)-phase, 349
  boundary, 250
  structure, 224
Equilibrium configurations (interface intersections), 31
Equilibrium, shapes, 3-6
  of crystalline particles, 3-6
  of liquid drops (metal), 8
  of polyhedra, 7, 175
Etch pits, at grain boundaries, 189, 190
  chemical reaction, 275
Euler-Poincaré relation, 31
Eutectic, growth, 318-320
  interfacial free energy, 154, 319
  phase equilibrium, 74
Evaporation-condensation, 43, 44, 291

Facets (and faceting), 7, 174, 314
Fick's first law, 291
Field evaporation, 7, 208
Field-ion microscopy, 7-9
  atom-probe, 241
  in surface structure observations, 170-172, 176, 177
  of coherent twin boundary (figure), 221
  of contact phenomena, 290
  of grain boundary (figures), 13, 209, 220
Fischer-Tropsch synthesis, 272, 273
Flotation, 90
Flow, plastic, 340-342
  stress, 340-342
Fracture, intergranular, 308
  work of, 303
Frank's formula, 212
Free energy ratios, see Energy ratios
Free volume, grain boundary, 218, 219
Friction, 284-290
  and surface structure, 285-288
  coefficient of, 286

force, 286
rolling, 286-288

Gamma (γ)-plot, 5, 6, 51-53, 70
  cusps, in, 54
  experimental, 53
Gibbs adsorption isotherm, 7, 15
  binary alloys (general), 17
  equation, 260
  liquid alloys, 112
  solid alloys, 127
Gibbs-Curie theorem, 3
Gibbs dividing surface, 11, 17, 260
Gibbs-Duhem equation, 11, 15, 22
Gibbs-Thomson theorem, 6, 29
Gibbs-Wulff construction, see γ-plot
Grain boundary, as source of dislocations, 338-340
  asymmetry, 40
  corrosion, 313, 314
  crack nucleation, 309-311
  creep, 335
  effect of impurities and inclusions on, 328
  entropy (calculation of), 248, (table), 131
  hardening, 343, 344
  influence on mechanical properties, 338-344
  migration, 322
    effects of misorientation on, 325
    mechanisms of, 330, 331
    rate, 327, 328
  mobility, 328-330
  precipitates, 25
  serrations, 42, see also Grain boundary structure
  shear (heterogeneous), 204, 205
  sliding, 322-336
    in polycrystals, 335
    mechanisms of, 331-334
  steps, 330, 331
  stress corrosion cracking, 313, 314
  thickness, 12, 13
  torque, 33, 41, 56
  triple junctions, 37, 38, 64
Grain boundary dislocations, 200-205
  Burgers vector for, 201, (tables), 202, 203
  coalescence of to form ledges, 204, 205
  in grain boundary sliding, 333-335
  ledges, 204, 205
  movement of, 334
  pile up of, 334, 335, 345
Grain boundary free energy, effects of composition on, 132
  isotherms of, 133
  measurement of, 130, (table), 131
  theoretical calculation of, 245-248
Grain boundary geometrical conventions, 40
Grain boundary intersections, 35
  with a free surface, 43, 74
  with a precipitate, 71
  with a solid-liquid surface, 74
  with a twin boundary, 54, 59, 60
    equilibrium configurations, 60, 74
Grain boundary ledge, 203-208
  as a dislocation source, 338-342
  Burgers vector of, 204
  crack initiation from, 309-311
  creation of, 204 (figure), 205
  density, 205, 207, 342
  effect of misorientation on, 204, 207
  electron micrographs of (figure), 206
  height, 205
  structure, 340, 345
  see also Grain boundary structure
Grain boundary misorientation, 40, 189
  histograms of (figure), 217
  theoretical curves (figure), 247
  versus energy (figure), 193, 194
Grain boundary models, see Grain boundary structure
Grain boundary segregation, 16, 135-138
  measurement of, 135
Grain boundary structure, 188-218
  boundary faults in, 225
  coincidence-site models of, 196-200
  disclination models of, 214-216
  dislocation model of, 189-194
  free volume in, 218, 219
  interpretation and description of, 208-219
  ledges in, 194, 203-208
  low-angle, 191-196
  O-lattice interpretation of, 210-212
  periodic, 212-214
  plane-matching theory of, 212-214
  practical description of, 216-219
  serrations, 194
  three-dimensional model of, 210, 211
  tilt and twist, 189-194
  units of (structural units), 209
Grain growth, 323-328
Griffith equation, 305
Groove profile, angle, 43
  at grain boundary-surface intersection, 43, 46
  at twin boundary-surface intersection, 48, 49
  formation, 44
  thermal, 80
  interferograms of, 46, 49
  scanning electron micrograph of, 46
  transmission electron micrograph of, 46
Guinier-Preston zones, 237

Hall-Petch relation, 340-342
Heat of adsorption, 129, 260-262
Heat of fusion, 123-126, 157, 182, 318

Heat of vaporization, 112
Helmholtz surface free energy, 1-3
Henry's law, 18, 20

Inclusion (interfacial) structure, 237
Interface phase, concept, 10-13
Interfacial, energy of contacting surfaces, 285
   energy ratios, see Energy ratios
   free energy calculation (for solid interfaces), 244-248
   structure in liquid metals, 167, 168
   structure in solid metals and alloys, 187-227
Interference microscopy, of grain boundary grooves, 45
Interferometric measurements, 80
Interphase interface, as dislocation source, 338-341
   energy measurements, 153-156
   free energy of metals and alloys (table), 155
   heterogeneous systems, 21, 23
   influence of on mechanical properties, 338
   multicomponent structure, 234, 235
   strengthening, 344
   structure, 227-240
   transformation structure, 235-236
Interstitial alloys, 19
Ion mass spectrometry, in grain boundary segregation analysis, 136

Jellium model, of surface structure, 185

Kinks, 172-175
Kronberg-Wilson boundaries, 196

Langmuir equation, 261
Laplace's fundamental equation, 87
Ledge, see Grain boundary ledge
Lennard-Jones potential, 53, 246
Liquid drops, equilibrium shapes of, 67, 69
   on solid substrate, 67, 69
Liquid metal, eutectic growth, 318
   interfacial energy, 168
   interfacial structure, 167, 168
   surface adsorption, 180, 181
   surface energy, 187
   surface reaction, 180, 181
   surface structure, 167, 168
   surface tension (table), 101
   see also Surface tension
Liquid–solid interface, see Solid–liquid interface
Low-energy electron diffraction (LEED), 130, 135
Lubrication, 289, 290

Martensitic, phase equilibrium, 74
   transformation, 349
Melting, grain boundary, 321, 322
   of metals and alloys, 320
   theory of, 320
Meniscus, 89
Metallography, quantitative, 36
Misfit lines, in grain boundary, 212-214
Misorientation angle, definition, 40
   versus grain boundary diffusion coefficient (figure), 296
   versus grain boundary energy, 193, 194 (calculated), 247
   versus grain boundary ledge density (figure), 207
Modulus of elasticity, variation with solid surface free energy (figure), 127
Moire pattern in grain boundary structure, 213, 214
Morse potential, 53, 246
Multiphase equilibria, 66
Multiphase interfacial free energy ratios (table), 82
Multiplicity, (coincidence lattice ratio), 196-200, (table), 197

Node measurements, 142-145, 152
Nodes, extended dislocation, 142-145, 152, 154
Nucleation, heterogeneous, 276
   homogeneous, 276
   theory, 276-280

O-lattice, 210-212
Order-disorder phenomena, 228
Oxidation, effect of surface defects on, 269, 270
   of metals, 268-270
   surface, 267

Parachor, equation, 108
Particle equilibrium, in a solid matrix, 70
   liquid structure, 169
   on a free surface, 67, 69
   solid structure, 169
Pendant drop, experimental observation of, 95
   parameters for surface tension calculation (table), 96
   see also Surface tension
Phase boundary structure, 320-340
Physisorption, 259, 260
Plane matching theory, 212-214
Precipitate, coherent structure, 237, 238
   equilibrium of at grain boundary, 71
   interfacial structure, 237-240
   ledge structure of, 239, 240
   noncoherent structure, 238

SUBJECT INDEX 375

Precipitation, at interfaces, 242-245
  driving force equation, 243
Protrusions, grain boundary, 208
Pseudoequilibrium, of polycrystalline grain structure, 34, 63
Pseudomorphic growth, 232, 233
Pseudopotential model, 185, 186

Quantitative metallography, 36

Recovery, grain boundary, 323
Recrystallization, 323-325

Scanning electron microscopy, observations of particle shapes, 8, 9, 69
  observations of reaction kinetics, 274
  of fracture surfaces, 307, 309
  of grain boundary grooves, 46
  of grain structure, 34
Segregation, at solid-solid interfaces, 16, 135, 240, 242-244
  embrittlement by, 306
  grain boundary, 16, 134, 135
    measurement of, 135
  of solute at interfaces, 243-245
  solid surface, 130
    measurement of, 130
  Suzuki, 153
  to stacking faults, 153, 154
Sessile drop, 91
  experimental observation of, 94
  parameters for surface tension measurements (table), 93
Shape anisotropy, 68
Shear-structure boundaries, 229, 230
Shock loading, explosive, 348-350
Single crystal structure, 168-170
Sintering, diagrams, 295, 302
  geometry (particle), 297-300, 301
  growth-rate equations (table), 299
  mechanisms (table), 299
  of three spheres or wires (cylinders), 300
  of two spheres, 297
  rate, 298
  stages, 301, 302
Solidification, 314, 317
  kinetics, 316
  mechanisms (of growth), 316
Solid interfaces, interface phase concept, 10
  thermodynamics of, 1
Solid interfacial free energy ratios (table), 76
Solid-liquid interface, 74
  energy calculation, 187 (table), 188
  energy of, 157, 319
  structure, 177 179
Solid-melt interface, see Solid-liquid interface

Solid-solution alloys, 20
Solute segregation, 16
Spinodal phase boundary structure, 235
Spirals, growth, see Surface structure
Stacking fault, 65, extrinsic, 249
  interface structure, 219-221
  intrinsic, 249
  nature, 148-150
  segregation, 153, 154, 244
  strengthening, 346
Stacking-fault free energy, 142-148
  effect of temperature and composition on, 150-153
  effect on mechanical properties, 346, 347
  measurements (table), 145
  temperature coefficients of (table), 145
  theoretical calculation of, 249
  twin boundary free energy relationship, 149
Steps, grain boundary, 330, 331
  see also Surface structure
  surface, see Surface structure
Stereometric microscopy, 36
Structural units, see Grain boundary structure
Subgrain refinement, 351
Substitutional alloys, 19
Superlattice structure, 227-230
Superplasticity, 336-338
Surface energy, calculation of from surface tension values, 123
  empirical calculation of, 182-184
  tables, 183, 185
  isotherms for solid alloys, 127, 128
  of solid metals and alloys, 116-130
  measurement of, 116-122
    zero-creep methods, 116-121
    table, 124
  theoretical calculation of, 184-187
    table, 186
Surface excess, 7, 50, 114, 115
  see also Segregation
Surface mobility, 291
Surface molecule, concept, 263-266
Surface reaction kinetics; catalysis, 270-276
Surface strain, 2
Surface stresses, solid, 1, 2
Surface structure, adatoms, 172
  adsorption, 179-181, 259-266
  and catalysis, 273-275
  atomically smooth solid, 9
  free-electron model of, see Jellium model
  jellium model, 185
  kinks, 172-175
  of liquid metals, and alloys, 165-167
  of solid metals and alloys, 168-181
  oxidation, 179-181, 267-270
  reactions, 179-181

Surface structure (cont.)
 spirals (dislocation), 175-177
 steps, 172-177
 terraces, 174, 175
 vacancies, 172
Surface tension (liquid metal), calculations, 181-187
 (tables), 183, 185, 186
 isotherms for alloy solutions, 112
 measurement of, 90-106
  drop weight method, 96
  maximum bubble pressure method, 98
  pendant drop method, 95
  sessile drop method, 91
  table of values, 101
 negative, 2
 of solid metals and alloys (table), 101
 periodic arrangement at melting point (figure), 109
 variation with temperature coefficient (figure), 107
 versus temperature coefficient (figure), 107
  see also Surface energy of solids

Temperature coefficients, of coherent twin boundary free energy (table), 138 (figure), 139
 of grain boundary free energy (table), 131
 of liquid metal and alloy surface energy (tension), 100 (table), 101
 of solid metal and alloy surface free energy, 122, 129, (table), 124
Terrace, see Surface structure
Thermal grooves, 80
Thin film structure, 281-284
Three-phase system, 66, 71
Tilt boundary, see Dislocation
Torque, at phase intersections, 74
 at twin boundary-surface intersections, 49, 50
 at twin-grain boundary intersections, 56-58
 grain boundary, 33, 41
 net effective, 41
 ratios, 42, 54, 55, 58, 61
 resultant, 57
Transformation boundary structure, 235
Transport phenomena, 291-295
 see also Diffusion
 vapor (see Evaporation-condensation)
Twin, coherent annealing (fcc), 45
 deformation (interface structure), 221-223, 347
 dissociation, 65
 fault structure, 221-223

index of (order), 47
in fcc metals and alloys, 45
in grain corner, 59, 60, 62, 63
interface structure, 219-226
noncoherent boundary structure, 224-226
origin of (in fcc metals), 62
residual strengthening of, 347-349
Twin boundary energy, coherent fcc (table), 138
 noncoherent, 140-142
  measurement of, 140 (table), 142
 temperature coefficient of, 139 (table), 138
 theoretical calculation of, 249, 250
Twin boundary intersections, configurational equilibrium, 55, 60
 torque, 49
 with a free surface, 47-50
 with a grain boundary, 54-61
Twin boundary migration, 62
Twist boundary, see Dislocation
Two-phase system, 70
 equilibrium of, 70, 71, 73

Undercooling, 315, 318, 321

Vacancies, in solid surfaces, 172
Vacancy adsorption, grain boundary, 14
Vacancy diffusion, see Diffusion
Vapor deposition, 176, 177, 276
 physical, 276-284
 vacuum, 280-284
Vapor growth, see Vapor deposition
Vaporization, see Heat of

Weak-beam, technique, 144, 145
Wear, surface, 287-289
Work function, alteration, 266, 267
 electron, 183, 184
Wulff, condition, 4
 constant, 6, 68
 construction (γ-plot), 6, 53
 theorem, 5

X-ray, Laue back reflection, 54
 nondispersive analysis, 135

Yield strength, 285
Young's modulus, in surface energy calculation, 181
 relationship to hardness, 285
 variation with solid surface energy (figure), 127

Zero creep, of polycrystalline thin foils, 116
 of wires, 120